"SQUARING THE CIRCLE"

A HISTORY OF THE PROBLEM

"SQUARING THE CIRCLE"

A HISTORY OF THE PROBLEM

BY

E. W. HOBSON, Sc.D, LL.D., F.R.S.

SADLEIRIAN PROFESSOR OF PURE MATHEMATICS, AND FELLOW OF CHRIST'S COLLEGE, IN THE UNIVERSITY OF CAMBRIDGE

PREFACE

IN the Easter Term of the present year I delivered a short course of six Professorial Lectures on the history of the problem of the quadrature of the circle, in the hope that a short account of the fortunes of this celebrated problem might not only prove interesting in itself, but might also act as a stimulant of interest in the more general history of Mathematics. It has occurred to me that, by the publication of the Lectures, they might perhaps be of use, in the same way, to a larger circle of students of Mathematics.

The account of the problem here given is not the result of any independent historical research, but the facts have been taken from the writings of those authors who have investigated various parts of the history of the problem.

The works to which I am most indebted are the very interesting book by Prof. F. Rudio entitled "Archimedes, Huygens, Lambert, Legendre. Vier Abhandlungen über die Kreismessung" (Leipzig, 1892), and Sir T. L. Heath's treatise "The works of Archimedes" (Cambridge, 1897). I have also made use of Cantor's "Geschichte der Mathematik," of Vahlen's "Konstruktionen und Approximationen" (Leipzig, 1911), of Yoshio Mikami's treatise "The development of Mathematics in China and Japan" (Leipzig, 1913), of the translation by T. J. McCormack (Chicago, 1898) of H. Schubert's "Mathematical Essays and Recreations," and of the article "The history and transcendence of π" written by Prof. D. E. Smith which appeared in the "Monographs on Modern Mathematics" edited by Prof. J. W. A. Young. On special points I have consulted various other writings.

E. W. H.

CHRIST'S COLLEGE, CAMBRIDGE.
October, 1913.

CONTENTS

CHAPTER I

GENERAL ACCOUNT OF THE PROBLEM

A GENERAL survey of the history of thought reveals to us the fact of the existence of various questions that have occupied the almost continuous attention of the thinking part of mankind for long series of centuries. Certain fundamental questions presented themselves to the human mind at the dawn of the history of speculative thought, and have maintained their substantial identity throughout the centuries, although the precise terms in which such questions have been stated have varied from age to age in accordance with the ever varying attitude of mankind towards fundamentals. In general, it may be maintained that, to such questions, even after thousands of years of discussion, no answers have been given that have permanently satisfied the thinking world, or that have been generally accepted as final solutions of the matters concerned. It has been said that those problems that have the longest history are the insoluble ones.

If the contemplation of this kind of relative failure of the efforts of the human mind is calculated to produce a certain sense of depression, it may be a relief to turn to certain problems, albeit in a more restricted domain, that have occupied the minds of men for thousands of years, but which have at last, in the course of the nineteenth century, received solutions that we have reasons of overwhelming cogency to regard as final. Success, even in a comparatively limited field, is some compensation for failure in a wider field of endeavour. Our legitimate satisfaction at such exceptional success is but slightly qualified by the fact that the answers ultimately reached are in a certain sense of a negative character. We may rest contented with proofs that these problems, in their original somewhat narrow form, are insoluble, provided we attain, as is actually the case in some celebrated instances, to a complete comprehension of the grounds, resting upon a thoroughly established theoretical basis, upon which our final conviction of the insolubility of the problems is founded.

The three celebrated problems of the quadrature of the circle, the trisection of an angle, and the duplication of the cube, although all of them are somewhat special in character, have one great advantage for the purposes of historical study, *viz.* that their complete history as scientific problems lies, in a completed form, before us. Taking the first of these problems, which will be here our special subject of study, we possess indications of its origin in remote antiquity, we are able to follow the lines on which the treatment of the problem proceeded and changed from age to age in accordance with the progressive development of general Mathematical Science, on which it exercised a noticeable reaction. We are also able to see how the progress of endeavours towards a solution was affected by the intervention of some of the greatest Mathematical thinkers that the world has seen, such men as Archimedes, Huyghens, Euler, and Hermite. Lastly, we know when and how the resources of modern Mathematical Science became sufficiently powerful to make possible that resolution of the problem which, although negative, in that the impossibility of the problem subject to the implied restrictions was proved, is far from being a mere negation, in that the true grounds of the impossibility have been set forth with a finality and completeness which is somewhat rare in the history of Science.

If the question be raised, why such an apparently special problem, as that of the quadrature of the circle, is deserving of the sustained interest which has attached to it, and which it still possesses, the answer is only to be found in a scrutiny of the history of the problem, and especially in the closeness of the connection of that history with the general history of Mathematical Science. It would be difficult to select another special problem, an account of the history of which would afford so good an opportunity of obtaining a glimpse of so many of the main phases of the development of general Mathematics; and it is for that reason, even more than on account of the intrinsic interest of the problem, that I have selected it as appropriate for treatment in a short course of lectures.

Apart from, and alongside of, the scientific history of the problem, it has a history of another kind, due to the fact that, at all times, and almost as much at the present time as formerly, it has attracted the attention of a class of persons who have, usually with a very inadequate equipment of knowledge of the true nature of the problem or of its history, devoted their attention to it, often with passionate enthusiasm. Such persons have very frequently maintained, in the face of all efforts at refutation made by genuine Mathematicians, that they had obtained a solution of the problem. The solutions propounded by the circle squarer exhibit every grade of skill, varying from the most futile attempts, in which the writers shew an utter lack of power to reason correctly, up to approximate solutions the construction of which required much ingenuity on the part of their inventor. In some cases it requires an effort of sustained attention to find out the precise point in the demonstration at which the error occurs, or in which an approximate determination is made to do duty for a theoretically exact one. The psychology of the scientific crank is a subject with which the officials of every Scientific Society have some practical acquaintance. Every Scientific Society still receives from time to time communications from the circle squarer and the trisector of angles, who often make amusing attempts to disguise the real character of their essays. The solutions propounded by such persons usually involve some misunderstanding as to the nature of the conditions under which the problems are to be solved, and ignore the difference between an approximate construction and the solution of the ideal problem. It is a common occurrence that such a person sends his solution to the authorities of a foreign University or Scientific Society, accompanied by a statement that the men of Science of the writer's own country have entered into a conspiracy to suppress his work, owing to jealousy, and that he hopes to receive fairer treatment abroad. The statement is not infrequently accompanied with directions as to the forwarding of any prize of which the writer may be found worthy by the University or Scientific Society addressed, and usually indicates no lack of confidence that the bestowal of such a prize has been amply deserved as the fit reward for the final solution of a problem which has baffled the efforts of a great multitude of predecessors in all ages. A very interesting detailed account of the peculiarities of the circle squarer, and of the futility of attempts on the part of Mathematicians to convince him of his errors, will be found in Augustus De Morgan's *Budget of Paradoxes*. As early as the time of the Greek Mathematicians circle-squaring occupied the attention of non-Mathematicians; in fact the Greeks had a special word to denote this kind of activity, *viz.* τετραγωνίζειν, which means to occupy oneself with the quadrature. It is interesting to remark that, in the year 1775, the Paris Academy found it necessary to protect its officials against the waste of time and energy involved in examining the efforts of circle squarers. It passed a resolution, which appears in the Minutes of the Academy*, that no more solutions were to be examined of the problems of the duplication of the cube, the trisection of the angle, the quadrature of the circle, and that the same resolution should apply to machines for exhibiting perpetual motion. An account of the reasons which led to the adoption of this resolution, drawn up by Condorcet, who was then the perpetual Secretary of the Academy, is appended. It is interesting to remark the strength of the conviction of Mathematicians that the solution of the problem is impossible, more than a century before an irrefutable proof of the correctness of that conviction was discovered.

The popularity of the problem among non-Mathematicians may seem to require some explanation. No doubt, the fact of its comparative obviousness explains in part at least its popularity; unlike many Mathematical problems, its nature can in some sense be understood by anyone; although, as we shall presently see, the very terms in which it is usually stated tend to suggest an imperfect apprehension of its precise import. The accumulated celebrity which the problem attained, as one of proverbial difficulty, makes it an irresistible attraction to men with a certain kind of mentality. An exaggerated notion of the gain which would accrue to mankind by a solution of the problem has at various times been a factor in stimulating the efforts of men with more zeal than knowledge. The man of mystical tendencies has been attracted to the problem by a vague idea that its solution would, in some dimly discerned manner, prove a key to a knowledge of the inner connections of things far beyond those with which the problem is immediately connected.

Statement of the problem

The fact was well known to the Greek Geometers that the problems of the quadrature and the rectification of the circle are equivalent problems. It was in fact at an early time established that the ratio of the length of a complete circle to the diameter has a definite value equal to that of the area of the circle to that of a square of which the radius is side. Since the time of Euler this ratio has always been denoted by the familiar notation π. The problem of "squaring the circle" is roughly that of constructing a square of which the area is equal to that enclosed by the circle. This is then equivalent to the problem of the rectification of the circle, *i.e.* of the determination of a

* *Histoire de l'Académie royale*, année 1775, p. 61.

straight line, of which the length is equal to that of the circumference of the circle. But a problem of this kind becomes definite only when it is specified what means are to be at our disposal for the purpose of making the required construction or determination; accordingly, in order to present the statement of our problem in a precise form, it is necessary to give some preliminary explanations as to the nature of the postulations which underlie all geometrical procedure.

The Science of Geometry has two sides; on the one side, that of practical or physical Geometry, it is a physical Science concerned with the actual spatial relations of the extended bodies which we perceive in the physical world. It was in connection with our interests, of a practical character, in the physical world, that Geometry took its origin. Herodotus ascribes its origin in Egypt to the necessity of measuring the areas of estates of which the boundaries had been obliterated by the inundations of the Nile, the inhabitants being compelled, in order to settle disputes, to compare the areas of fields of different shapes. On this side of Geometry, the objects spoken of, such as points, lines, &c., are physical objects; a point is a very small object of scarcely perceptible and practically negligible dimensions; a line is an object of small, and for some purposes negligible, thickness; and so on. The constructions of figures consisting of points, straight lines, circles, &c., which we draw, are constructions of actual physical objects. In this domain, the possibility of making a particular construction is dependent upon the instruments which we have at our disposal.

On the other side of the subject, Geometry is an abstract or rational Science which deals with the relations of objects that are no longer physical objects, although these ideal objects, points, straight lines, circles, &c., are called by the same names by which we denote their physical counterparts. At the base of this rational Science there lies a set of definitions and postulations which specify the nature of the relations between the ideal objects with which the Science deals. These postulations and definitions were suggested by our actual spatial perceptions, but they contain an element of absolute exactness which is wanting in the rough data provided by our senses. The objects of abstract Geometry possess in absolute precision properties which are only approximately realized in the corresponding objects of physical Geometry. In every department of Science there exists in a greater or less degree this distinction between the abstract or rational side and the physical or concrete side; and the progress of each department of Science involves a continually increasing amount of rationalization. In Geometry the passage from a purely empirical treatment to the setting up of a rational Science proceeded by much more rapid stages than in other cases. We have in the Greek Geometry, known to us all through the presentation of it given in that oldest of all scientific text books, Euclid's *Elements of Geometry*, a treatment of the subject in which the process of rationalization has already reached an advanced stage. The possibility of solving a particular problem of determination, such as the one we are contemplating, as a problem of rational Geometry, depends upon the postulations that are made as to the allowable modes of determination of new geometrical elements by means of assigned ones. The restriction in practical Geometry to the use of specified instruments has its counterpart in theoretical Geometry in restrictions as to the mode in which new elements are to be determined by means of given ones. As regards the postulations of rational Geometry in this respect there is a certain arbitrariness corresponding to the more or less arbitrary restriction in practical Geometry to the use of specified instruments.

The ordinary obliteration of the distinction between abstract and physical Geometry is furthered by the fact that we all of us, habitually and almost necessarily, consider both aspects of the subject at the same time. We may be thinking out a chain of reasoning in abstract Geometry, but if we draw a figure, as we usually must do in order to fix our ideas and prevent our attention from wandering owing to the difficulty of keeping a long chain of syllogisms in our minds, it is excusable if we are apt to forget that we are not in reality reasoning about the objects in the figure, but about objects which are their idealizations, and of which the objects in the figure are only an imperfect representation. Even if we only visualize, we see the images of more or less gross physical objects, in which various qualities irrelevant for our specific purpose are not entirely absent, and which are at best only approximate images of those objects about which we are reasoning.

It is usually stated that the problem of squaring the circle, or the equivalent one of rectifying it, is that of constructing a square of an area equal to that of the circle, or in the latter case of constructing a straight line of length equal to that of the circumference, by a method which involves the use only of the compass and of the ruler as a single straight-edge. This mode of statement, although it indicates roughly the true statement of the problem, is decidedly defective in that it entirely leaves out of account the fundamental distinction between the two aspects of Geometry to which allusion has been made above. The compass and the straight-edge are physical objects by the use of which other objects can be constructed, *viz.* circles of small thickness, and lines which are approximately straight and very thin, made of ink or other material. Such instruments can clearly have no direct relation to theoretical Geometry, in which circles and straight lines are ideal objects possessing in absolute precision properties that are only approximately realized in the circles and straight lines that can be constructed by compasses and rulers. In theoretical Geometry, a restriction to the use of rulers and compasses, or of other instruments, must be replaced by corresponding postulations as to the allowable modes of determination of geometrical objects. We will see what these postulations really are in the case of Euclidean Geometry. Every Euclidean problem of construction, or as it would be preferable to say, every problem of determination, really consists in the determination of one or more points which shall satisfy prescribed conditions. We have here to consider the fundamental modes in which, when a number of points are regarded as given, or already determined, a new point is allowed to be determined.

Two of the fundamental postulations of Euclidean Geometry are that, having given two points A and B, then (1) a unique straight line (A, B) (the whole straight line, and not merely the segment between A and B) is determined such that A and B are incident on it, and (2) that a unique circle $A\,(B)$, of which A is centre and on which B is incident, is determined. The determinancy or assumption of existence of such straight lines and circles is in theoretical Geometry sufficient for the purposes of the subject. When we know that these objects, having known properties, exist, we may reason about them and employ them for the purposes of our further procedure; and that is sufficient for our purpose. The notion of drawing or constructing them by means of a straight-edge or compass has no relevance to abstract Geometry, but is borrowed from the language of practical Geometry.

A new point is determined in Euclidean Geometry exclusively in one of the three following ways:

Having given four points A, B, C, D, not all incident on the same straight line, then

(1) Whenever a point P exists which is incident both on (A, B) and on (C, D), that point is regarded as determinate.

(2) Whenever a point P exists which is incident both on the straight line (A, B) and on the circle $C(D)$, that point is regarded as determinate.

(3) Whenever a point P exists which is incident on both the circles $A(B)$, $C(D)$, that point is regarded as determinate.

The cardinal points of any figure determined by a Euclidean construction are always found by means of a finite number of successive applications of some or all of these rules (1), (2) and (3). Whenever one of these rules is applied it must be shewn that it does not fail to determine the point. Euclid's own treatment is sometimes defective as regards this requisite; as for example in the first proposition of his first book, in which it is not shewn that the circles intersect one another.

In order to make the practical constructions which correspond to these three Euclidean modes of determination, corresponding to (1) the ruler is required, corresponding to (2) both the ruler and the compass, and corresponding to (3) the compass only.

As Euclidean plane Geometry is concerned with the relations of points, straight lines, and circles only, it is clear that the above system of postulations, although arbitrary in appearance, is the system that the exigencies of the subject would naturally suggest. It may, however, be remarked that it is possible to develop Euclidean Geometry with a more restricted set of postulations. For example it can be shewn that all Euclidean constructions can be carried out by means of (3) alone*, without employing (1) or (2).

Having made these preliminary explanations we are now in a position to state in a precise form the ideal problem of "squaring the circle," or the equivalent one of the rectification of the circle.

The historical problem of "squaring the circle" is that of determining a square of which the area shall equal that of a given circle, by a method such that the determination of the corners of the square is to be made by means of the above rules (1), (2), (3), each of which may be applied any finite number of times. In other words, each new point successively determined in the process of construction is to be obtained as the intersection of two straight lines already determined, or as an intersection of a straight line and a circle already determined, or as an intersection of two circles already determined. A

* See for example the *Mathematical Gazette* for March 1913, where I have treated this point in detail in the Presidential Address to the Mathematical Association.

similar statement applies to the equivalent problem of the rectification of the circle.

This mode of determination of the required figure we may speak of shortly as a Euclidean determination.

Corresponding to any problem of Euclidean determination there is a practical problem of physical Geometry to be carried out by actual construction of straight lines and circles by the use of ruler and compasses. Whenever an ideal problem is soluble as one of Euclidean determination the corresponding practical problem is also a feasible one. The ideal problem has then a solution which is ideally perfect; the practical problem has a solution which is an approximation limited only by the imperfections of the instruments used, the ruler and the compass; and this approximation may be so great that there is no perceptible defect in the result. But it is an error which accounts I think, in large measure, for the aberrations of the circle squarer and the trisector of angles, to assume the converse that, when a practical problem is soluble by the use of the instruments in such a way that the error is negligible or imperceptible, the corresponding ideal problem is also soluble. This is very far from being necessarily the case. It may happen that in the case of a particular ideal problem no solution is obtainable by a finite number of successive Euclidean determinations, and yet that such a finite set gives an approximation to the solution which may be made as close as we please by taking the process far enough. In this case, although the ideal problem is insoluble by the means which are permitted, the practical problem is soluble in the sense that a solution may be obtained in which the error is negligible or imperceptible, whatever standard of possible perceptions we may employ. As we have seen, a Euclidean problem of construction is reducible to the determination of one or more points which satisfy prescribed conditions. Let P be one such point; then it may be possible to determine in Euclidean fashion each point of a set $P_1, P_2, \dots P_n, \dots$ of points which converge to P as limiting point, and yet the point P may be incapable of determination by Euclidean procedure. This is what we now know to be the state of things in the case of our special problem of the quadrature of the circle by Euclidean determination. As an ideal problem it is not capable of solution, but the corresponding practical problem is capable of solution with an accuracy bounded only by the limitations of our perceptions and the imperfections of the instruments employed. Ideally we can actually determine by Euclidean methods a square of which the area differs from that of a given circle by less than an arbitrarily prescribed magnitude, although we cannot pass to the limit. We can obtain solutions of the corresponding physical problem which leave nothing to be desired from the practical point of view. Such is the answer which has been obtained to the question raised in this celebrated historical problem of Geometry. I propose to consider in some detail the various modes in which the problem has been attacked by people of various races, and through many centuries; how the modes of attack have been modified by the progressive development of Mathematical tools, and how the final answer, the nature of which had been long anticipated by all competent Mathematicians, was at last found and placed on a firm basis.

General survey of the history of the problem

The history of our problem is typical as exhibiting in a remarkable degree many of the phenomena that are characteristic of the history of Mathematical Science in general. We notice the early attempts at an empirical solution of the problem conceived in a vague and sometimes confused manner; the gradual transition to a clearer notion of the problem as one to be solved subject to precise conditions. We observe also the intimate relation which the mode of regarding the problem in any age had with the state then reached by Mathematical Science in its wider aspect; the essential dependence of the mode of treatment of the problem on the powers of the existing tools. We observe the fact that, as in Mathematics in general, the really great advances, embodying new ideas of far-reaching fruitfulness, have been due to an exceedingly small number of great men; and how a great advance has often been followed by a period in which only comparatively small improvements in, and detailed developments of, the new ideas have been accomplished by a series of men of lesser rank. We observe that there have been periods when for a long series of centuries no advance was made; when the results obtained in a more enlightened age have been forgotten. We observe the times of revival, when the older learning has been rediscovered, and when the results of the progress made in distant countries have been made available as the starting points of new efforts and of a fresh period of activity.

The history of our problem falls into three periods marked out by fundamentally distinct differences in respect of method, of immediate aims, and of equipment in the possession of intellectual tools. The first period embraces the time between the first records of empirical determinations of the ratio of the circumference to the diameter of a circle until the invention of the Differential and Integral Calculus, in the middle of the seventeenth century. This period, in which the ideal of an exact construction was never entirely lost sight of, and was occasionally supposed to have been attained, was the geometrical period, in which the main activity consisted in the approximate determination of π by calculation of the sides or areas of regular polygons in- and circum-scribed to the circle. The theoretical groundwork of the method was the Greek method of Exhaustions. In the earlier part of the period the work of approximation was much hampered by the backward condition of arithmetic due to the fact that our present system of numerical notation had not yet been invented; but the closeness of the approximations obtained in spite of this great obstacle are truly surprising. In the later part of this first period methods were devised by which approximations to the value of π were obtained which required only a fraction of the labour involved in the earlier calculations. At the end of the period the method was developed to so high a degree of perfection that no further advance could be hoped for on the lines laid down by the Greek Mathematicians; for further progress more powerful methods were requisite.

The second period, which commenced in the middle of the seventeenth century, and lasted for about a century, was characterized by the application of the powerful analytical methods provided by the new Analysis to the determination of analytical expressions for the number π in the form of convergent series, products, and continued fractions. The older geometrical forms of investigation gave way to analytical processes in which the functional relationship as applied to the trigonometrical functions became prominent. The new methods of systematic representation gave rise to a race of calculators of π, who, in their consciousness of the vastly enhanced means of calculation placed in their hands by the new Analysis, proceeded to apply the formulae to obtain numerical approximations to π to ever larger numbers of places of decimals, although their efforts were quite useless for the purpose of throwing light upon the true nature of that number. At the end of this period no knowledge had been obtained as regards the number π of a kind likely to throw light upon the possibility or impossibility of the old historical problem of the ideal construction; it was not even definitely known whether the number is rational or irrational. However, one great discovery, destined to furnish the clue to the solution of the problem, was made at this time; that of the relation between the two numbers π and e, as a particular case of those exponential expressions for the trigonometrical functions which form one of the most fundamentally important of the analytical weapons forged during this period.

In the third period, which lasted from the middle of the eighteenth century until late in the nineteenth century, attention was turned to critical investigations of the true nature of the number π itself, considered independently of mere analytical representations. The number was first studied in respect of its rationality or irrationality, and it was shewn to be really irrational. When the discovery was made of the fundamental distinction between algebraic and transcendental numbers, *i.e.* between those numbers which can be, and those numbers which cannot be, roots of an algebraical equation with rational coefficients, the question arose to which of these categories the number π belongs. It was finally established by a method which involved the use of some of the most modern devices of analytical investigation that the number π is transcendental. When this result was combined with the results of a critical investigation of the possibilities of a Euclidean determination, the inference could be made that the number π, being transcendental, does not admit of construction either by a Euclidean determination, or even by a determination in which the use of other algebraic curves besides the straight line and the circle is permitted. The answer to the original question thus obtained is of a conclusively negative character; but it is one in which a clear account is given of the fundamental reasons upon which that negative answer rests.

We have here a record of human effort persisting throughout the best part of four thousand years, in which the goal to be attained was seldom wholly lost sight of. When we look back, in the light of the completed history of the problem, we are able to appreciate the difficulties which in each age restricted the progress which could be made within limits which could not be surpassed by the means then available; we see how, when new weapons became available, a new race of thinkers turned to the further consideration of the problem with a new outlook.

The quality of the human mind, considered in its collective aspect, which most strikes us, in surveying this record, is its colossal patience.

CHAPTER II

THE FIRST PERIOD

Earliest traces of the problem

THE earliest traces of a determination of π are to be found in the Papyrus *Rhind* which is preserved in the British Museum and was translated and explained* by Eisenlohr. It was copied by a clerk, named Ahmes, of the king Raaus, probably about 1700 B.C., and contains an account of older Egyptian writings on Mathematics. It is there stated that the area of a circle is equal to that of a square whose side is the diameter diminished by one ninth; thus $A=(\frac{8}{9})^2 d^2$, or comparing with the formula $A=\frac{1}{4}\pi d^2$, this would give

$$\pi = \frac{256}{81} = 3{\cdot}1604\ldots.$$

No account is given of the means by which this, the earliest determination of π, was obtained; but it was probably found empirically.

The approximation $\pi = 3$, less accurate than the Egyptian one, was known to the Babylonians, and was probably connected with their discovery that a regular hexagon inscribed in a circle has its side equal to the radius, and with the division of the circumference into $6 \times 60 \equiv 360$ equal parts.

This assumption ($\pi = 3$) was current for many centuries; it is implied in the Old Testament, 1 Kings vii. 23, and in 2 Chronicles iv. 2, where the following statement occurs:

"Also he made a molten sea of ten cubits from brim to brim, round in compass, and five cubits the height thereof; and a line of thirty cubits did compass it round about."

The same assumption is to be found in the Talmud, where the statement is made "that which in circumference is three hands broad is one hand broad."

* Eisenlohr, *Ein mathematisches Handbuch der alten Ägypter* (Leipzig, 1877).

The earlier Greek Mathematicians

It is to the Greek Mathematicians, the originators of Geometry as an abstract Science, that we owe the first systematic treatment of the problems of the quadrature and rectification of the circle. The oldest of the Greek Mathematicians, Thales of Miletus (640—548 B.C.) and Pythagoras of Samos (580—500 B.C.), probably introduced the Egyptian Geometry to the Greeks, but it is not known whether they dealt with the quadrature of the circle. According to Plutarch (in *De exilio*), Anaxagoras of Clazomene (500—428 B.C.) employed his time during an incarceration in prison on Mathematical speculations, and constructed the quadrature of the circle. He probably made an approximate construction of an equal square, and was of opinion that he had obtained an exact solution. At all events, from this time the problem received continuous consideration.

About the year 420 B.C. Hippias of Elis invented a curve known as the τετραγωνίζουσα or Quadratrix, which is usually connected with the name of Dinostratus (second half of the fourth century) who studied the curve carefully, and who shewed that the use of the curve gives a construction for π.

This curve may be described as follows, using modern notation.

Let a point Q starting at A describe the circular quadrant AB with uniform velocity, and let a point R starting at O describe the radius OB with uniform velocity, and so that if Q and R start simultaneously they will reach the point B simultaneously. Let the point P be the intersection of OQ with a line perpendicular to OB drawn from R. The locus of P is the quadratrix. Letting $\angle QOA = \theta$, and $OR = y$, the ratio y/θ is constant, and equal to $2a/\pi$, where a denotes the radius of the circle. We have

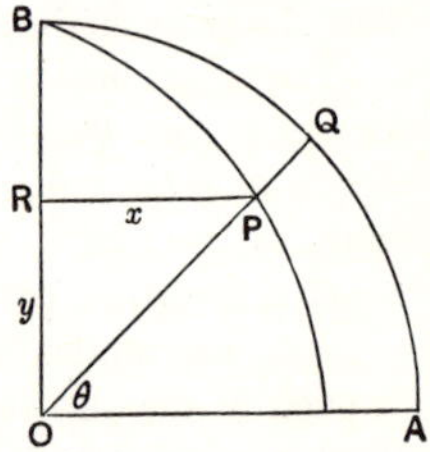

FIG. 1.

$$x = y \cot\theta, \text{ or } x = y \cot\frac{\pi y}{2a},$$

the equation of the curve in rectangular coordinates. The curve will intersect the x axis at the point

$$x = \lim_{y=0}\left(y \cot\frac{\pi y}{2a}\right) = 2a/\pi.$$

If the curve could be constructed, we should have a construction for the length $2a/\pi$, and thence one for π. It was at once seen that the construction of the curve itself involves the same difficulty as that of π.

The problem was considered by some of the Sophists, who made futile attempts to connect it with the discovery of "cyclical square numbers," *i.e.* such square numbers as end with the same cipher as the number itself, as for example $25 = 5^2$, $36 = 6^2$; but the right path to a real treatment of the problem was discovered by Antiphon and further developed by Bryson, both of them contemporaries of Socrates (469—399 B.C.). Antiphon inscribed a square in the circle and passed on to an octagon, 16agon, &c., and thought that by proceeding far enough a polygon would be obtained of which the sides would be so small that they would coincide with the circle. Since a square can always be described so as to be equal to a rectilineal polygon, and a circle can be replaced by a polygon of equal area, the quadrature of the circle would be performed. That this procedure would give only an approximate solution he overlooked. The important improvement was introduced by Bryson of considering circumscribed as well as inscribed polygons; in this procedure he foreshadowed the notion of upper and lower limits in a limiting process. He thought that the area of the circle could be found by taking the mean of the areas of corresponding in- and circum-scribed polygons.

Hippocrates of Chios who lived in Athens in the second half of the fifth century B.C., and wrote the first text book on Geometry, was the first to give examples of curvilinear areas which admit of exact quadrature. These figures are the menisci or lunulae of Hippocrates.

If on the sides of a right-angled triangle ACB semi-circles are described on the same side, the sum of the areas of the two lunes AEC, BDC is equal to that of the triangle ACB. If the right-angled triangle is isosceles, the two lunes are equal, and each of them is half the area of the triangle. Thus the area of a lunula is found.

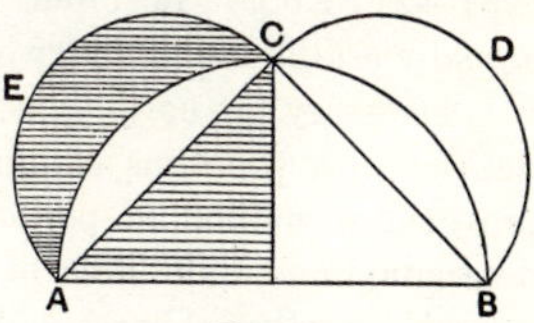

FIG. 2.

If $AC = CD = DB =$ radius OA (see Fig. 3), the semi-circle ACE is $\frac{1}{4}$ of the semi-circle $ACDB$. We have now

$$\text{D}AB - 3\text{D}AC = ACDB - 3\,.\,\text{meniscus } ACE,$$

and each of these expressions is $\frac{1}{4}\text{D}AB$ or half the circle on $\frac{1}{2}AB$ as diameter. If then the meniscus AEC were quadrable so also would be the circle on $\frac{1}{2}AB$ as diameter. Hippocrates recognized the fact that the meniscus is not quadrable, and he made attempts to find other quadrable lunulae in order to make the quadrature of the circle depend on that of such quadrable lunulae. The question of the existence of various kinds of quadrable lunulae was taken up by Th. Clausen* in 1840, who discovered four other quadrable lunulae in addition to the one mentioned above. The question was considered in a general manner by Professor Landau † of Göttingen in 1890, who pointed out that two of the four lunulae which Clausen supposed to be new were already known to Hippocrates.

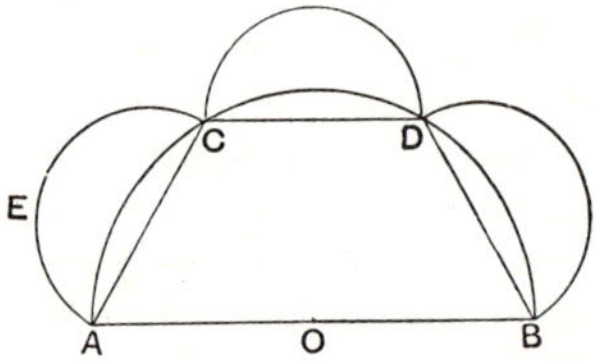

FIG. 3.

From the time of Plato (429—348 B.C.), who emphasized the distinction between Geometry which deals with incorporeal things or images of pure thought and Mechanics which is concerned with things in the external world, the idea became prevalent that problems such as that with which we are concerned should be solved by Euclidean determination only, equivalent on the practical side to the use of two instruments only, the ruler and the compass.

The work of Archimedes

The first really scientific treatment of the problem was undertaken by the greatest of all the Mathematicians of antiquity, Archimedes (287—212 B.C.). In order to understand the mode in which he actually established his very important approximation to the value of π it is necessary for us to consider in some detail the Greek method of dealing with problems of limits, which in the hands of Archimedes provided a method of performing genuine integrations, such as his determination of the area of a segment of a parabola, and of a considerable number of areas and volumes.

This method is that known as the method of exhaustions, and rests on a principle stated in the enunciation of Euclid X. 1, as follows:

"Two unequal magnitudes being set out, if from the greater there be subtracted a magnitude greater than its half, and from that which

* *Journal für Mathematik*, vol. 21, p. 375.

† *Archiv Math. Physik* (3) 4 (1903).

is left a magnitude greater than its half, and if this process be repeated continually, there will be left some magnitude which will be less than the lesser magnitude set out."

This principle is deduced by Euclid from the axiom that, if there are two magnitudes of the same kind, then a multiple of the smaller one can be found which will exceed the greater one. This latter axiom is given by Euclid in the form of a definition of ratio (Book v. def. 4), and is now known as the axiom of Archimedes, although, as Archimedes himself states in the introduction to his work on the quadrature of the parabola, it was known and had been already employed by earlier Geometers. The importance of this so-called axiom of Archimedes, now generally considered as a postulate, has been widely recognized in connection with the modern views as to the arithmetic continuum and the theory of continuous magnitude. The attention of Mathematicians was directed to it by O. Stolz*, who shewed that it was a consequence of Dedekind's postulate relating to "sections." The possibility of dealing with systems of numbers or of magnitudes for which the principle does not hold has been considered by Veronese and other Mathematicians, who contemplate non-Archimedean systems, *i.e.* systems for which this postulate does not hold. The acceptance of the postulate is equivalent to the ruling out of infinite and of infinitesimal magnitudes or numbers as existent in any system of magnitudes or of numbers for which the truth of the postulate is accepted.

The example of the use of the method of exhaustions which is most familiar to us is contained in the proof given in Euclid XII. 2, that the areas of two circles are to one another as the squares on their diameters. This theorem which is a presupposition of the reduction of the problem of squaring the circle to that of the determination of a definite ratio π is said to have been proved by Hippocrates, and the proof given by Euclid is pretty certainly due to Eudoxus, to whom various other applications of the method of Exhaustions are specifically attributed by Archimedes. Euclid shews that the circle can be "exhausted" by the inscription of a sequence of regular polygons each of which has twice as many sides as the preceding one. He shews that the area of the inscribed square exceeds half the area of the circle; he then passes to an octagon by bisecting the arcs bounded by the sides of the square. He shews that the excess of the area of the circle over that of the octagon is less than half what is left of the circle when the square is removed from it, and so on through the further stages of the process.

* See *Math. Annalen*, vol. 22, p. 504, and vol. 39, p. 107.

The truth of the theorem is then inferred by shewing that a contrary assumption leads to a contradiction.

A study of the works of Archimedes, now rendered easily accessible to us in Sir T. L. Heath's critical edition, is of the greatest interest not merely from the historical point of view but also as affording a very instructive methodological study of rigorous treatment of problems of determination of limits. The method by which Archimedes and other Greek Mathematicians contemplated limit problems impresses one, apart from the geometrical form, with its essentially modern way of regarding such problems. In the application of the method of exhaustions and its extensions no use is made of the ideas of the infinite or the infinitesimal; there is no jumping to the limit as the supposed end of an essentially endless process, to be reached by some inscrutable *saltus*. This passage to the limit is always evaded by substituting a proof in the form of a *reductio ad absurdum*, involving the use of inequalities such as we have in recent times again adopted as appropriate to a rigorous treatment of such matters. Thus the Greeks, who were however thoroughly familiar with all the difficulties as to infinite divisibility, continuity, &c., in their mathematical proofs of limit theorems never involved themselves in the morass of indivisibles, indiscernibles, infinitesimals, &c., in which the Calculus after its invention by Newton and Leibnitz became involved, and from which our own text books are not yet completely free.

The essential rigour of the processes employed by Archimedes, with such fruitful results, leaves, according to our modern views, one point open to criticism. The Greeks never doubted that a circle has a definite area in the same sense that a rectangle has one; nor did they doubt that a circle has a length in the same sense that a straight line has one. They had not contemplated the notion of non-rectifiable curves, or non-quadrable areas; to them the existence of areas and lengths as definite magnitudes was obvious from intuition. At the present time we take only the length of a segment of a straight line, the area of a rectangle, and the volume of a rectangular parallelepiped as primary notions, and other lengths, areas, and volumes we regard as derivative, the actual existence of which in accordance with certain definitions requires to be established in each individual case or in particular classes of cases. For example, the measure of the length of a circle is defined thus: A sequence of inscribed polygons is taken so that the number of sides increases indefinitely as the sequence proceeds, and such that the length of the greatest side of the polygon diminishes indefinitely, then if the numbers which represent the perimeters of the successive polygons form a convergent sequence, of which the arithmetical limit is one and the same number for all sequences of polygons which satisfy the prescribed conditions, the circle has a length represented by this limit. It must be proved that this limit exists and is independent of the particular sequence employed, before we are entitled to regard the circle as rectifiable.

In his work *κύκλου μέτρησις*, the measurement of a circle, Archimedes proves the following three theorems.

(1) The area of any circle is equal to a right-angled triangle in which one of the sides about the right angle is equal to the radius, and the other to the circumference, of the circle.

(2) The area of the circle is to the square on its diameter as 11 to 14.

(3) The ratio of the circumference of any circle to its diameter is less than $3\frac{1}{7}$ but greater than $3\frac{10}{71}$.

It is clear that (2) must be regarded as entirely subordinate to (3). In order to estimate the accuracy of the statement in (3), we observe that

$$3\tfrac{1}{7} = 3{\cdot}14285\ldots, \quad 3\tfrac{10}{71} = 3{\cdot}14084\ldots, \quad \pi = 3{\cdot}14159\ldots.$$

In order to form some idea of the wonderful power displayed by Archimedes in obtaining these results with the very limited means at his disposal, it is necessary to describe briefly the details of the method he employed.

His first theorem is established by using sequences of in- and circum-scribed polygons and a *reductio ad absurdum*, as in Euclid XII. 2, by the method already referred to above.

In order to establish the first part of (3), Archimedes considers a regular hexagon circumscribed to the circle.

In the figure, AC is half one of the sides of this hexagon. Then

$$\frac{OA}{AC} = \sqrt{3} > \frac{265}{153}.$$

Bisecting the angle AOC, we obtain AD half the side of a regular circumscribed 12agon. It is then shewn that $\dfrac{OD}{DA} > \dfrac{591\frac{1}{8}}{153}$. If OE is the bisector of the angle DOA, AE is half the side of a circumscribed 24agon, and it is then shewn that $\dfrac{OE}{EA} > \dfrac{1172\frac{1}{8}}{153}$. Next, bisecting EOA, we obtain AF the half side of a 48agon, and it is shewn that $\frac{OF}{FA} > \frac{2339\frac{1}{4}}{153}$. Lastly if OG (not shewn in the figure) be the bisector of FOA, AG is the half side of a regular 96agon circumscribing the circle, and it is shewn that $\frac{OA}{AG} > \frac{4673\frac{1}{2}}{153}$, and thence that the ratio of the diameter to the perimeter of the 96agon

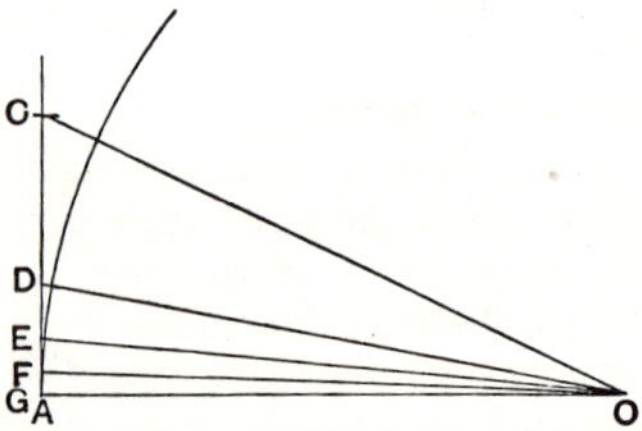

FIG. 4.

is $> \frac{4673\frac{1}{2}}{14688}$, and it is deduced that the circumference of the circle, which is less than the perimeter of the polygon, is $< 3\frac{1}{7}$ of the diameter. The second part of the theorem is obtained in a similar manner by determination of the side of a regular 96agon inscribed in the circle.

In the course of his work, Archimedes assumes and employs, without explanation as to how the approximations were obtained, the following estimates of the values of square roots of numbers:

$$\tfrac{1351}{780} > \sqrt{3} > \tfrac{265}{153},\quad 3013\tfrac{3}{4} > \sqrt{9082321},\quad 1838\tfrac{9}{11} > \sqrt{3380929},$$

$$1009\tfrac{1}{6} > \sqrt{1018405},\quad 2017\tfrac{1}{4} > \sqrt{4069284\tfrac{1}{36}},\quad 591\tfrac{1}{8} < \sqrt{349450},$$

$$1172\tfrac{1}{8} < \sqrt{1373943\tfrac{33}{64}},\quad 2339\tfrac{1}{4} < \sqrt{5472132\tfrac{1}{16}}.$$

In order to appreciate the nature of the difficulties in the way of obtaining these approximations we must remember the backward condition of Arithmetic with the Greeks, owing to the fact that they possessed a system of notation which was exceedingly inconvenient for the purpose of performing arithmetical calculations.

The letters of the alphabet together with three additional signs were employed, each letter being provided with an accent or with a short horizontal stroke; thus the nine integers

1, 2, 3, 4, 5, 6, 7, 8, 9 were denoted by α′, β′, γ′, δ′, ε′, ϛ′, ζ′, η′, θ′,

the multiples of 10,

10, 20, 30, ... 90 were denoted by ι′, κ′, λ′, μ′, ν′, ξ′, ο′, π′, ϙ′, the multiples of 100,

100, 200, ... 900 by ρ′, σ′, τ′, υ′, φ′, χ′, ψ′, ω′, ϡ′.

The intermediate numbers were expressed by juxtaposition, representing here addition, the largest number being placed on the left, the next largest following, and so on in order. There was no sign for zero. Thousands were represented by the same letters as the first nine integers but with a small dash in front and below the line; thus for example ͵δ′ was 4000, and 1913 was expressed by ͵αϡιγ′ or $\overline{\text{͵αϡιγ}}$. 10000 and higher numbers were expressed by using the ordinary numerals with M or Mυ as an abbreviation for the word μυριάς; the number of myriads, or the multiple of 10000, was generally written over the abbreviation, thus 349450 was $\overset{\lambda\delta}{\mathrm{M}}$͵θων′. A variety of devices were employed for the representation of fractions*.

The determinations of square roots such as $\sqrt{3}$ by Archimedes were much closer than those of earlier Greek writers. There has been much speculation as to the method he must have employed in their determination. There is reason to believe that he was acquainted with the method of approximation that we should denote by

$$a \pm \frac{b}{2a} > \sqrt{a^2 \pm b} > a \pm \frac{b}{2a \pm 1}.$$

Various alternative explanations have been suggested; some of these suggest that a method equivalent to the use of approximation by continued fractions was employed.

A full discussion of this matter will be found in Sir T. L. Heath's work on Archimedes.

The treatise of Archimedes on the measurement of the circle must be regarded as the one really great step made by the Greeks towards the solution of the problem; in fact no essentially new mode of attack was made until the invention of the Calculus provided Mathematicians with new weapons. In a later writing which has been lost, but which is mentioned by Hero, Archimedes found a still closer approximation to π.

The essential points of the method of Archimedes, when generalized and expressed in modern notation, consist of the following theorems:

(1) The inequalities $\sin\theta < \theta < \tan\theta$.

* For an interesting account of the Arithmetic of Archimedes, see Heath's *Works of Archimedes*, Chapter IV.

(2) The relations for the successive calculation of the perimeters and areas of polygons inscribed and circumscribed to a circle.

Denoting by p_n, a_n the perimeter and area of an inscribed regular polygon of n sides, and by P_n, A_n the perimeter and area of a circumscribed regular polygon of n sides, these relations are

$$p_{2n} = \sqrt{p_n P_{2n}}, \quad a_{2n} = \sqrt{a_n A_n},$$

$$P_{2n} = \frac{2p_n P_n}{p_n + P_n}, \quad A_{2n} = \frac{2a_{2n} A_n}{a_{2n} + A_n}.$$

Thus the two series of magnitudes

$$P_n, p_n, \; P_{2n}, p_{2n}, P_{4n}, p_{4n}, \ldots,$$

$$A_n, a_{2n}, A_{2n}, a_{4n}, A_{4n}, a_{8n}, \ldots,$$

are calculated successively in accordance with the same law. In each case any element is calculated from the two preceding ones by taking alternately their harmonic and geometric means. This system of formulae is known as the Archimedean Algorithm; by means of it the chords and tangents of the angles at the centre of such polygons as are constructible can be calculated. By methods essentially equivalent to the use of this algorithm the sines and tangents of small angles were obtained to a tolerably close approximation. For example, Aristarchus (250 B.C.) obtained the limits $\frac{1}{45}$ and $\frac{1}{60}$ for sin 1°.

The work of the later Greeks

Among the later Greeks, Hipparchus (180—125 B.C.) calculated the first table of chords of a circle and thus founded the science of Trigonometry. But the greatest step in this direction was made by Ptolemy (87—165 A.D.) who calculated a table of chords in which the chords of all angles at intervals of $\frac{1}{2}$° from 0° to 180° are contained, and thus constructed a trigonometry that was not surpassed for 1000 years. He was the first to obtain an approximation to π more exact than that of Archimedes; this was expressed in sexagesimal measure by 3° 8′ 30″ which is equivalent to

$$3 + \tfrac{8}{60} + \tfrac{30}{3600} \text{ or } 3\tfrac{17}{120} \equiv 3{\cdot}14166\ldots.$$

The work of the Indians

We have now to pass over to the Indian Mathematicians. Áryabhatta (about 500 A.D.) knew the value

$$\tfrac{62832}{20000} \equiv 3{\cdot}1416 \text{ for } \pi.$$

The same value in the form $\frac{3927}{1250}$ was given by Bhâskara (born 1114 A.D.)

in his work *The crowning of the system*; and he describes this value as exact, in contrast with the inexact value $\frac{22}{7}$. His commentator Gancea explains that this result was obtained by calculating the perimeters of polygons of 12, 24, 48, 96, 192, and 384 sides, by the use of the formula

$$a_{2n} = \sqrt{2 - \sqrt{4 - a_n^2}}$$

connecting the sides of inscribed polygons of $2n$ and n sides respectively, the radius being taken as unity. If the diameter is 100, the side of an inscribed 384agon is $\sqrt{98694}$ which leads to the above value* given by Áryabhatta. Brahmagupta (born 598 A.D.) gave as the exact value $\pi = \sqrt{10}$. Hankel has suggested that this was obtained as the supposed limit ($\sqrt{1000}$) of $\sqrt{965}$, $\sqrt{981}$, $\sqrt{986}$, $\sqrt{987}$ (diameter 10), the perimeters of polygons of 12, 24, 48, 96 sides, but this explanation is doubtful. It has also been suggested that it was obtained by the approximate formula

$$\sqrt{a^2 + x} = a + \frac{1}{2a + x},$$

which gives $\sqrt{10} = 3 + \frac{1}{7}$.

The work of the Chinese Mathematicians

The earliest Chinese Mathematicians, from the time of Chou-Kong who lived in the 12th century B.C., employed the approximation $\pi = 3$. Some of those who used this approximation were mathematicians of considerable attainments in other respects.

According to the Sui-shu, or *Records of the Sui dynasty*, there were a large number of circle-squarers, who calculated the length of the circular circumference, obtaining however divergent results.

Chang Híng, who died in 139 A.D., gave the rule

$$\frac{(\text{circumference})^2}{(\text{perimeter of circumscribed square})^2} = \frac{5}{8},$$

which is equivalent to $\pi = \sqrt{10}$.

Wang Fau made the statement that if the circumference of a circle is 142 the diameter is 45; this is equivalent to $\pi = 3{\cdot}1555\ldots$. No record has been found of the method by which this result was obtained.

* See Colebrooke's *Algebra with arithmetic and mensuration, from the Sanscrit of Brahmagupta and Bhâskara*, London, 1817.

Liu Hui published in 263 A.D. an *Arithmetic in nine sections* which contains a determination of π. Starting with an inscribed regular hexagon, he proceeds to the inscribed dodecagon, 24agon, and so on, and finds the ratio of the circumference to the diameter to be 157 : 50, which is equivalent to $\pi = 3{\cdot}14$.

By far the most interesting Chinese determination was that of the great Astronomer *Tsu Ch'ung-chih* (born 430 A.D.). He found the two values $\frac{22}{7}$ and $\frac{355}{113}$ $(=3{\cdot}1415929\ldots)$. In fact he proved that 10π lies between $31{\cdot}415927$ and $31{\cdot}415926$, and deduced the value $\frac{355}{113}$.

The value $\frac{22}{7}$ which is that of Archimedes he spoke of as the "inaccurate" value, and $\frac{355}{113}$ as the "accurate value." This latter value was not obtained either by the Greeks or the Hindoos, and was only rediscovered in Europe more than a thousand years later, by Adriaen Anthonisz. The later Chinese Mathematicians employed for the most part the "inaccurate" value, but the "accurate" value was rediscovered by Chang Yu-chin, who employed an inscribed polygon with 2^{14} sides.

The work of the Arabs

In the middle ages a knowledge of Greek and Indian mathematics was introduced into Europe by the Arabs, largely by means of Arabic translations of Euclid's elements, Ptolemy's σύνταξις, and treatises by Appollonius and Archimedes, including the treatise of Archimedes on the measurement of the circle.

The first Arabic Mathematician Muhammed ibn Mûsâ Alchwarizmi, at the beginning of the ninth century, gave the Greek value $\pi = 3\frac{1}{7}$, and the Indian values $\pi = \sqrt{10}$, $\pi = \frac{62832}{20000}$, which he states to be of Indian origin. He introduced the Indian system of numerals which was spread in Europe at the beginning of the 13th century by Leonardo Pisano, called Fibonacci.

The time of the Renaissance

The greatest Christian Mathematician of medieval times, Leonardo Pisano (born at Pisa at the end of the 12th century), wrote a work entitled *Practica geometriae*, in 1220, in which he improved on the results of Archimedes, using the same method of employing the in- and circum-scribed 96agons. His limits are $\dfrac{1440}{458\frac{1}{5}} = 3{\cdot}1427$ and

$\frac{1440}{458\frac{4}{9}} = 3\cdot1410\ldots$, whereas $3\frac{1}{7} = 3\cdot1428$, $3\frac{10}{71} = 3\cdot1408\ldots$ were the values given by Archimedes. From these limits he chose $\frac{1440}{458\frac{1}{3}}$ or $\pi = 3\cdot1418\ldots$ as the mean result.

During the period of the Renaissance no further progress in the problem was made beyond that due to *Leonardo Pisano*; some later writers still thought that $3\frac{1}{7}$ was the exact value of π. George Purbach (1423—1461), who constructed a new and more exact table of sines of angles at intervals of 10′, was acquainted with the Archimedean and Indian values, which he fully recognized to be approximations only. He expressed doubts as to whether an exact value exists. Cardinal Nicholas of Cusa (1401—1464) obtained $\pi = 3\cdot1423$ which he thought to be the exact value. His approximations and methods were criticized by Regiomontanus (Johannes Müller, 1436—1476), a great mathematician who was the first to shew how to calculate the sides of a spherical triangle from the angles, and who calculated extensive tables of sines and tangents, employing for the first time the decimal instead of the sexagesimal notation.

The fifteenth and sixteenth centuries

In the fifteenth and sixteenth centuries great improvements in trigonometry were introduced by Copernicus (1473—1543), Rheticus (1514—1576), Pitiscus (1561—1613), and Johannes Kepler (1571—1630).

These improvements are of importance in relation to our problem, as forming a necessary part of the preparation for the analytical developments of the second period.

In this period Leonardo da Vinci (1452—1519) and Albrecht Dürer (1471—1528) should be mentioned, on account of their celebrity, as occupying themselves with our subject, without however adding anything to the knowledge of it.

Orontius Finaeus (1494—1555) in a work *De rebus mathematicis hactenus desiratis*, published after his death, gave two theorems which were later established by Huyghens, and employed them to obtain the limits $\frac{22}{7}$, $\frac{245}{78}$ for π; he appears to have asserted that $\frac{245}{78}$ is the exact value. His theorems when generalized are expressed in our notation by the fact that θ is approximately equal to $(\sin^2\theta \tan\theta)^{\frac{1}{3}}$.

The development of the theory of equations which later became of fundamental importance in relation to our problem was due to the Italian Mathematicians of the 16th century, Tartaglia (1506—1559), Cardano (1501—1576), and Ferrari (1522—1565).

The first to obtain a more exact value of π than those hitherto known in Europe was Adriaen Anthonisz (1527—1607) who rediscovered the Chinese value $\pi = \frac{355}{113} = 3{\cdot}1415929\ldots$, which is correct to 6 decimal places. His son Adriaen who took the name of Metius (1571—1635), published this value in 1625, and explained that his father had obtained the approximations $\frac{333}{106} < \pi < \frac{377}{120}$ by the method of Archimedes, and had then taken the mean of the numerators and denominators, thus obtaining his value.

The first explicit expression for π by an infinite sequence of operations was obtained by Vieta (François Viète, 1540—1603). He proved that, if two regular polygons are inscribed in a circle, the first having half the number of sides of the second, then the area of the first is to that of the second as the supplementary chord of a side of the first polygon is to the diameter of the circle. Taking a square, an octagon, then polygons of 16, 32, ... sides, he expressed the supplementary chord of the side of each, and thus obtained the ratio of the area of each polygon to that of the next. He found that, if the diameter be taken as unity, the area of the circle is

$$2\,\frac{1}{\sqrt{\tfrac{1}{2}}\sqrt{\tfrac{1}{2}+\tfrac{1}{2}\sqrt{\tfrac{1}{2}}}\sqrt{\tfrac{1}{2}+\tfrac{1}{2}\sqrt{\tfrac{1}{2}+\tfrac{1}{2}\sqrt{\tfrac{1}{2}}}}\cdots},$$

from which we obtain

$$\frac{\pi}{2} = \frac{1}{\sqrt{\tfrac{1}{2}}\sqrt{\tfrac{1}{2}+\tfrac{1}{2}\sqrt{\tfrac{1}{2}}}\cdots}.$$

It may be observed that this expression is obtainable from the formula

$$\theta = \frac{\sin\theta}{\cos\frac{\theta}{2}\cos\frac{\theta}{4}\cos\frac{\theta}{8}\cdots} \qquad (\theta < \pi)$$

afterwards obtained by Euler, by taking $\theta = \frac{\pi}{2}$.

Applying the method of Archimedes, starting with a hexagon and proceeding to a polygon of $2^{16}.6$ sides, Vieta shewed that, if the diameter of the circle be 100000, the circumference is $> 314159\frac{26535}{100000}$ and is $< 3{\cdot}14159\frac{26537}{100000}$; he thus obtained π correct to 9 places of decimals.

Adrianus Romanus (Adriaen van Rooman, born in Lyons, 1561—1615) by the help of a $15 \cdot 2^{24}$agon calculated π to 15 places of decimals.

Ludolf van Ceulen (Cologne) (1539—1610), after whom the number π is still called in Germany "Ludolph's number," is said to have calculated π to 35 places of decimals. According to his wish the value was engraved on his tombstone which has been lost. In his writing *Van den Cirkel* (Delft, 1596) he explained how, by employing the method of Archimedes, using in- and circum-scribed polygons up to the $60 \cdot 2^{29}$agon, he obtained π to 20 decimal places. Later, in his work *De Arithmetische en Geometrische fondamenten* he obtained the limits given by

$$3\,\frac{14159265358979323846264338327950}{100000000000000000000000000000000},$$

and the same expression with 1 instead of 0 in the last place of the numerator.

The work of Snellius and Huyghens

In a work *Cyclometricus*, published in 1621, Willebrod Snellius (1580—1626) shewed how narrower limits can be determined, without increasing the number of sides of the polygons, than in the method of Archimedes. The two theorems, equivalent to the approximations

$$\tfrac{1}{3}(2\sin\theta + \tan\theta) < \theta < 3/(2\operatorname{cosec}\theta + \cot\theta),$$

by which he attained this result were not strictly proved by him, and were afterwards established by Huyghens; the approximate formula $\theta = \dfrac{3\sin\theta}{2+\cos\theta}$ had been already obtained by Nicholas of Cusa (1401—1464). Using in- and circum-scribed hexagons the limits 3 and 3·464 are obtained by the method of Archimedes, but Snellius obtained from the hexagons the limits 3·14022 and 3·14160, closer than those obtained by Archimedes from the 96agon. With the 96agon he found the limits 3·1415926272 and 3·1415928320. Finally he verified Ludolf's determination with a great saving of labour, obtaining 34 places with the 2^{30}agon, by which Ludolf had only obtained 14 places. Grunberger* calculated 39 places by the help of the formulae of Snellius.

The extreme limit of what can be obtained on the geometrical lines laid down by Archimedes was reached in the work of Christian Huyghens (1629—1665). In his work† *De circuli magnitudine inventa*,

* *Elementa Trigonometriae*, Rome, 1630.

† A study of the German Translation by Rudio will repay the trouble.

which is a model of geometrical reasoning, he undertakes by improved methods to make a careful determination of the area of a circle. He establishes sixteen theorems by geometrical processes, and shews that by means of his theorems three times as many places of decimals can be obtained as by the older method. The determination made by Archimedes he can get from the triangle alone. The hexagon gives him the limits 3·1415926533 and 3·1415926538.

The following are the theorems proved by Huyghens:

I. If ABC is the greatest triangle in a segment less than a semi-circle, then

$$\triangle ABC < 4\,(\triangle AEB + \triangle BFC),$$

where AEB, BFC are the greatest triangles in the segments AB, BC.

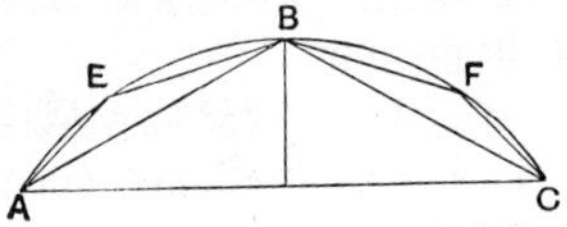

FIG. 5.

II. $\triangle FEG > \frac{1}{2}\triangle ABC$,

where ABC is the greatest triangle in the segment.

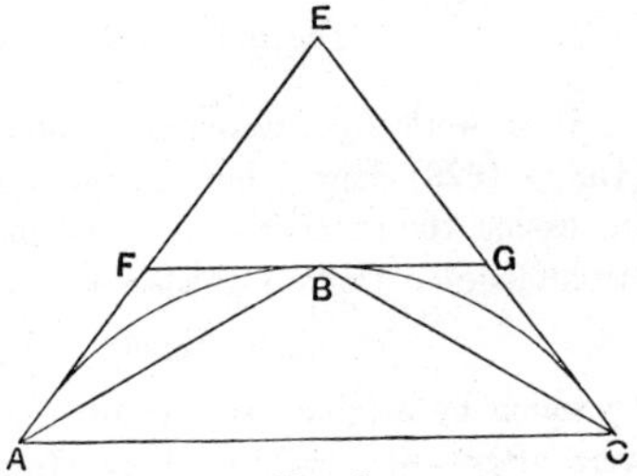

FIG. 6.

III. $$\frac{\text{segment } ACB}{\triangle ACB} > \tfrac{4}{3},$$

provided the segment is less than the semi-circle.

This theorem had already been given by Hero.

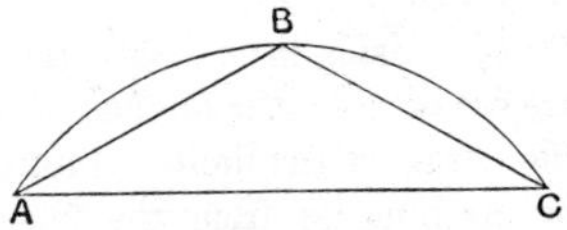

FIG. 7.

IV. $$\frac{\text{segment } ACB}{\triangle ATC} < \tfrac{2}{3}.$$

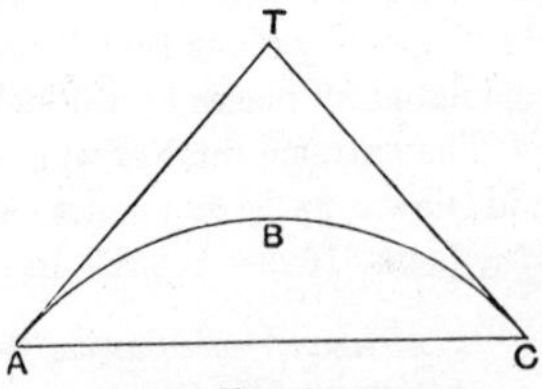

FIG. 8.

V. If A_n is the area of an inscribed regular polygon of n sides, and S the area of the circle, then $S > A_{2n} + \frac{1}{3}(A_{2n} - A_n)$.

VI. If A_n' is the area of the circumscribed regular polygon of n sides, then $S < \frac{2}{3}A_n' + \frac{1}{3}A_n$.

VII. If C_n denotes the perimeter of the inscribed polygon, and C the circumference of the circle, then $C > C_{2n} + \frac{1}{3}(C_{2n} - C_n)$.

VIII. $\frac{2}{3}CD + \frac{1}{3}EF >$ arc CE, where E is any point on the circle.

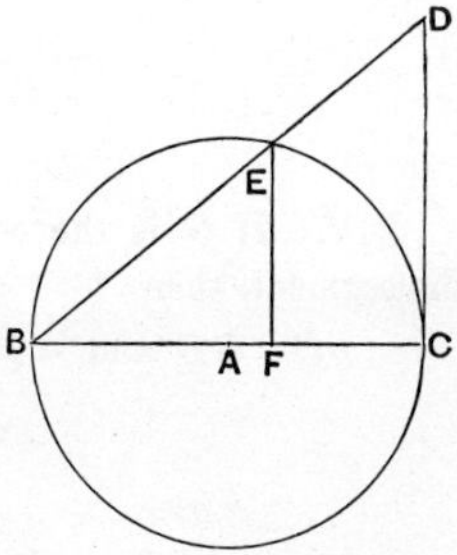

Fig. 9.

IX. $$C < \tfrac{2}{3}C_n + \tfrac{1}{3}C_n',$$
where C_n' is the perimeter of the circumscribed polygon of n sides.

X. If a_n, a_n' denote the sides of the in- and circum-scribed polygons, then $a_{2n}{}^2 = a_{2n}' \cdot \frac{1}{2}a_n$.

XI. $C <$ the smaller of the two mean proportionals between C_n and C_n'.

$S <$ the similar polygon whose perimeter is the larger of the two mean proportionals.

XII. If ED equals the radius of the circle, then $BG >$ arc BF.

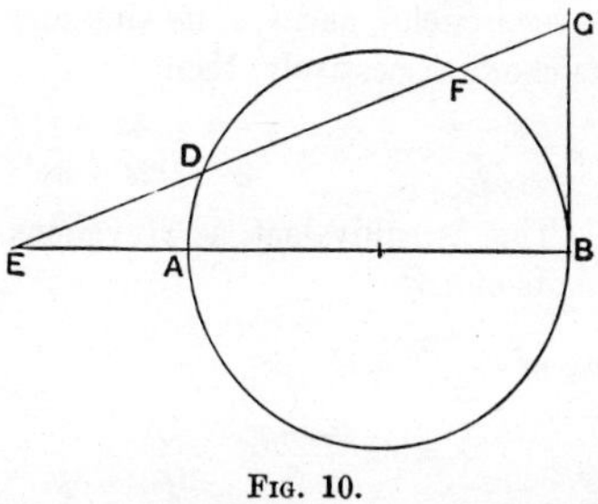

Fig. 10.

XIII. If AC = radius of the circle, then $BL <$ arc BE.

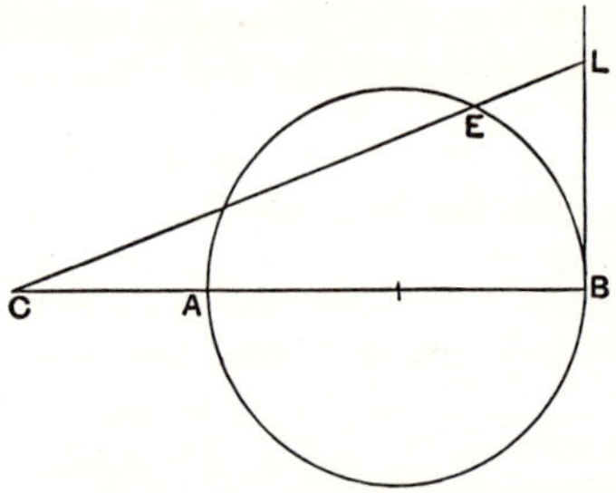

FIG. 11.

XIV. If G is the centroid of the segment, then

$$BG > GD \text{ and } < \tfrac{3}{2}GD.$$

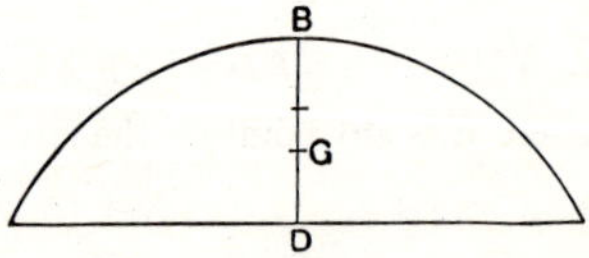

FIG. 12.

XV. $$\frac{\text{segment } ABC}{\triangle ABC} > \frac{4}{3}$$

and $$< 3\tfrac{1}{3} \cdot \frac{B'D}{BB' + 3OD}.$$

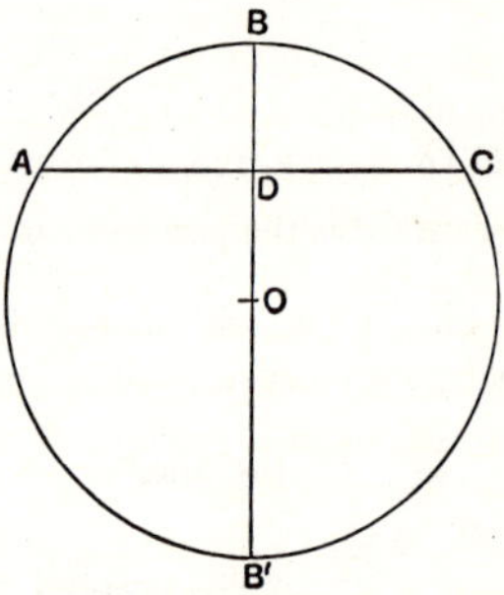

FIG. 13.

XVI. If a denote the arc ($<$ semi-circle), and s, s' its sine and its chord respectively, then

$$s' + \frac{s'-s}{3} < a < s' + \frac{s'-s}{3} \cdot \frac{4s'+s}{2s'+3s}.$$

This is equivalent, as Huyghens points out, to

$$p_{2n} + \frac{p_{2n}-p_n}{3} < C$$
$$< p_{2n} + \frac{p_{2n}-p_n}{3} \cdot \frac{4p_{2n}+p_n}{2p_{2n}+3p_n},$$

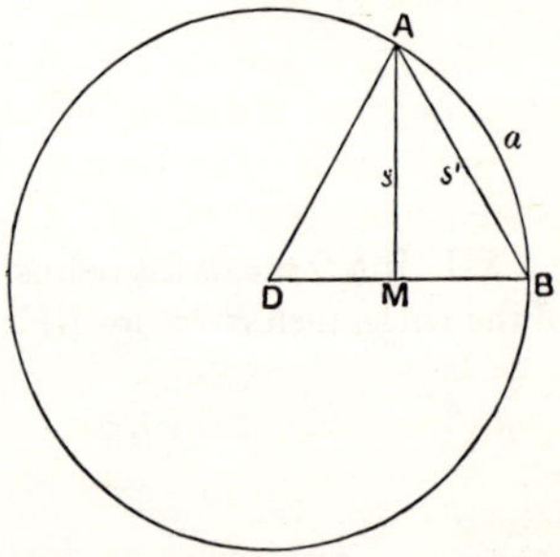

FIG. 14.

where p_n is the perimeter of a regular inscribed polygon of n sides, and C is the circumference of the circle.

The work of Gregory

The last Mathematician to be mentioned in connection with the development of the method of Archimedes is James Gregory (1638—1675), Professor in the Universities of St Andrews and Edinburgh, whose important work in connection with the development of the new Analysis we shall have to refer to later. Instead of employing the perimeters of successive polygons, he calculated their areas, using the formulae

$$A_{2n}' = \frac{2A_n A_n'}{A_n + A_{2n}} = \frac{2A_n' A_{2n}}{A_n' + A_{2n}};$$

where A_n, A_n' denote the areas of in- and circum-scribed regular n-agons; he also employed the formula $A_{2n} = \sqrt{A_n A_n'}$ which had been obtained by Snellius. In his work *Exercitationes geometricae* published in 1668, he gave a whole series of formulae for approximations on the lines of Archimedes. But the most interesting step which Gregory took in connection with the problem was his attempt to prove, by means of the Archimedean algorithm, that the quadrature of the circle is impossible. This is contained in his work *Vera circuli et hyperbolae quadratura* which is reprinted in the works of Huyghens (*Opera varia* I, pp. 315—328) who gave a refutation of Gregory's proof. Huyghens expressed his own conviction of the impossibility of the quadrature, and in his controversy with Wallis remarked that it was not even decided whether the area of the circle and the square of the diameter are commensurable or not. In default of a theory of the distinction between algebraic and transcendental numbers, the failure of Gregory's proof was inevitable. Other such attempts were made by Lagny (*Paris Mém.* 1727, p. 124), Saurin (*Paris Mém.* 1720), Newton (*Principia* I, 6, Lemma 28), and Waring (*Proprietates algebraicarum curvarum*) who maintained that no algebraical oval is quadrable. Euler also made some attempts in the same direction (*Considerationes cyclometricae, Novi Comm. Acad. Petrop.* XVI, 1771); he observed that the irrationality of π must first be established, but that this would not of itself be sufficient to prove the impossibility of the quadrature. Even as early as 1544, Michael Stifel, in his *Arithmetic integra*, expressed the opinion that the construction is impossible. He emphasized the distinction between a theoretical and a practical construction.

The work of Descartes

The great Philosopher and Mathematician René Descartes (1596—1650), of immortal fame as the inventor of coordinate geometry, regarded the problem from a new point of view. A given straight line being taken as equal to the circumference of a circle he proposed to determine the diameter by the following construction:

Take AB one quarter of the given straight line. On AB describe the square $ABCD$; by a known process a point C_1 on AC produced, can be so determined that the rectangle $BC_1 = \frac{1}{4}ABCD$. Again C_2 can be so determined that rect. $B_1C_2 = \frac{1}{4}BC_1$; and so on indefinitely. The diameter required is given by AB_ω, where B_ω is the limit to which $B, B_1, B_2, \ldots$ converge. To see the reason of this, we can shew that AB is the diameter of the circle inscribed in $ABCD$, that AB_1 is the diameter of the circle circumscribed by the regular octagon having the same perimeter as the square; and generally that AB_n is the diameter of the regular 2^{n+2}-agon having the same perimeter as the square. To verify this, let

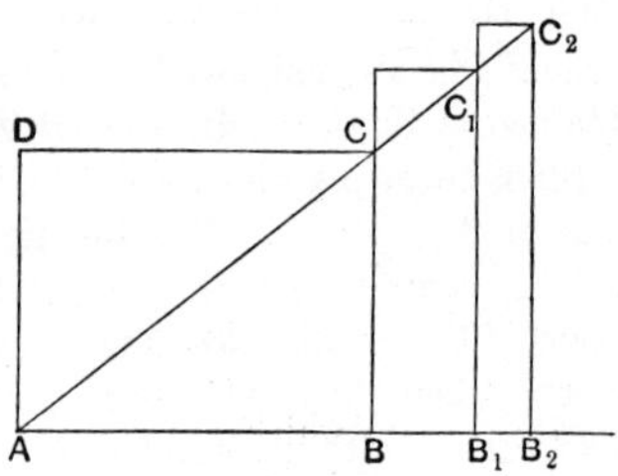

FIG. 15.

$$x_n = AB_n,\ x_0 = AB;$$

then by the construction,

$$x_n(x_n - x_{n-1}) = \frac{1}{4^n}x_0^2,$$

and this is satisfied by $x_n = \dfrac{4x_0}{2^n}\cot\dfrac{\pi}{2^n}$; thus

$$\lim x_n = \frac{4x_0}{\pi} = \text{diameter of the circle.}$$

This process was considered later by Schwab (Gergonne's *Annales de Math.* vol. VI), and is known as the process of isometers.

This method is equivalent to the use of the infinite series

$$\frac{4}{\pi} = \tan\frac{\pi}{4} + \frac{1}{2}\tan\frac{\pi}{8} + \frac{1}{4}\tan\frac{\pi}{16} + \ldots,$$

which is a particular case of the formula

$$\frac{1}{x} - \cot x = \frac{1}{2}\tan\frac{x}{2} + \frac{1}{4}\tan\frac{x}{4} + \frac{1}{8}\tan\frac{x}{8} + \ldots,$$

due to Euler.

The discovery of logarithms

One great invention made early in the seventeenth century must be specially referred to; that of logarithms by John Napier (1550—1617). The special importance of this invention in relation to our subject is due to the fact of that essential connection between the numbers π and e which, after its discovery in the eighteenth century, dominated the later theory of the number π. The first announcement of the discovery was made in Napier's *Mirifici logarithmorum canonis descriptio* (Edinburgh, 1614), which contains an account of the nature of logarithms, and a table giving natural sines and their logarithms for every minute of the quadrant to seven or eight figures. These logarithms are not what are now called Napierian or natural logarithms (*i.e.* logarithms to the base e), although the former are closely related with the latter. The connection between the two is

$$L = 10^7 \log_e 10^7 - 10^7 . l, \text{ or } e^l = 10^7 e^{-\frac{L}{10^7}},$$

where l denotes the logarithm to the base e, and L denotes Napier's logarithm. It should be observed that in Napier's original theory of logarithms, their connection with the number e did not explicitly appear. The logarithm was not defined as the inverse of an exponential function; indeed the exponential function and even the exponential notation were not generally used by mathematicians till long afterwards.

Approximate constructions

A large number of approximate constructions for the rectification and quadrature of the circle have been given, some of which give very close approximations. It will suffice to give here a few examples of such constructions.

(1) The following construction for the approximate rectification of the circle was given by Kochansky (*Acta Eruditorum*, 1685).

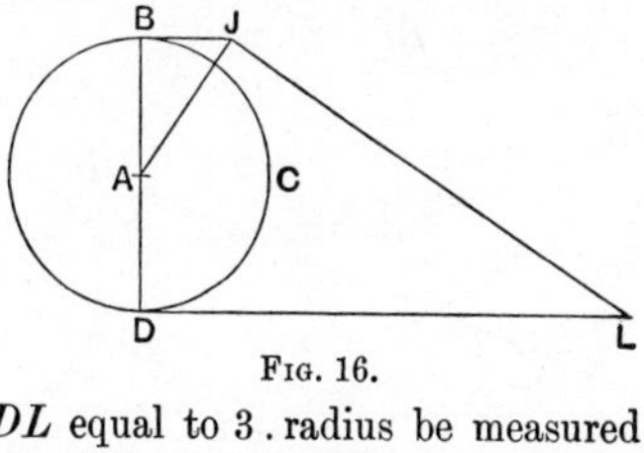

FIG. 16.

Let a length DL equal to 3 . radius be measured off on a tangent to the circle; let DAB be the diameter perpendicular to DL.

Let J be on the tangent at B, and such that $\angle BAJ = 30^\circ$. Then JL is approximately equal to the semi-circular arc BCD. Taking the radius as unity, it can easily be proved that

$$JL = \sqrt{\frac{40}{3} - \sqrt{12}} = 3{\cdot}141533 \ldots$$

the correct value to four places of decimals.

(2) The value $\frac{355}{113} = 3{\cdot}141592 \ldots$ is correct to six decimal places. Since $\dfrac{355}{113} = 3 + \dfrac{4^2}{7^2 + 8^2}$, it can easily be constructed.

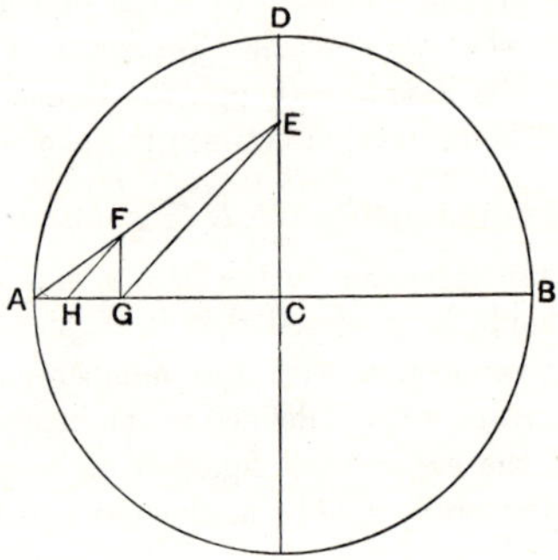

Fig. 17.

Let $CD = 1$, $CE = \frac{7}{8}$, $AF = \frac{1}{2}$; and let FG be parallel to CD and FH to EG; then $AH = \dfrac{4^2}{7^2 + 8^2}$.

This construction was given by Jakob de Gelder (*Grünert's Archiv*, vol. 7, 1849).

(3) At A make $AB = (2 + \frac{1}{5})$ radius on the tangent at A and let

$$BC = \tfrac{1}{5}\,.\text{ radius.}$$

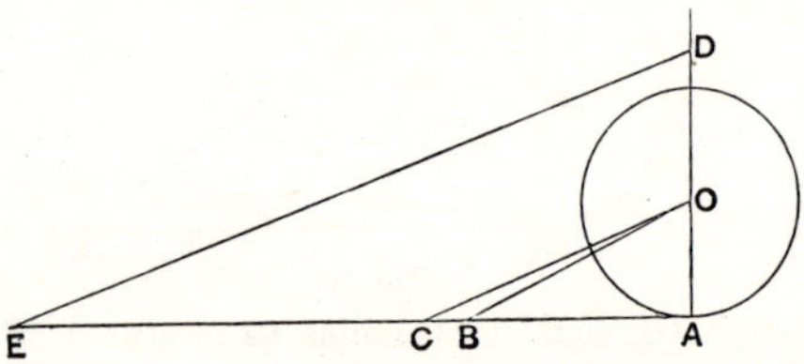

Fig. 18.

On the diameter through A take $AD=OB$, and draw DE parallel to OC. Then

$$\frac{AE}{AD}=\frac{AC}{AO}=\frac{13}{5}\text{; therefore } AE=r\,.\,\frac{13}{5}\sqrt{1+\left(\frac{11}{5}\right)^2}=r\,.\,\frac{13}{25}\sqrt{146}\,;$$

thus $AE=r\,.\,6{\cdot}2831839\ldots$, so that AE is less than the circumference of the circle by less than two millionths of the radius.

The rectangle with sides equal to AE and half the radius r has very approximately its area equal to that of the circle. This construction was given by Specht (*Crelle's Journal*, vol. 3, p. 83).

(4) Let AOB be the diameter of a given circle. Let

$$OD=\tfrac{3}{5}r,\quad OF=\tfrac{3}{2}r,\quad OE=\tfrac{1}{2}r.$$

Describe the semi-circles DGE, AHF with DE and AF as diameters; and let the perpendicular to AB through O cut them in G and H respectively. The square of which the side is GH is approximately of area equal to that of the circle.

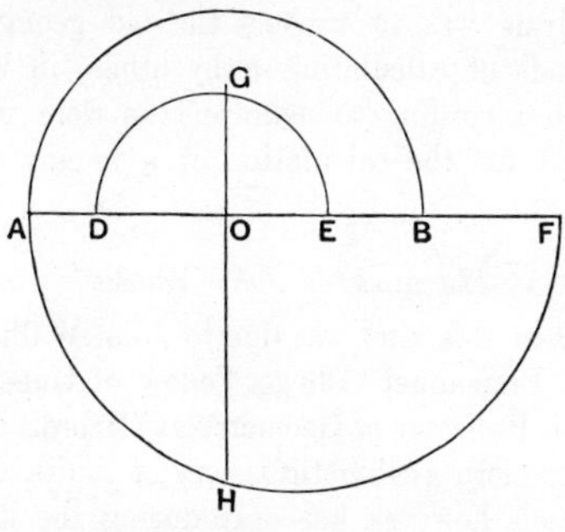

FIG. 19.

We find that $GH=r\,.\,1{\cdot}77246\ldots$, and since $\sqrt{\pi}=1{\cdot}77245$ we see that GH is greater than the side of the square whose area is equal to that of the circle by less than two hundred thousandths of the radius.

CHAPTER III

THE SECOND PERIOD

The new Analysis

THE foundations of the new Analysis were laid in the second half of the seventeenth century when Newton (1642—1727) and Leibnitz (1646—1716) founded the Differential and Integral Calculus, the ground having been to some extent prepared by the labours of Huyghens, Fermat, Wallis, and others. By this great invention of Newton and Leibnitz, and with the help of the brothers James Bernouilli (1654—1705) and John Bernouilli (1667—1748), the ideas and methods of Mathematicians underwent a radical transformation which naturally had a profound effect upon our problem. The first effect of the new Analysis was to replace the old geometrical or semi-geometrical methods of calculating π by others in which analytical expressions formed according to definite laws were used, and which could be employed for the calculation of π to any assigned degree of approximation.

The work of John Wallis

The first result of this kind was due to John Wallis (1616—1703), Undergraduate at Emmanuel College, Fellow of Queens' College, and afterwards Savilian Professor of Geometry at Oxford. He was the first to formulate the modern arithmetic theory of limits, the fundamental importance of which, however, has only during the last half century received its due recognition; it is now regarded as lying at the very foundation of Analysis. Wallis gave in his *Arithmetica Infinitorum* the expression

$$\frac{\pi}{2}=\frac{2}{1}\cdot\frac{2}{3}\cdot\frac{4}{3}\cdot\frac{4}{5}\cdot\frac{6}{5}\cdot\frac{6}{7}\cdot\frac{8}{7}\cdot\frac{8}{9}\cdots$$

for π as an infinite product, and he shewed that the approximation obtained by stopping at any fraction in the expression on the right is in defect or in excess of the value $\frac{\pi}{2}$ according as the fraction is proper or improper. This expression was obtained by an ingenious method depending upon the expression for $\frac{\pi}{8}$ the area of a semi-circle of diameter 1 as the definite integral $\int_0^1 \sqrt{x-x^2}\,dx$. The expression has the advantage over that of Vieta that the operations required by it are all rational ones.

Lord Brouncker (1620—1684), the first President of the Royal Society, communicated without proof to Wallis the expression

$$\frac{4}{\pi}=1+\frac{1}{2+}\frac{9}{2+}\frac{25}{2+}\frac{49}{2+}\cdots,$$

a proof of which was given by Wallis in his *Arithmetica Infinitorum*. It was afterwards shewn by Euler that Wallis' formula could be obtained from the development of the sine and cosine in infinite products, and that Brouncker's expression is a particular case of much more general theorems.

The calculation of π by series

The expression from which most of the practical methods of calculating π have been obtained is the series which, as we now write it, is given by

$$\tan^{-1}x=x-\tfrac{1}{3}x^3+\tfrac{1}{5}x^5-\dots \qquad (-1\leqq x\leqq 1).$$

This series was discovered by Gregory (1670) and afterwards independently by Leibnitz (1673). In Gregory's time the series was written as

$$a=t-\frac{t^3}{3r^2}+\frac{t^5}{5r^4}-\dots,$$

where a, t, r denote the length of an arc, the length of a tangent at one extremity of the arc, and the radius of the circle; the definition of the tangent as a ratio had not yet been introduced.

The particular case

$$\frac{\pi}{4}=1-\frac{1}{3}+\frac{1}{5}-\dots$$

is known as Leibnitz's series; he discovered it in 1674 and published it in 1682, with investigations relating to the representation of π, in his work "De vera proportione circuli ad quadratum circumscriptum in numeris rationalibus." The series was, however, known previously to Newton and Gregory.

By substituting the values $\frac{\pi}{6}, \frac{\pi}{8}, \frac{\pi}{10}, \frac{\pi}{12}$ in Gregory's series, the calculation of π up to 72 places was carried out by Abraham Sharp under instructions from Halley (Sherwin's *Mathematical Tables*, 1705, 1706).

The more quickly convergent series

$$\sin^{-1}x = x + \frac{1}{2}\frac{x^3}{3} + \frac{1\,.\,3}{2\,.\,4}\frac{x^5}{5} + \dots,$$

discovered by Newton, is troublesome for purposes of calculation, owing to the form of the coefficients. By taking $x = \frac{1}{2}$, Newton himself calculated π to 14 places of decimals.

Euler and others occupied themselves in deducing from Gregory's series formulae by which π could be calculated by means of rapidly converging series.

Euler, in 1737, employed special cases of the formula

$$\tan^{-1}\frac{1}{p} = \tan^{-1}\frac{1}{p+q} + \tan^{-1}\frac{q}{p^2+pq+1},$$

and gave the general expression

$$\tan^{-1}\frac{x}{y} = \tan^{-1}\frac{ax-y}{ay+x} + \tan^{-1}\frac{b-a}{ab+1} + \tan^{-1}\frac{c-b}{cb+1} + \dots,$$

from which more such formulae could be obtained. As an example, we have, if $a, b, c, \dots$ are taken to be the uneven numbers, and $\frac{x}{y} = 1$,

$$\frac{\pi}{4} = \tan^{-1}\frac{1}{2} + \tan^{-1}\frac{1}{2\,.\,4} + \tan^{-1}\frac{1}{2\,.\,9} + \dots.$$

In the year 1706, Machin (1680—1752), Professor of Astronomy in London, employed the series

$$\frac{\pi}{4} = 4\left(\frac{1}{5} - \frac{1}{3\,.\,5^3} + \frac{1}{5\,.\,5^5} - \frac{1}{7\,.\,5^7} + \dots\right)$$
$$- \left(\frac{1}{239} - \frac{1}{3\,.\,239^3} + \frac{1}{5\,.\,239^5} - \frac{1}{7\,.\,239^7} + \dots\right),$$

which follows from the relation

$$\frac{\pi}{4} = 4\tan^{-1}\frac{1}{5} - \tan^{-1}\frac{1}{239},$$

to calculate π to 100 places of decimals. This is a very convenient expression, because in the first series $\frac{1}{5}, \frac{1}{5^3}, \dots$ can be replaced by $\frac{4}{100}, \frac{64}{1000000}$, &c., and the second series is very rapidly convergent.

In 1719, de Lagny (1660—1734), of Paris, determined in two different ways the value of π up to 127 decimal places. Vega (1754—1802) calculated π to 140 places, by means of the formulae

$$\frac{\pi}{4} = 5\tan^{-1}\frac{1}{7} + 2\tan^{-1}\frac{3}{79} = 2\tan^{-1}\frac{1}{3} + \tan^{-1}\frac{1}{7},$$

due to Euler, and shewed that de Lagny's determination was correct with the exception of the 113th place, which should be 8 instead of 7.

Clausen calculated in 1847, 248 places of decimals by the use of Machin's formula and the formula

$$\frac{\pi}{4} = 2\tan^{-1}\frac{1}{3} + \tan^{-1}\frac{1}{7}.$$

In 1841, 208 places, of which 152 are correct, were calculated by Rutherford by means of the formula

$$\frac{\pi}{4} = 4\tan^{-1}\frac{1}{5} - \tan^{-1}\frac{1}{70} + \tan^{-1}\frac{1}{99}.$$

In 1844 an expert reckoner, Zacharias Dase, employed the formula

$$\frac{\pi}{4} = \tan^{-1}\frac{1}{2} + \tan^{-1}\frac{1}{5} + \tan^{-1}\frac{1}{8},$$

supplied to him by Prof. Schultz, of Vienna, to calculate π to 200 places of decimals, a feat which he performed in two months.

In 1853 Rutherford gave 440 places of decimals, and in the same year W. Shanks gave first 530 and then 607 places (*Proc. R. S.*, 1853).

Richter, working independently, gave in 1853 and 1855, first 333, then 400 and finally 500 places.

Finally, W. Shanks, working with Machin's formula, gave (1873—74) 707 places of decimals.

Another series which has also been employed for the calculation of π is the series

$$\tan^{-1} t = \frac{t}{1+t^2}\left\{1 + \frac{2}{3}\frac{t^2}{1+t^2} + \frac{2 \cdot 4}{3 \cdot 5}\left(\frac{t^2}{1+t^2}\right)^2 + \frac{2 \cdot 4 \cdot 6}{3 \cdot 5 \cdot 7}\left(\frac{t^2}{1+t^2}\right)^3 + \dots\right\}.$$

This was given in the year 1755 by Euler, who, applying it in the formula

$$\pi = 20\tan^{-1}\tfrac{1}{7} + 8\tan^{-1}\tfrac{3}{79},$$

calculated π to 20 places, in one hour as he states. The same series was also discovered independently by Ch. Hutton (*Phil. Trans.*, 1776). It was later rediscovered by J. Thomson and by De Morgan.

An expression for π given by Euler may here be noticed; taking the identity

$$\tan^{-1}\frac{x}{2-x} = 2\int_0^x \frac{dx}{4+x^4} + 2\int_0^x \frac{x\,dx}{4+x^4} + \int_0^x \frac{x^2\,dx}{4+x^4},$$

he developed the integrals in series, then put $x = \frac{1}{2}$, $x = \frac{1}{4}$, obtaining series for $\tan^{-1}\frac{1}{3}$, $\tan^{-1}\frac{1}{7}$, which he substituted in the formula

$$\frac{\pi}{4} = 2\tan^{-1}\frac{1}{3} + \tan^{-1}\frac{1}{7}.$$

In China a work was published by Imperial order in 1713 which contained a chapter on the quadrature of the circle where the first 19 figures in the value of π are given.

At the beginning of the eighteenth century, analytical methods were introduced into China by Tu Tê-mei (Pierre Jartoux) a French missionary; it is, however, not known how much of his work is original, or whether he borrowed the formulae he gave directly from European Mathematicians.

One of his series

$$\pi = 3\left(1 + \frac{1^2}{4.6} + \frac{1^2.3^2}{4.6.8.10} + \frac{1^2.3^2.5^2}{4.6.8.10.12.14} + \dots\right)$$

was employed at the beginning of the nineteenth century by Chu-Hung for the calculation of π. By this means 25 correct figures were obtained.

Tsêng Chi-hung, who died in 1877, published values of π and $1/\pi$ to 100 places. He is said to have obtained his value of π in a month, by means of the formula

$$\frac{\pi}{4} = \tan^{-1}\frac{1}{2} + \tan^{-1}\frac{1}{3}$$

and Gregory's series.

In Japan, where a considerable school of Mathematics was developed in the eighteenth century, π was calculated by Takebe in 1722 to 41 places, by employment of the regular 1024agon. It was calculated by Matsunaga in 1739 to 50 places by means of the same series as had been employed by Chu-Hung.

The rational values $\pi = \frac{5419351}{1725033}$, $\pi = \frac{428224593349304}{136308121570117}$, correct to 12 and 30 decimal places respectively, were given by Arima in 1766.

Kurushima Yoshita (died 1757) gave for π^2 the approximate values $\frac{227}{13}$, $\frac{10748}{1089}$, $\frac{10975}{1112}$, $\frac{98548}{9885}$.

Tanyem Shŏkei published in 1728 the series

$$\pi^2 = 8\left(1 + \frac{1}{6} + \frac{1.4}{6.15} + \frac{1.4.9}{6.15.28} + \dots\right),$$

$$\pi^2 = 4\left(1 + \frac{2}{6} + \frac{2.8}{6.15} + \frac{2.8.18}{6.15.28} + \dots\right)$$

due to Takebe, and ultimately to Jartoux.

The following series published in 1739 by Matsunaga may be mentioned:

$$\pi^2 = 9\left(1 + \frac{1^2}{3.4} + \frac{1^2.2^2}{3.4.5.6} + \frac{1^2.2^2.3^2}{3.4.5.6.7.8} + \dots\right).$$

The work of Euler

Developments of the most far-reaching importance in connection with our subject were made by Leonhard Euler, one of the greatest Analysts of all time, who was born at Basel in 1707 and died at St Petersburg in 1783. With his vast influence on the development of Mathematical Analysis in general it is impossible here to deal, but some account must be given of those of his discoveries which come into relation with our problem.

The very form of modern Trigonometry is due to Euler. He introduced the practice of denoting each of the sides and angles of a triangle by a single letter, and he introduced the short designation of the trigonometrical ratios by $\sin \alpha$, $\cos \alpha$, $\tan \alpha$, &c. Before Euler's time there was great prolixity in the statement of propositions, owing to the custom of denoting these expressions by words, or by letters specially introduced in the statement. The habit of denoting the ratio of the circumference to the diameter of a circle by the letter π, and the base of the natural system of logarithms by e, is due to the influence of the works of Euler, although the notation π appears as early as 1706, when it was used by William Jones in the *Synopsis palmariorum Matheseos*. In Euler's earlier work he frequently used p instead of π, but by about 1740 the letter π was used not only by Euler but by other Mathematicians with whom he was in correspondence.

A most important improvement which had a great effect not only upon the form of Trigonometry but also on Analysis in general was the introduction by Euler of the definition of the trigonometrical ratios in order to replace the old sine, cosine, tangent, &c., which were the lengths of straight lines connected with the circular arc. Thus these trigonometrical ratios became functions of an angular magnitude, and therefore numbers, instead of lengths of lines related by equations with the radius of the circle. This very important improvement was not generally introduced into our text books until the latter half of the nineteenth century.

This mode of regarding the trigonometrical ratios as analytic functions led Euler to one of his greatest discoveries, the connection of these functions with the exponential function. On the basis of the definition of e^z by means of the series

$$1 + z + \frac{z^2}{2\,!} + \frac{z^3}{3\,!} + \dots,$$

he set up the relations

$$\cos x = \frac{e^{ix} + e^{-ix}}{2}, \quad i \sin x = \frac{e^{ix} - e^{-ix}}{2},$$

which can also be written

$$e^{ix} = \cos x + i \sin x, \quad e^{-ix} = \cos x - i \sin x.$$

The relation $e^{i\pi} = -1$, which Euler obtained by putting $x = \pi$, is the fundamental relation between the two numbers π and e which was indispensable later on in making out the true nature of the number π.

In his very numerous memoirs, and especially in his great work, *Introductio in analysin infinitorum* (1748), Euler displayed the most wonderful skill in obtaining a rich harvest of results of great interest, largely dependent on his theory of the exponential function. Hardly any other work in the history of Mathematical Science gives to the reader so strong an impression of the genius of the author as the *Introductio*. Many of the results given in that work are obtained by bold generalizations, in default of proofs which would now be regarded as completely rigorous; but this it has in common with a large part of all Mathematical discoveries, which are often due to a species of divining intuition, the rigorous demonstrations and the necessary restrictions coming later. In particular there may be mentioned the expressions for the sine and cosine functions as infinite products, and a great number of series and products deduced from these expressions; also a number of expressions relating the number e with continued fractions which were afterwards used in connection with the investigation of the nature of that number.

Great as the progress thus made was, regarded as preparatory to a [illegible]lution of our problem, nothing definite as to the true nature of the number π was as yet established, although Mathematicians were convinced that e and π are not roots of algebraic equations. Euler himself gave expression to the conviction that this is the case. Somewhat later, Legendre gave even more distinct expression to this view in his *Éléments de Géométrie* (1794), where he writes: "It is probable that the number π is not even contained among the algebraical irrationalities, *i.e.* that it cannot be a root of an algebraical equation with a finite number of terms, whose coefficients are rational. But it seems to be very difficult to prove this strictly."

CHAPTER IV

THE THIRD PERIOD

The irrationality of π and e

THE third and final period in the history of the problem is concerned with the investigation of the real nature of the number π. Owing to the close connection of this number with the number e, the base of natural logarithms, the investigation of the nature of the two numbers was to a large extent carried out at the same time.

The first investigation, of fundamental importance, was that of J. H. Lambert (1728—1777), who in his "Mémoire sur quelques propriétés remarquables des quantités transcendentes circulaires et logarithmiques" (*Hist. de l'Acad. de Berlin,* 1761, printed in 1768), proved that e and π are irrational numbers. His investigations are given also in his treatise *Vorläufige Kenntnisse für die, so die Quadratur und Rektification des Zirkels suchen*, published in 1766.

He obtained the two continued fractions

$$\frac{e^x-1}{e^x+1}=\frac{1}{2/x+}\,\frac{1}{6/x+}\,\frac{1}{10/x+}\,\frac{1}{14/x+\ldots},$$

$$\tan x=\frac{1}{1/x-}\,\frac{1}{3/x-}\,\frac{1}{5/x-}\,\frac{1}{7/x-\ldots},$$

which are closely related with continued fractions obtained by Euler, but the convergence of which Euler had not established. As the result of an investigation of the properties of these continued fractions, Lambert established the following theorems :

(1) If x is a rational number, different from zero, e^x cannot be a rational number.

(2) If x is a rational number, different from zero, $\tan x$ cannot be a rational number.

If $x=\frac{1}{4}\pi$, we have $\tan x=1$, and therefore $\frac{1}{4}\pi$, or π, cannot be a rational number.

It has frequently been stated that the first rigorous proof of Lambert's results is due to Legendre (1752—1833), who proved these theorems in his *Éléments de Géométrie* (1794), by the same method, and added a proof that π^2 is an irrational number. The essential rigour of Lambert's proof has however been pointed out by Pringsheim (*Münch. Akad. Ber.*, Kl. 28, 1898), who has supplemented the investigation in respect of the convergence.

A proof of the irrationality of π and π^2 due to Hermite (*Crelle's Journal*, vol. 76, 1873) is of interest, both in relation to the proof of Lambert, and as containing the germ of the later proof of the transcendency of e and π.

A simple proof of the irrationality of e was given by Fourier (Stainville, *Mélanges d'analyse*, 1815), by means of the series

$$1+\frac{1}{1!}+\frac{1}{2!}+\frac{1}{3!}+\dots$$

which represents the number. This proof can be extended to shew that e^2 is also irrational. On the same lines it was proved by Liouville (1809—1882) (*Liouville's Journal*, vol. 5, 1840) that neither e nor e^2 can be a root of a quadratic equation with rational coefficients. This last theorem is of importance as forming the first step in the proof that e and π cannot be roots of any algebraic equation with rational coefficients. The probability had been already recognized by Legendre that there exist numbers which have this property.

Existence of transcendental numbers

The confirmation of this surmised existence of such numbers was obtained by Liouville in 1840, who by an investigation of the properties of the convergents of a continued fraction which represents a root of an algebraical equation, and also by another method, proved that numbers can be defined which cannot be the root of any algebraical equation with rational coefficients.

The simpler of Liouville's methods of proving the existence of such numbers will be here given.

Let x be a real root of the algebraic equation

$$ax^n+bx^{n-1}+cx^{n-2}+\dots=0,$$

with coefficients which are all positive or negative integers. We shall assume that this equation has all its roots unequal; if it had equal roots we might suppose it to be cleared of them in the usual manner.

Let the other roots be denoted by $x_1, x_2, \dots x_{n-1}$; these may be real or complex. If $\frac{p}{q}$ be any rational fraction, we have

$$\frac{p}{q} - x = \frac{ap^n + bp^{n-1}q + cp^{n-2}q^2 + \dots}{q^n \cdot a\left(\frac{p}{q} - x_1\right)\left(\frac{p}{q} - x_2\right)\dots\left(\frac{p}{q} - x_n\right)}.$$

If now we have a sequence of rational fractions converging to the value x as limit, but none of them equal to x, and if $\frac{p}{q}$ be one of these fractions,

$$\left(\frac{p}{q} - x_1\right)\left(\frac{p}{q} - x_2\right) \dots \left(\frac{p}{q} - x_n\right)$$

approximates to the fixed number

$$(x - x_1)(x - x_2)\dots(x - x_n).$$

We may therefore suppose that for all the fractions $\frac{p}{q}$,

$$a\left(\frac{p}{q} - x_1\right)\left(\frac{p}{q} - x_2\right) \dots \left(\frac{p}{q} - x_n\right)$$

is numerically less than some fixed positive number A. Also

$$ap^n + bp^{n-1}q + \dots$$

is an integer numerically $\geqq 1$; therefore

$$\left|\frac{p}{q} - x\right| > \frac{1}{Aq^n}.$$

This must hold for all the fractions $\frac{p}{q}$ of such a sequence, from and after some fixed element of the sequence, for some fixed number A. If now a number x can be so defined such that, however far we go in the sequence of fractions $\frac{p}{q}$, and however A be chosen, there exist fractions belonging to the sequence for which $\left|\frac{p}{q} - x\right| < \frac{1}{Aq^n}$, it may be concluded that x cannot be a root of an equation of degree n with integral coefficients. Moreover, if we can shew that this is the case whatever value n may have, we conclude that x cannot be a root of any algebraic equation with rational coefficients.

Consider a number

$$x = \frac{k_1}{r^{1!}} + \frac{k_2}{r^{2!}} + \dots + \frac{k_m}{r^{m!}} + \dots,$$

where the integers $k_1, k_2, \ldots k_m, \ldots$ are all less than the integer r, and do not all vanish from and after a fixed value of m.

Let
$$\frac{p}{q} = \frac{k_1}{r^{1!}} + \frac{k_2}{r^{2!}} + \ldots + \frac{k_m}{r^{m!}},$$

then $\frac{p}{q}$ continually approaches x as m is increased. We have

$$x - \frac{p}{q} = \frac{k_{m+1}}{r^{(m+1)!}} + \frac{k_{m+2}}{r^{(m+2)!}} + \ldots$$

$$< r\left(\frac{1}{r^{(m+1)!}} + \frac{1}{r^{(m+2)!}} + \ldots\right)$$

$$< \frac{2r}{q^{m+1}}, \text{ since } q = r^{m!}.$$

It is clear that, whatever values A and n may have, if m, and therefore q, is large enough, we have $\frac{2r}{q^{m+1}} < \frac{1}{Aq^n}$; and thus the relation $\left|\frac{p}{q} - x\right| > \frac{1}{Aq^n}$ is not satisfied for all the fractions $\frac{p}{q}$. The numbers x so defined are therefore transcendental. If we take $r = 10$, we see how to define transcendental numbers that are expressed as decimals.

This important result provided a complete justification of the division of numbers into two classes, algebraical numbers, and transcendental numbers; the latter being characterized by the property that such a number cannot be a root of an algebraical equation of any degree whatever, of which the coefficients are rational numbers.

A proof of this fundamentally important distinction, depending on entirely different principles, was given by G. Cantor (*Crelle's Journal*, vol. 77, 1874) who shewed that the algebraical numbers form an enumerable aggregate, that is to say that they are capable of being counted by means of the integer sequence 1, 2, 3, ..., whereas the aggregate of all real numbers is not enumerable. He shewed how numbers can be defined which certainly do not belong to the sequence of algebraic numbers, and are therefore transcendental.

This distinction between algebraic and transcendental numbers being recognized, the question now arose, as regards any particular number defined in an analytical manner, to which of the two classes it belongs; in particular whether π and e are algebraic or transcendental. The difficulty of answering such a question arises from the fact that the recognition of the distinction between the two classes of numbers does not of itself provide a readily applicable criterion by the use of which the question may be answered in respect of a particular number.

The scope of Euclidean determinations

Before proceeding to describe the manner in which it was finally shewn that the number π is a transcendental number, it is desirable to explain in what way this result is connected with the problems of the quadrature and rectification of the circle by means of Euclidean determinations.

The development of Analytical Geometry has made it possible to replace every geometrical problem by a corresponding analytical one which involves only numbers and their relations. As we have already remarked, every Euclidean problem of what is called construction consists essentially in the determination of one or more points which shall satisfy certain prescribed relations with regard to a certain finite number of assigned points, the data of the problem. Such a problem has as its analytical counterpart the determination of a number, or a finite set of numbers, which shall satisfy certain prescribed relations relatively to a given set of numbers. The determination of the required numbers is always made by means of a set of algebraical equations.

The development of the theory of algebraical equations, especially that due to Abel, Gauss, and Galois, led the Mathematicians of the last century to scrutinize with care the limits of the possibility of solving geometrical problems subject to prescribed limitations as to the nature of the geometrical operations regarded as admissible. In particular, it has been ascertained what classes of geometrical problems are capable of solution when operations equivalent in practical geometry to the use of certain instruments are admitted*. The investigations have led to the discovery of cases such as that of inscribing a regular polygon of 17 sides in a circle, in which a problem, not previously known to be capable of solution by Euclidean means, has been shewn to be so.

We shall here give an account of as much of the theory of this subject as is necessary for the purpose of application to the theory of the quadrature and rectification of the circle.

In the first place we observe that, having given two or more points in a plane, a Cartesian set of axes can be constructed by means of a

* An interesting detailed account of investigations of this kind will be found in Enriques' *Questions of Elementary Geometry*, German Edition, 1907.

Euclidean construction, for example by bisecting the segment of the line on which two of the given points are incident, and then determining a perpendicular to that segment. We may therefore assume that a given set of points, the data of a Euclidean problem, are specified by means of a set of numbers, the coordinates of these points.

The determination of a required point P is, in a Euclidean problem, made by means of a finite number of applications of the three processes, (1) of determining a new point as the intersection of straight lines given each by a pair of points already determined, (2) of determining a new point as an intersection of a straight line given by two points and a circle given by its centre and one point on the circumference, all four points having been already determined, and (3) of determining a new point as an intersection of two circles which are determined by four points already determined.

In the analytical interpretation we have an original set of numbers $a_1, a_2, \dots a_{2r}$ given, the coordinates of the r given points; $(r \geqq 2)$. At each successive stage of the geometrical process we determine two new numbers, the coordinates of a fresh point.

When a certain stage of the process has been completed, the data for the next step consist of numbers $(a_1, a_2, \dots a_{2n})$ containing the original data and those numbers which have been already ascertained by the successive stages of the process already carried out.

If (1) is employed for the next step of the geometrical process, the new point determined by that step corresponds to numbers determined by two equations

$$Ax + By + C = 0, \quad A'x + B'y + C' = 0,$$

where A, B, C, A', B', C' are rational functions of eight of the numbers $(a_1, a_2, \dots a_{2n})$. Therefore x, y the coordinates of the new point determined by this step are rational functions of $a_1, a_2, \dots a_{2n}$.

In order to get the data for the next step afterwards, we have only to add to $a_1, a_2, \dots a_{2n}$ these two rational functions of eight of them.

If case (2) is employed, the next point is determined by two equations of the form

$$(x - a_p)^2 + (y - a_q)^2 = (a_r - a_s)^2 + (a_t - a_u)^2,$$
$$y = mx + n,$$

where m, n are rational functions of four of the numbers $a_1, a_2, \dots a_{2n}$. On elimination of y, we have a quadratic equation for x; and thus x is determined as a quadratic irrational function of $(a_1, a_2, \dots a_{2n})$, of the form $A \pm \sqrt{B}$, where A and B are rational functions; it is clear that y will be determined in a similar way.

If, in the new step, (3) is employed, the equations for determining (x, y) consist of two equations of the form

$$(x-a_p)^2+(y-a_q)^2=(a_r-a_s)^2+(a_t-a_u)^2;$$

on subtracting these equations, we obtain a linear equation, and thus it is clear that this case is essentially similar to that in which (2) is employed, so far as the form of x, y is concerned.

Since the determination of a required point P is to be made by a finite number of such steps, we see that the coordinates of P are determined by means of a finite succession of operations on

$$(a_1, a_2, \dots a_{2r}),$$

the coordinates of the points; each of these operations consists either of a rational operation, or of one involving the process of taking a square root of a rational function as well as a rational operation.

We have now established the following result:

In order that a point P can be determined by the Euclidean mode it is necessary and sufficient that its coordinates can be expressed as such functions of the coordinates $(a_1, a_2, \dots a_{2r})$ of the given points of the problem, as involve the successive performance, a finite number of times, of operations which are either rational or involve taking a square root of a rational function of the elements already determined.

That the condition stated in this theorem is necessary has been proved above; that it is sufficient is seen from the fact that a single rational operation, and the single operation of taking a square root of a number already known, are both operations which correspond to possible Euclidean determinations.

The condition stated in the result just obtained may be put in another form more immediately available for application. The expression for a coordinate x of the point P may, by the ordinary processes for the simplification of surd expressions, by getting rid of surds from the denominators of fractions, be reduced to the form

$$x=a+b\sqrt{c_1\pm\sqrt{c_2\pm\sqrt{c_3+\dots}}}+b'\sqrt{c_1'\pm\sqrt{c_2'\pm\sqrt{c_3'+\dots}}}+\dots,$$

where all the numbers

$$a, b, c_1, c_2, \dots, b', c_1', c_2', \dots$$

are rational functions of the given numbers $(a_1, a_2, \dots a_{2r})$, and the number of successive square roots is in every term finite. Let m be the greatest number of successive square roots in any term of x; this may be called the rank of x. We may then write

$$x = a + b\sqrt{B} + b'\sqrt{B'} + \dots,$$

where $B, B', \dots$ are all of rank not greater than $m-1$. We can form an equation which x satisfies, and such that all its coefficients are rational functions of $a, b, b', B, B' \dots$; for $\sqrt{B}$ may be eliminated by taking $(x-a-b'\sqrt{B'}-\dots)^2 = b^2B$, and this is of the form

$$P_2 + \sqrt{B'}P_2' = 0,$$

from which we form the biquadratic

$$P_2^2 - B'P_2'^2 = 0,$$

in which $\sqrt{B'}$ does not occur. Proceeding in this way we obtain an equation in x of degree some power of 2, and of which the coefficients are rational functions of $a, b, B, B', \dots$, and are therefore of rank $\leqq m-1$. This equation is of the form

$$L_1x^{2^s} + L_2x^{2^s-1} + \dots = 0,$$

where $L_1, L_2, \dots$ are at most of rank $m-1$. If $L_1, L_2, \dots$ involve a radical $\sqrt{K}$, the equation is of the form

$$\sqrt{K}(b_1x^{2^s} + \dots) + (b_1'x^{2^s} + \dots) = 0,$$

and we can as before reduce this to an equation of degree 2^{s+1} in which $\sqrt{K}$ does not occur; by repeating the process for each radical like $\sqrt{K}$, we may eliminate them all, and finally obtain an equation such that the rank of every coefficient is $\leqq m-2$. By continual repetition of this procedure we ultimately reach an equation, such that the coefficients are all of rank zero, *i.e.* rational functions of $(a_1, a_2, \dots a_{2r})$. We now see that the following result has been established:

In order that a point P may be determinable by Euclidean procedure it is necessary that each of its coordinates be a root of an equation of some degree, a power of 2, *of which the coefficients are rational functions of* $(a_1, a_2, \dots a_{2r})$, *the coordinates of the points given in the data of the problem.*

From our investigation it is clear that only those algebraic equations which are obtainable by elimination from a sequence of linear and quadratic equations correspond to possible Euclidean problems.

The quadratic equations must consist of sets, those in the first set having coefficients which are rational functions of the given numbers, those in the second set having coefficients of rank at most 1; in the next set the coefficients have rank at most 2, and so on.

The criterion thus obtained is sufficient, whenever it can be applied, to determine whether a proposed Euclidean problem is a possible one or not.

In the case of the rectification of the circle, we may assume that the data of the problem consist simply of the two points (0, 0) and (1, 0), and that the point to be determined has the coordinates (π, 0). This will, in accordance with the criterion obtained, be a possible problem only if π is a root of an algebraic equation with rational coefficients, of that special class which has roots expressible by means of rational numbers and numbers obtainable by successive operations of taking the square roots. The investigations of Abel have shewn that this is only a special class of algebraic equations.

As we shall see, it is now known that π, being transcendental, is not a root of any algebraic equation at all, and therefore in accordance with the criterion is not determinable by Euclidean construction. The problems of duplication of the cube, and of the trisection of an angle, although they lead to algebraic equations, are not soluble by Euclidean constructions, because the equations to which they lead are not in general of the class referred to in the above criterion.

The transcendence of π

In 1873 Ch. Hermite* succeeded in proving that the number e is transcendental, that is that no equation of the form

$$ae^m + be^n + ce^r + \ldots = 0$$

can subsist, where $m, n, r, \ldots a, b, c, \ldots$ are whole numbers. In 1882, the more general theorem was stated by Lindemann that such an equation cannot hold, when $m, n, r, \ldots a, b, c, \ldots$ are algebraic numbers, not necessarily real; and the particular case that $e^{ix} + 1 = 0$ cannot be satisfied by an algebraic number x, and therefore that π is not algebraic, was completely proved by Lindemann†.

Lindemann's general theorem may be stated in the following precise form:

If $x_1, x_2, \ldots x_n$ *are any real or complex algebraical numbers, all distinct, and* $p_1, p_2, \ldots p_n$ *are n algebraical numbers at least one of which is different from zero, then the sum*

$$p_1 e^{x_1} + p_2 e^{x_2} + \ldots + p_n e^{x_n}$$

is certainly different from zero.

* "Sur la fonction exponentielle," *Comptes Rendus*, vol. 77, 1873.

† *Ber. Akad. Berlin*, 1882.

The particular case of this theorem in which

$$n=2, \quad x_1=ix, \quad x_2=0, \quad p_1=p_2=1,$$

shews that $e^{ix}+1$ cannot be zero if x is an algebraic number, and thus that, since $e^{i\pi}+1=0$, it follows that *the number π is transcendental.*

From the general theorem there follow also the following important results:

(1) Let $n=2$, $p_1=1$, $p_2=-a$, $x_1=x$, $x_2=0$; then the equation $e^x-a=0$ cannot hold if x and a are both algebraic numbers and x is different from zero. Hence *the exponential e^x is transcendent if x is an algebraic number different from zero. In particular e is transcendent.* Further, *the natural logarithm of an algebraic number different from zero is a transcendental number. The transcendence of $i\pi$ and therefore of π is a particular case of this theorem.*

(2) Let $n=3$, $p_1=-i$, $p_2=i$, $p_3=-2a$, $x_1=ix$, $x_2=-ix$, $x_3=0$; it then follows that the equation $\sin x=a$ cannot be satisfied if a and x are both algebraic numbers different from zero. Hence, *if $\sin x$ is algebraic, x cannot be algebraic, unless $x=0$, and if a is algebraic, $\sin^{-1} a$ cannot be algebraic, unless $a=0$.*

It is easily seen that a similar theorem holds for the cosine and the other trigonometrical functions.

The fact that π is a transcendental number, combined with what has been established above as regards the possibility of Euclidean constructions or determinations with given data, affords the final answer to the question whether the quadrature or the rectification of the circle can be carried out in the Euclidean manner.

The quadrature and the rectification of a circle whose diameter is given are impossible, as problems to be solved by the processes of Euclidean Geometry, in which straight lines and circles are alone employed in the constructions.

It appears, however, that the transcendence of π establishes the fact that *the quadrature or the rectification of a circle whose diameter is given are impossible by a construction in which the use only of algebraic curves is allowed.*

The special case (2) of Lindemann's theorem throws light on the interesting problems of the rectification of arcs of circles and of the quadrature of sectors of circles. If we take the radius of a circle to be unity then $2\sin\frac{1}{2}x$ is the length of the chord of an arc of which the length is x. It has been shewn that $2\sin\frac{1}{2}x$ and x cannot both be algebraic, unless $x=0$. We have therefore the following result:

If the chord of a circle bears to the diameter a ratio which is algebraic, then the corresponding arc is not rectifiable by any construction in which algebraic curves alone are employed; neither can the quadrature of the corresponding sector of the circle be carried out by such a construction.

The method employed by Hermite and Lindemann was of a complicated character, involving the use of complex integration. The method was very considerably simplified by Weierstrass*, who gave a complete proof of Lindemann's general theorem.

Proofs of the transcendence of e and π, progressively simple in character, were given by Stieltjes†, Hilbert, Hurwitz and Gordan‡, Mertens§, and Vahlen‖.

All these proofs consist of a demonstration that an equation which is linear in a number of exponential functions, such that the coefficients are whole numbers, and the exponents algebraic numbers, is impossible. By choosing a multiplier of the equation of such a character that its employment reduces the given equation to the equation of the sum of a non-vanishing integer and a number proved to lie numerically between 0 and 1 to zero, the impossibility is established.

Simplified presentations of the proofs will be found in Weber's *Algebra*, in Enriques' *Questions of Elementary Geometry* (German Edition, 1907), in Hobson's *Plane Trigonometry* (second edition, 1911), and in Art. IX. of the "Monographs on Modern Mathematics," edited by J. W. A. Young.

Proof of the transcendence of π

The proof of the transcendence of π which will here be given is founded upon that of Gordan.

(1) Let us assume that, if possible, π is a root of an algebraical equation with integral coefficients; then $i\pi$ is also a root of such an equation.

Assume that $i\pi$ is a root of the equation

$$C(x-\alpha_1)(x-\alpha_2)\dots(x-\alpha_s)=0,$$

where all the coefficients

$$C,\ C\Sigma\alpha,\ C\Sigma\alpha_r\alpha_s,\ \dots,\ C\alpha_1\alpha_2\dots\alpha_s$$

* *Ber. Akad. Berlin*, 1885.

† *Comptes Rendus*, Paris Acad. 1890.

‡ These proofs are to be found in the *Math. Annalen*, vol. 43 (1896), by Hilbert, Hurwitz and Gordan.

§ *Wiener Ber.* Kl. cv. II*a* (1896).

‖ *Math. Annalen*, vol. 53 (1900).

are positive or negative integers (including zero); thus one of the numbers $\alpha_1, \dots \alpha_s$ is $i\pi$.

From Euler's equation $e^{i\pi}+1=0$, we see that the relation

$$(1+e^{\alpha_1})(1+e^{\alpha_2})\dots(1+e^{\alpha_s})=0$$

must hold, since one of the factors vanishes. If we multiply out the factors in this equation, it clearly takes the form

$$A+e^{\beta_1}+e^{\beta_2}+\dots+e^{\beta_n}=0,$$

where A is some positive integer ($\geqq 1$), being made up of 1 together with those terms, if any, which are of the form $e^{\alpha_p+\alpha_q+\cdots}$, where

$$\alpha_p+\alpha_q+\dots=0.$$

(2) A symmetrical function consisting of the sum of the products taken in every possible way, of a fixed number of the numbers $C\alpha_1, C\alpha_2, \dots C\alpha_s$, is an integer. It will be proved that the symmetrical functions of $C\beta_1, C\beta_2, \dots C\beta_n$ have the same property. In order to prove this we have need of the following lemma:

A symmetrical function consisting of the sums of the products taken p together of $\alpha+\beta+\gamma+\dots$ letters

$$x_1, x_2, \dots x_\alpha; \quad y_1, y_2, \dots y_\beta; \quad z_1, z_2, \dots z_\gamma; \text{ \&c.},$$

belonging to any number of separate sets, can be expressed in terms of symmetrical functions of the letters in the separate sets.

It will be sufficient to prove this in the case in which there are only two sets of letters, the extension to the general case being then obvious.

Denote by $\sum\limits_p P(x, y)$ the sum of the products which we require to express; and denote by $\sum\limits_r P(x)$ the sum of the products of r dimensions of the letters $x_1, x_2, \dots x_\alpha$ only. In case $p \leqq \alpha$, we see that

$$\sum_p P(x,y)=\sum_p P(x)+\sum_1 P(y)\sum_{p-1}P(x)+\sum_2 P(y)\sum_{p-2}P(x)+\dots;$$

in case $p>\alpha$, we see that

$$\sum_p P(x,y)$$
$$=\sum_\alpha P(x)\sum_{p-\alpha}P(y)+\sum_{\alpha-1}P(x)\sum_{p-\alpha+1}P(y)+\sum_{\alpha-2}P(x)\sum_{p-\alpha+2}P(y)+\dots;$$

and the terms on the right-hand side involve in each case only symmetrical functions of the letters of the two separate sets; thus the lemma is established.

To apply this lemma, we observe that the numbers β fall into separate sets, according to the way they are formed from the letters α.

The general value of β consists of the sum of r of the letters $a_1, a_2, \dots a_s$; and we consider those values of β that correspond to a fixed value of r to belong to one set. It is clear that a symmetrical function of those letters β which belong to one and the same set is expressible as a symmetrical function of $a_1, a_2, \dots a_s$; therefore a symmetrical function of the products $C\beta$, where all the β's belong to one and the same set, is in virtue of what has been established in (1) an integer. Applying the above lemma to all the n numbers $C\beta$, we see that the symmetrical products formed by all the numbers $C\beta$ are integral, or zero. We have supposed those of the numbers β which vanish to be suppressed and the corresponding exponentials to be absorbed in the integer A; whether this is done before or after the symmetrical functions of $C\beta$ are formed makes no difference, so that the above reasoning applies to the numbers $C\beta$ when those of them which vanish are removed.

(3) Let p be a prime number greater than all the numbers A, n, C $|C^n\beta_1\beta_2\dots\beta_n|$; and let

$$\phi(x)=\frac{x^{p-1}}{(p-1)!}C^{np+p-1}\{(x-\beta_1)(x-\beta_2)\dots(x-\beta_n)\}^p.$$

We observe that $\phi(x)$ is of the form

$$\frac{(Cx)^{p-1}}{(p-1)!}[(Cx)^n-q_1(Cx)^{n-1}+q_2(Cx)^{n-2}-\dots+(-1)^n q_n]^p,$$

where $q_1, q_2, \dots q_n$ are integers. The function $\phi(x)$ may be expressed in the form

$$\phi(x)=c_{p-1}x^{p-1}+c_px^p+\dots+c_{np+p-1}x^{np+p-1},$$

where $c_{p-1}(p-1)!$, $c_pp!$, ... are integral.

We see that $\phi^{p-1}(0)=(-1)^{np}C^{p-1}q_n{}^p$, which is an integer not divisible by p.

Also $\phi^p(0)$ is the value when $x=0$ of

$$pC^{p-1}\frac{d}{dx}[(Cx)^n-q_1(Cx)^{n-1}+\dots]^p$$

and is clearly an integer divisible by p. We see also that

$$\phi^{(p+1)}(0),\quad \phi^{(p+2)}(0),\ \dots\ \phi^{np+p-1}(0)$$

are all multiples of p.

Further if $m\leqq n$, $\phi(\beta_m)$, $\phi'(\beta_m)$, ... $\phi^{(p-1)}(\beta_m)$ all vanish, and

$$\sum_{m=1}^{m=n}\phi^{(p)}(\beta_m),\quad \sum_{m=1}^{m=n}\phi^{(p+1)}(\beta_m),\ \dots\ \sum_{m=1}^{m=n}\phi^{(np+p-1)}(\beta_m)$$

are all integers divisible by p. This follows from the fact that $\sum_{m=1}^{m=n} (C\beta_m)^r$ is expressible in terms of those symmetrical functions which consist of the sums of products of the numbers $C\beta_1$, $C\beta_2$, ...; and these expressions have integral values.

(4) Let K_p denote the integer

$$(p-1)!\,c_{p-1} + p!\,c_p + \ldots + (np+p-1)!\,c_{np+p-1}$$

which may be written in the form

$$\phi^{(p-1)}(0) + \phi^{(p)}(0) + \ldots + \phi^{(np+p-1)}(0).$$

In virtue of what has been established in (3) as to the values of

$$\phi^{(p-1)}(0), \quad \phi^{(p)}(0), \ldots$$

we see that $K_p A$ is not a multiple of p.

We examine the form to which the equation

$$A + e^{\beta_1} + e^{\beta_2} + \ldots + e^{\beta_n} = 0$$

is reduced by multiplying all the terms by K_p.

We have

$$K_p e^{\beta_m} = \sum_{r=p-1}^{r=np+p-1} c_r \left\{ \beta_m{}^r + r\beta_m{}^{r-1} + r(r-1)\beta_m{}^{r-2} + \ldots + r! + \frac{\beta_m{}^{r+1}}{r+1} + \frac{\beta_m{}^{r+2}}{(r+1)(r+2)} + \ldots \right\}$$

$$= \phi(\beta_m) + \phi'(\beta_m) + \ldots + \phi^{np+p-1}(\beta_m) + \sum_{r=p-1}^{r=np+p-1} c_r\beta_m{}^r \left\{ \frac{\beta_m}{r+1} + \frac{\beta_m{}^2}{(r+1)(r+2)} + \ldots \right\}.$$

The modulus of the sum of the series

$$\frac{\beta_m}{r+1} + \frac{\beta_m^2}{(r+1)(r+2)} + \ldots$$

does not exceed

$$\frac{|\beta_m|}{r+1} + \frac{|\beta_m|^2}{(r+1)(r+2)} + \ldots,$$

and this is less than $e^{|\beta_m|}$; hence we have

$$c_r\beta_m{}^r \left\{ \frac{\beta_m}{r+1} + \frac{\beta_m{}^2}{(r+1)(r+2)} + \ldots \right\} = \theta_r |c_r\beta_m{}^r| e^{|\beta_m|},$$

where θ_r is some number whose modulus is between 0 and 1.

The modulus of

$$\sum_{r=p-1}^{r=np+p-1} \theta_r |c_r\beta_m{}^r| e^{|\beta_m|} \text{ is less than } e^{|\beta_m|} \sum_{r=p-1}^{r=np+p-1} |c_r\beta_m{}^r|,$$

or than

$$e^{|\beta_m|}\frac{|\beta_m|^{p-1}}{(p-1)!}C^{np+p-1}\{(|\beta_m|+|\beta_1|)(|\beta_m|+|\beta_2|)\dots(|\beta_m|+|\beta_n|)\}^p,$$

or than

$$e^{\bar{\beta}}\frac{\bar{\beta}^{p-1}}{(p-1)!}C^{np+p-1}\{(\bar{\beta}+|\beta_1|)(\bar{\beta}+|\beta_2|)\dots(\bar{\beta}+|\beta_n|)\}^p;$$

where $\bar{\beta}$ denotes the greatest of the numbers $|\beta_1|, |\beta_2|, \dots |\beta_n|$.

It thus appears that the modulus of

$$\sum_{r=p-1}^{r=np+p-1} c_r\beta_m{}^r\left\{\frac{\beta_m}{r+1}+\frac{\beta_m{}^2}{(r+1)(r+2)}+\dots\right\}$$

is less than a number of the form $PQ^p/(p-1)!$, where P and Q are independent of p and of m.

We have now

$$K_p(A+\sum_{m=1}^{m=n}e^{\beta_m})=K_pA+\sum_{m=1}^{m=n}\{\phi^{(p)}(\beta_m)+\dots+\phi^{(np+p-1)}(\beta_m)\}+L,$$

where K_pA is not a multiple of p, the second term is an integer divisible by p, and L is less than $nPQ^p/(p-1)!$. The prime p may be chosen so large that $nPQ^p/(p-1)!$ is numerically less than unity. Since $K_p(A+\sum_{m=1}^{m=n}e^{\beta_m})$ is expressed as the sum of an integer which does not vanish and of a number numerically less than unity, it is impossible that it can vanish. Having now shewn that no such equation as

$$A+e^{\beta_1}+e^{\beta_2}+\dots+e^{\beta_n}=0$$

can subsist, we see that πi cannot be a root of an algebraic equation with integral coefficients, and thus that π is transcendental.

It has thus been proved that π is a transcendental number, and hence, taking into account the theorem proved on page 50, the impossibility of "squaring the circle" has been effectively established.

RULER AND COMPASSES

RULER & COMPASSES

BY

HILDA P. HUDSON

M.A., Sc.D.

WITH DIAGRAMS

CONTENTS

REFERENCES.

Enriques, Fragen der Elementargeometrie. Deutsche Aufgabe von Fleischer. II Teil. Leipzig, 1907.

Vahlen, Konstruktionen und Approximationen. Leipzig, 1911.

Klein, Ausgewählte Fragen der Elementargeometrie. Göttingen, 1895.

Hilbert, Grundlagen der Geometrie. Leipzig, 2te Aufl., 1903.

Richmond, To construct a regular polygon of 17 sides. Math. Ann. LXVII. p. 459, 1909.

Gérard, Construction du polygone régulier de 17 côtés au moyen du seul compas. Math. Ann. XLVIII. p. 390, 1896.

Mathews, Projective geometry. London, 1914.

Zühlke, Ausführung elementargeometrischer Konstruktionen bei ungünstigen Lageverhältnissen. Leipzig, 1906.

Lemoine, Géométrographie ou art des constructions géométriques. Paris, 1902.

Poncelet, Traité des propriétés projectives des figures. Paris, 1822.

Steiner, Die geometrischen Konstructionen ausgeführt mittlest der geraden Linie und Eines festen Kreises. Berlin, 1833.

Mascheroni, Geometria del Compasso. Pavia, 1797.

Adler, Zur Theorie der Mascheronischen Constructionen. Wiener Sitzungsber., 1890.

Hobson, On geometrical constructions by means of the compass. Math. Gazette, March 1913.

These works will be cited by the name of the author only. Valuable help has also been given by Mr. C. S. Jackson and Dr. F. S. Macaulay, to whom the writer offers her thanks.

CHAPTER I.

INTRODUCTION.

At the beginning of his *Elements*, Euclid places his three Postulates: " Let it be granted

(i) that a straight line may be drawn from any one point to any other point;

(ii) that a terminated straight line may be produced to any length in a straight line;

(iii) that a circle may be described from any centre, at any distance from that centre";

and all the constructions used in the first six books are built up from these three operations only. The first two tell us what Euclid could do with his ruler or straight edge. It can have had no graduations, for he does not use it to carry a distance from one position to another, but only to draw straight lines and produce them. The first postulate gives us that part of the straight line AB which lies between the given points A, B; and the second gives us the parts lying beyond A and beyond B; so that together they give the power to draw the whole of the straight line which is determined by the two given points, or rather as much of it as may be required for any problem in hand.

The last postulate tells us what Euclid could do with his compasses. Again, he does not use them to carry distance, except from one radius to another of the same circle; his instrument, whatever it was, must have collapsed in some way as soon as the centre was shifted, or either point left the plane (see p. 70). The three postulates then amount to granting the use of ruler and compasses, in order to draw a straight line through two given points, and to describe a circle with a given centre to pass through a given point ; and these two operations carry us through all the plane constructions of the *Elements*. The term *Euclidean construction* is used of any construction, whether contained in his works or not, which can be carried out with Euclid's two operations repeated any finite number of times.

In fact, Euclid gives only very few of the constructions which can be carried out with ruler and compasses, and probably every student of geometry has at some time or other constructed a figure which no one else had ever made before. But from very early times there were certain figures which everyone tried to make with ruler and compasses, and no one succeeded. The most famous of these baffling figures are the square equal in area to a given circle, and the angle equal to the third part of a given angle ; and it has at last been proved that neither of these can possibly be constructed by a finite series of Euclidean operations.

The set of figures which it is possible to construct with ruler and compasses is thus on the one hand infinite, and on the other hand limited. It is easy to see that it is infinite : even if we consider only the very simple type of figure consisting of a set of points at equal distances on a straight line, which can certainly be constructed with ruler and compasses, the figure may contain either three or four or a greater number of points without any upper limit, so that there are an infinite number of figures even of this one simple type ; much more is the whole set of possible figures infinite. And yet the set is limited, for many figures can be thought of which do not belong to it, and require apparatus other than ruler and compasses for their construction ; besides those mentioned above, there are for example the regular heptagon and nonagon ; or the ellipse, which can be drawn as a continuous curve with the help of two pins and a thread, but of which we can only obtain an unlimited number of separate points by Euclidean constructions.

The first question treated in this book is the one which naturally arises here : what constructions can be built upon Euclid's postulates, and what cannot ? or, in other words, what problems can be solved by ruler and compasses only ? For centuries, vain attempts were made to square the circle and trisect an angle by Euclidean constructions, and these attempts were often of use in other ways, though their immediate object failed; but it was only through the growth of analysis that it was proved once for all that they must fail. The ancient or classical geometry lends itself curiously little to any general treatment; and even modern geometry lacks a notation or calculus by which to examine its own powers and limitations. Occasionally we can be sure beforehand that a certain class of problem will yield to a certain type of method, but as a rule each problem has to be taken on its own merits and a separate method invented for it. There can be endless variety in the methods, and it might seem a hopeless task to sum them up, and impossible to say of any one problem that no ruler and compass construction for it ever will be found by some ingenious person yet unborn. Yet this is just what can be said in the case of the circle-squarers, and the final word came not from a geometer but from an analyst. The methods of coordinate geometry allow us to translate any geometrical statement into the language of algebra, and though this language is less elegant, it has a larger vocabulary; it can discuss problems in general as well as in particular, and it can give us the complete answers to the questions: what constructions are possible with ruler only, or with ruler and compasses?

In the next chapter we shall show how each step of a ruler and compass construction is equivalent to a certain analytical process; it is found that the power to use a ruler corresponds exactly to the power to solve linear equations, and the power to use compasses to the power to solve quadratics. For this reason, problems that can be solved with ruler only are called *linear problems*, and those that can be solved with ruler and compasses are called *quadratic problems*. Since each step of a ruler and compass construction is equivalent to the solution of an equation of the first or second degree, we consider what these algebraic processes can lead to, when combined in every possible way, and that enables us to answer the question before us and say (p. 19) that those problems and those problems alone can be solved by ruler only, which can be made to depend on a linear equation, whose root can be calculated by carrying out rational operations only; and (pp. 24, 30) that those problems and those problems alone can be solved by ruler and compasses, which can be made to depend on an algebraic equation, whose degree must be a power of 2, and whose roots can be calculated by carrying out rational operations together with the extraction of square roots only.

This is a complete answer to our question, but it is stated entirely in algebraic language, and in the general form in which it stands it cannot be translated into the language of pure geometry, for the words are lacking. But they are not really needed; for if a definite problem is before us, stated geometrically, we can always apply the test to the algebraic equivalent of the problem, though the application in some particular cases may not be easy; and if the test is satisfied, we can from the analysis deduce a Euclidean construction for the geometrical problem. As an example, we shall consider, at the end of chapter II, what regular polygons are within our powers of construction. It is a curious and unexpected result that the regular polygon of 17 sides is included, and a construction for this is given on p. 34.

When we have agreed that the set of possible Euclidean constructions is both infinite and limited, and when we have found out in some measure what its limitations are, it is natural to seek a clearer view of the set, and to ask what are the best ways of classification. The first main subdivision has already been brought to our notice; it consists of linear problems, which can be solved with ruler only, and chapter III is devoted to these. We try to show how the data of a problem control its construction, and how the properties and relations of the data fall into distinct classes, each of which allows a particular set of constructions to be carried out.

Now though Euclid's compasses can to some extent carry distances, we are not making use of compasses in the section referred to; and Euclid's ruler cannot carry distances at all. So we find that in general, that is, if the data of the problem do not have some exceptional relations, we are not able to carry a distance from one part of the figure to another, nor to compare the lengths of two segments, even of the same straight line, unless one is a part of the other, and then all that we can say is that the whole is greater than the part. We can never say that two segments are equal in length unless they coincide; for the very idea of comparing the lengths of two segments which have different positions, involves the idea of making them lie alongside of one another in order that we may compare them, and therefore of carrying one at least into another position; and this we cannot in general do with ruler only. Now the property of a parallelogram, that its opposite sides are equal in length, shows that if we can draw parallels, we can carry a distance from a straight line to a parallel straight line, so that there is a close connection between carrying distances and drawing parallels; and we shall show that, to a certain extent, if we can do either we can do both. But even so, it is only the parallel sides of a parallelogram that are equal, and not the adjacent sides, and drawing parallels does not help us to transfer the length of a segment into any direction other than that of the original segment. It is only in the special case in which the data allow us to construct a rhombus, whose adjacent sides are equal as well as its parallel sides, that we can obtain equal lengths on any of two different sets of parallel straight lines. We could hardly expect, with ruler only, to be able to turn a given length into any direction, for this is one of the chief uses of a pair of compasses.

In this way we get the usual classification of linear constructions according to the projective and metrical properties of the data, properties of length and properties of angle. The idea of cross-ratio is fundamental to them all, and when we have to compare the cross-ratios of two different ranges, we are led to the theories of homography and involution; but in connection with these we come upon several problems that require the common or double points; and the construction of these points is equivalent to the solution of a quadratic equation, and therefore impossible with ruler only.

In chapter IV therefore, in which we admit the use of compasses, the comparison is worked out between describing a circle and solving a quadratic, and then we are able to carry the treatment of cross-ratio and involution to a more satisfactory stage. A digression is introduced at this point (p. 70) to show that the ordinary modern instruments, dividers, parallel ruler and set-square, that are commonly used along with ruler and compasses, amount to short cuts in Euclidean constructions without extending the range of soluble problems. We also consider which ways of using these other instruments can completely replace the use of compasses.

But instead of classifying the data of a construction, we may classify the methods, and ask afterwards in what sort of problem each method is likely to be useful. Two fundamental ideas are put forward in chapter V, one or other or both of which are prominent in very many constructions: these are separation of properties and transformation. They give rise to half a dozen fairly well-marked lines of attack which are illustrated in that chapter. There is the method of loci (p. 78), when some of the conditions to be fulfilled convince us that a required point must lie on a certain locus, which must be made up of straight lines and circles if the method is to succeed, and the other conditions convince us that the same point must also lie on another such locus, so that it can only be a point of intersection of the two loci. There is the method of trial and error (p. 80), when from a finite number of unsuccessful attempts at a construction we are able to discover the way to begin which is bound to lead to success, if a solution exists; the method of projection (p. 90), and several of its particular cases, in which the two principles of separation of properties and of transformation are both present; the method of inversion (p. 92), which is very appropriate to ruler and compass constructions because of the way in which it relates circles and straight lines; and the method of reciprocation (p. 98), which rests upon the principle of duality.

Besides these intrinsic ways of classification, which come from considering the constructions themselves, there are others that come from considering them from some quite external point of view, according to how far they avoid certain draw-backs which are practical rather than mathematical, and arise from faults in our actual apparatus. Yet there is some theoretical interest also in devising constructions that shall as far as possible get over the difficulties of a small sheet of paper and a blunt pencil (chapter VI). The idea of the last section of the same chapter is to make a numerical estimate of the length of a construction, by reckoning up all the different operations with ruler and compasses that it requires, so as to be able to say which is the shortest of different solutions of the same problem. This plan of "giving marks" is little more than a pastime, and the scale of marking is very arbitrary; but Lemoine's book on Geometrography deserves to be better known, and some account of the matter is given here in the hope of introducing more English readers to his original work.

The subjects of the last two chapters are also now mere curiosities, though one at least arose as a practical point in machine construction. Any Euclidean problem can be solved by drawing only one circle and the requisite number of straight lines, usually a large number; or else by drawing the requisite number of circles and no straight line at all. This is proved, and examples of the methods are given, in chapters VII and VIII.

Thus the connecting link throughout the book is the idea of the whole set of ruler and compass constructions, its extent, its limitations and its divisions. But the matter of the following pages consists largely of examples, which have been brought in wherever possible, and it is hoped that those readers who are not attracted by general or analytical discussions, may yet find something to interest them among these problems and their geometrical solutions.

CHAPTER II.

POSSIBLE CONSTRUCTIONS.

BEFORE beginning to treat of any particular ruler and compass constructions, we shall discuss in this chapter the whole set of such constructions; and for this we need the help of analysis. It is only by finding the equivalent analytical process for each geometrical step, that we are able to tell what constructions can or cannot be carried out by using certain definite instruments in certain definite ways.

Coordinates.

We must first see how to express the data of a problem in arithmetical form, by means of some system of coordinates, and then see how each operation with ruler or compasses combines the numerical data in a certain way. The laws of algebra then show exactly what results can be obtained with either instrument or both.

It will simplify matters to suppose that all the geometrical data are points. This is convenient, and it is perfectly general. By Euclid's first two postulates, any straight line which is given ready drawn could have been constructed if, instead, two points lying on it had been given; so if there is a straight line among the data we can discard it, provided we take instead, as data, two points lying on the straight line. These must be definite points, not just "any" points. None of Euclid's postulates enable us to take any point on a straight line (see p. 13); a point must be either given, or constructed as the intersection of two lines. So unless two definite points on the straight line are among the other data, before we discard the straight line we must first intersect it by two other given straight lines, or by the joins of two pairs of given points; then we can take these two points of intersection as data, instead of the given straight line. This assumes that there are at least two straight lines or three points not in a straight line among the other data, or directly obtainable from the data; there are some trivial cases in which so few elements are given, that only a finite number of points and lines can be constructed from them (see p. 19). For example, if all that is given is one pair of straight lines, we obtain their point of intersection and nothing more; it is the only point which is completely determined. All the other points on the given lines are partially determined, but not completely; they form a definite class of points, distinguished from all the other points of the plane but not from one another, and therefore not sufficiently determined to be ready for use in further constructions. Such cases are excluded from the present discussion. In the same way, a circle can be replaced either by the centre and one point on the circumference, or by the centre and two other points whose distance apart is equal to the radius, or by three points on the circumference; again we have to exclude a few trivial cases. Also, any of the problems we are considering has for its object the construction of points, straight lines and circles having certain required properties; and it can be considered as solved when we have obtained the required points and also points from which the required straight lines and circles can be immediately constructed; so that not only what is given but also what is required may be taken to be a set of points.

We begin by referring all the points, given or required, to some axes of coordinates or other frame of reference; then every point has a set of coordinates, and every set of coordinates determines a point. These coordinates might be lengths, or areas, or angles, according to the system chosen; but we shall assume that they are pure numbers, the ratios of the geometrical quantities which determine the point to the corresponding units of the same kind. In a Cartesian system then, the coordinates are ratios of lengths, namely, of the abscissa to the unit of length on the axis of $\mathbf{x}$, and of the ordinate to the unit of length on the axis of $\mathbf{y}$; the two units may be the same or different. We shall use wherever possible an oblique Cartesian system, and we may take its frame of reference to consist of two axes together with a point on each at the corresponding unit distance from the origin. Later in the chapter we shall replace this by a more general system of projective coordinates (p. 19), for which the frame of reference consists of four points. In both systems, a point has two unique coordinates, and a straight line is represented by a linear equation between them.

Thus when a set of points is given, their coordinates are also given ; the whole set of data in any problem can be replaced by a set of numbers, and in the same way what is required can also be taken to be a set of numbers, the coordinates of the points which determine the points, straight lines and circles which are to be constructed. The question before us takes the form : what is the relation of this second set of numbers to the first, if the construction is one which can be carried out with ruler, or compasses, or both ?

I. Ruler Alone.

If the ruler alone is used, the only lines which can be drawn are the straight lines joining pairs of points already given or obtained ; and the only way of obtaining a new point is as the intersection of two such joins.

Let x_1, y_1 be the Cartesian coordinates of P_1, the point of intersection of the straight lines joining $A_1(a_1, b_1)$, $A_2(a_2, b_2)$ and A_3, A_4 respectively ; then x_1, y_1 satisfy both the equations of these two straight lines. The first of these equations is

$$\begin{vmatrix} x, & y, & 1 \\ a_1, & b_1, & 1 \\ a_2, & b_2, & 1 \end{vmatrix} = 0 ;$$

for it is formed from the general equation of a straight line,

$$lx + my + 1 = 0,$$

by putting for the undetermined coefficients l, m their values found from the two conditions that this line passes through the points A_1 and A_2,

$$la_1 + mb_1 + 1 = 0$$

and

$$la_2 + mb_2 + 1 = 0,$$

and the result of eliminating L, m between the last three equations is the determinantal equation first written.

So x_1, y_1 are given by the simultaneous equations

$$\begin{vmatrix} x, & y, & 1 \\ a_1, & b_1, & 1 \\ a_2, & b_2, & 1 \end{vmatrix} = 0, \qquad \begin{vmatrix} x, & y, & 1 \\ a_3, & b_3, & 1 \\ a_4, & b_4, & 1 \end{vmatrix} = 0,$$

that is,

$$x(b_1 - b_2) + y(a_2 - a_1) + a_1 b_2 - a_2 b_1 = 0,$$
$$x(b_3 - b_4) + y(a_4 - a_3) + a_3 b_4 - a_4 b_3 = 0,$$

whence

$$\frac{x_1}{\begin{array}{c}(a_2 - a_1)(a_3 b_4 - a_4 b_3) \\ -(a_4 - a_3)(a_1 b_2 - a_2 b_1)\end{array}} = \frac{y_1}{\begin{array}{c}(b_2 - b_1)(a_3 b_4 - a_4 b_3) \\ -(b_4 - b_3)(a_1 b_2 - a_2 b_1)\end{array}}$$

$$= \frac{1}{\begin{array}{c}(a_1 - a_2)(b_3 - b_4) \\ -(a_3 - a_4)(b_1 - b_2)\end{array}}.$$

What is important for our purpose is that x_1, y_1 are rational functions of each of the eight coordinates $a_1 \dots b_4$ of the four points A.

If then we have a set of given points (A), taking two pairs of them in every possible way, we can construct a new set of points (P) whose coordinates (x, y) are formed from the given coordinates (a, b) by equations of the above types. These points P can then be used in further constructions, and the numbers (x, y) can be added to the numerical data (a, b), and with these we can go on to construct fresh points and find fresh coordinates, at each step adding the fresh point to the data, and adding the fresh pair of coordinates to the set of numbers that may take the place of $a_1 \dots b_4$ in the above equations.

This process will terminate if we ever arrive at a figure in which every point obtained is already joined to every other by a straight line, as for example, a triangle, so that no new point or straight line can be obtained. We shall see later (p. 19) that this only happens in a few special cases; in general the process can be carried on indefinitely, and the number of points and straight lines which can be obtained is infinite.

At each step, the coordinates of the point just constructed are rational functions of those of the four points used in its construction. Hence, by successive substitution, the coordinates x and y of any point P, which can be constructed with ruler only from the given points $A(a, b)$, are rational functions of the set of numbers a, b. The steps of the substitution correspond exactly to the steps of the ruler construction taken in reverse order ; to the geometrical process of drawing the straight lines A_1A_2, A_3A_4 to meet in P_1 there corresponds the process of substituting for x_1, y_1 their values in terms of $a_1 \dots b_4$ given above.

For example, let there be four given points, $A(0, 0)$, $B(0, 1)$, $C(1, 0)$, $D(a, b)$.

(i) Join AB, CD to meet in $P(p, q)$, and join AC, BD to meet in $Q(r, s)$.

(ii) Join AD, PQ to meet in $X(x, y)$.

Here is a very simple construction in two stages: (i) from the four given points A, B, C, D we obtain two others, P, Q ; (ii) from these points, together with two of the given set, we obtain the final point X. Corresponding to the three pairs of straight lines drawn, there are three pairs of equations :

$$p=0,\ q=\frac{b}{1-a};\quad r=\frac{a}{1-b},\ s=0;\ \dots\dots\dots\dots\dots\text{(i)}$$

$$x=\frac{a(ps-qr)}{b(p-r)-a(q-s)},\quad y=\frac{b(ps-qr)}{b(p-r)-a(q-s)}.\ \dots\dots\text{(ii)}$$

In order to express the coordinates x, y of X in terms of those of A, B, C, D, we start with equations (ii), corresponding to the last step of the construction, which express x, y in terms of the coordinates of P, Q as well as of A, D. In these expressions we substitute for p, q, r, s from equations (i), which correspond to the preceding stage of the construction, and so obtain the final expressions of x, y in terms of the coordinates of A, B, C, D only, which reduce to

$$x=\frac{a}{2-a-b},\quad y=\frac{b}{2-a-b}.$$

This first general result may be briefly stated thus :

The points which can be constructed with ruler only from a given set of points have coordinates which are rational functions of the coordinates of the given points.

Indeterminate Constructions.

Often in the course of the solution of a problem we find instructions which are indeterminate, such as: "take any point" or "any straight line," "take any point upon a given straight line," "draw any straight line through a given point." These indeterminate operations are quite different from those we have been considering, in which each new point is completely determined by two straight lines already drawn, and each new straight line by two points. These vague operations are all equivalent to taking some arbitrary points $X(\xi, \eta)$ and carrying out definite operations upon them. Now these ill-defined points are of two classes: either the position of X does, or it does not, affect the position of the points to be finally constructed. If it does, the problem as it stands is indeterminate, and X must be regarded as belonging to the data, for the construction cannot be carried out until X has been chosen; and ξ, η must be included in the set of given coordinates $\mathbf{a}$, $\mathbf{b}$. If, on the other hand, X is a true auxiliary point, so that its position has no effect upon the final result, then the coordinates of the points required in the problem are the same wherever X may be chosen, and cannot involve ξ, η; in particular, they are the same as when X is chosen so that ξ, η are rational functions of $\mathbf{a}$, $\mathbf{b}$; as for example when X coincides with one of the given points A, or with one of the points P that can be immediately and definitely obtained from the points A. In this case, the final coordinates are certainly rational functions of the set of numbers $\mathbf{a}$, $\mathbf{b}$ only; therefore they are so, whatever the position of the auxiliary point X. It may quite well happen, when the given points are specially situated, that none of the points A or P will serve for X; as for example when all the given points lie on one straight line, and X has to be any point in the plane not lying on that line. If this line is taken to be the axis of $\mathbf{x}$, we have every $\mathbf{b}=\mathbf{0}$; but we can still take ξ, η to be rational functions of the $\mathbf{a}$, with the condition that η does not vanish. Strictly speaking, no ruler construction at all can be carried out on a set of given points all lying on one straight line; but by means of auxiliary points not lying on the line, we can construct other points, on the line, whose positions depend only on the given points, and which may therefore be said, in a slightly extended sense, to be constructed from the given points alone.

Converse Theory.

The next thing to do is to examine the converse theory, and determine whether all, or only some, of the points whose coordinates are rational functions of $\mathbf{a}$, $\mathbf{b}$ can be constructed with ruler only.

Now if our coordinates are Cartesian, the answer is that they cannot all be constructed. When a Cartesian frame of reference is given, the coordinates of a given point are two definite numbers; but though they are uniquely determined, they are defined by a hypothetical construction, namely by drawing a parallel to one of the axes; and if the data are a perfectly general set of points, we cannot draw parallels with ruler only. Though the coordinates are determinate, they cannot be constructed, and we cannot carry out an assigned rational operation on them. We must therefore eventually discard the Cartesian coordinates.

But we shall first show that, provided we can draw parallels, we can carry out any rational operation on the Cartesian coordinates. Since this is not so in general, we are now confining our attention to the particular case in which the data of the problem are not a perfectly general set of points, but are so specialized as to allow us to draw parallels. The simplest special property to assume is that among the data, or directly obtainable from them, are four points at the corners of a parallelogram; because when a parallelogram is given, we can with ruler only draw through any point a parallel to any straight line. This fundamental theorem is fully discussed in the next chapter, p. 51; for the present we take it for granted.

It is now convenient to choose a Cartesian frame as follows: let the origin O coincide with a given point A_1; let the axes pass through two other given points A_2, A_3, and let A_1A_2, A_1A_3 be taken as units of measurement along the respective axes, so that the points A, B at unit distances from O along the axes coincide with A_2, A_3. Then we must put

$$\mathbf{a}_1 = \mathbf{b}_1 = 0 ; \quad \mathbf{a}_2 = 1, \mathbf{b}_2 = 0 ; \quad \mathbf{a}_3 = 0, \mathbf{b}_3 = 1.$$

In the particular case which we are now considering, we can complete the parallelogram OAEB, where E is the unit point, whose coordinates are (**1**, **1**); and further, by drawing parallels through each of the given points to the sides and diagonals of this unit parallelogram, we can construct the points each of whose coordinates is one of the numbers $\mathbf{0}$, $\pm\mathbf{1}$, $\pm\mathbf{a}$, $\pm\mathbf{b}$.

First Four Rules.

Next, if P, Q are the given points ($\mathbf{a}$, $\mathbf{0}$), ($\mathbf{b}$, $\mathbf{0}$), we can construct the point R, whose coordinates are ($\mathbf{a}+\mathbf{b}$, $\mathbf{0}$), and hence, by repeating the process, all the points whose coordinates are linear functions of $\mathbf{a}$, $\mathbf{b}$, ... with positive or negative integral coefficients.

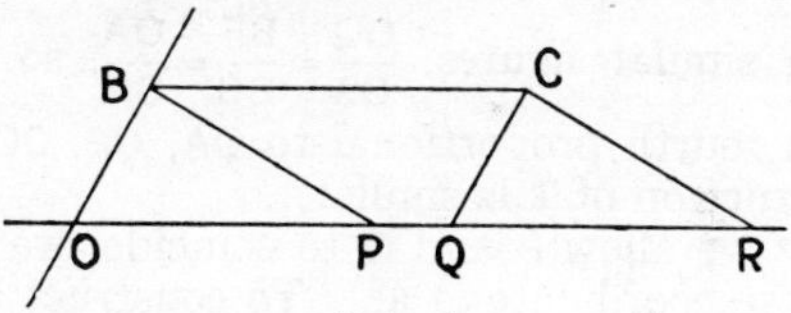

FIG. 1.

The construction of R is as follows:
Complete the parallelogram OQCB, and draw CR parallel to BP to meet OPQ in R, which is the required point.

For the triangles RCQ, PBO are congruent, and $QR = OP$;

$$\therefore\ OR = OQ + QR = OQ + OP = \mathbf{a} + \mathbf{b}.$$

If the three points O($\mathbf{0}$, $\mathbf{0}$), A($\mathbf{1}$, $\mathbf{0}$) and X($\mathbf{x}$, $\mathbf{0}$) are given, and we wish to construct the point whose coordinate is $\mathbf{7x}+\mathbf{3}$, we can lay off along the axis, beginning at X, six adjacent lengths XX_1, X_1X_2, ... X_5X_6, each equal to OX, and three more X_6Y_1, Y_1Y_2, Y_2Y_3, each equal to OA; then Y_3 is the point required.

We can also construct the points S($\mathbf{ab}$, $\mathbf{0}$) and T($\mathbf{a}/\mathbf{b}$, $\mathbf{0}$). These are such that OS is a fourth proportional to OA, OP, OQ, while OT is a fourth proportional to OQ, OP, OA. Then to construct the point whose coordinate is $\mathbf{3}/\mathbf{x}$, we need to find in succession points A_2, A_3, Y along the axis, where $AA_2 = A_2A_3 = OA$, and OY is a fourth proportional to OX, OA, OA_3.

The construction of S is as follows:

Take any point C on OB, for example, the point $(\mathbf{0}, \mathbf{a})$; join CA, CP to meet BE (or any other parallel to OPQ) in

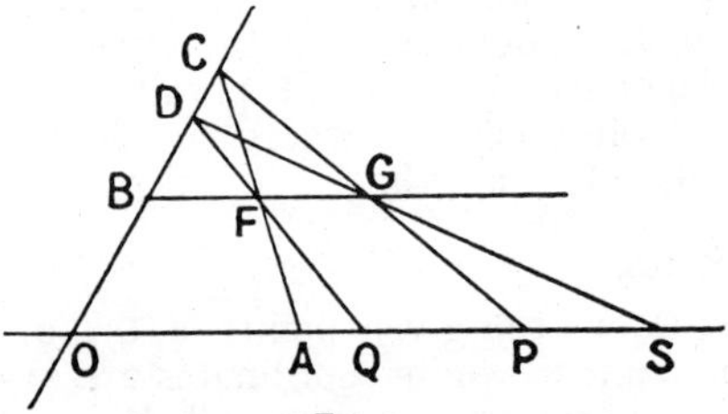

FIG. 2.

F, G; join QF to meet OC in D, and join DG to meet OPQ in S, which is the required point.

For, from similar figures, $\frac{OQ}{OS}=\frac{BF}{BG}=\frac{OA}{OP}$, so that OS is the required fourth proportional to OA, OP, OQ.

The construction of T is similar.

If in fig. 2 we allow P and Q to coincide, we obtain the point S, whose coordinate is $\mathbf{a}^2$. To construct the point Y whose coordinate is $\mathbf{x}^2+\mathbf{3x}+\mathbf{2}$, we can first find the point $S(\mathbf{x}^2, \mathbf{0})$, and then lay off from S along the axis three lengths equal to OX and two equal to OA; the end of the last segment is the point Y. Or it is shorter to construct the points H, K, whose coordinates are $\mathbf{x}+\mathbf{1}$, $\mathbf{x}+\mathbf{2}$, and then OY is a fourth proportional to OA, OH, OK.

Thus when once a parallelogram is drawn, we can find a ruler construction for any point whose coordinates are formed from those of points given or already obtained, by the processes of addition, subtraction, multiplication or division. But it is the definition of a rational function that it is formed from its arguments by a finite number of these four elementary operations. So that what we have just said is that we can find a ruler construction for any point whose coordinates are rational functions of the given coordinates. This set of functions includes all rational numbers, which are rational functions of the coordinates $(\mathbf{0}, \mathbf{1})$ of the single point A.

Combining this second general result with the first, we can now state that when a parallelogram is given it is possible to construct, with ruler only, all those and only those points whose Cartesian coordinates are rational functions of the coordinates of the given points.

Generalization by Projection.

So far we have only considered the case in which a parallelogram is given, and this allowed us to use Cartesian coordinates ; we have now to remove the restriction. This can be done by the method of projection. If the whole figure, with which we have been dealing hitherto, is the projection of one in another plane, the four corners of the parallelogram correspond to a quadrangle, or set of four points, which need have no special properties, and may be any four points in that plane, provided that no three of them lie on one straight line. For if we project a plane α on to a plane β from a vertex of projection V, then the straight line **fg**, in which α is cut by the plane through V parallel to β, projects into the straight line at infinity in β. Thus by properly choosing β, we can project any straight line of α to infinity, and then any two straight lines in α meeting on **fg** project into two straight lines in β meeting at infinity, that is, into two parallels. Now we choose **fg** to be the straight line joining the points of intersection **f**, **g** of two pairs of opposite sides of the quadrangle in α ; then the projection of the quadrangle has two pairs of parallel opposite sides, and is a parallelogram.

Then the join of any two points, or the intersection of any two straight lines, corresponds to the join or the intersection of their projections, so that ruler constructions in α correspond step by step to ruler constructions in β. If then, in any plane α in which points are given, four of these are chosen as a quadrangle of reference, we can project upon another plane β, choosing the projection so that the line at infinity in β corresponds to a diagonal of the quadrangle, and the quadrangle itself projects into a parallelogram. Then the points of α, which can be constructed with a ruler from the given points in α, project into the points of β which can be constructed with a ruler from the projections of the given points ; and since these given projections include the four corners of a parallelogram, we can choose as above a Cartesian frame of reference in β, such that the corners of the parallelogram have the coordinates $(\mathbf{0}, \mathbf{0})$, $(\mathbf{1}, \mathbf{0})$, $(\mathbf{0}, \mathbf{1})$, $(\mathbf{1}, \mathbf{1})$ respectively ;

and then we know that we can construct in β exactly those points whose coordinates are rational functions of the coordinates of the given projections; and hence we can construct in α exactly those points of which this set is the projection. We have therefore to find a meaning for these Cartesian coordinates in β, with reference only to the points of α and the quadrangle chosen from among them. This is done by expressing the coordinates as cross-ratios, which are not altered by projection (p. 38).

In the first plane α, let **oaeb** be the chosen quadrangle, and **fg** the one of its three diagonals which is projected into the line at infinity FG in β. Let capital letters denote the projections of the same small letters; then OAEB is

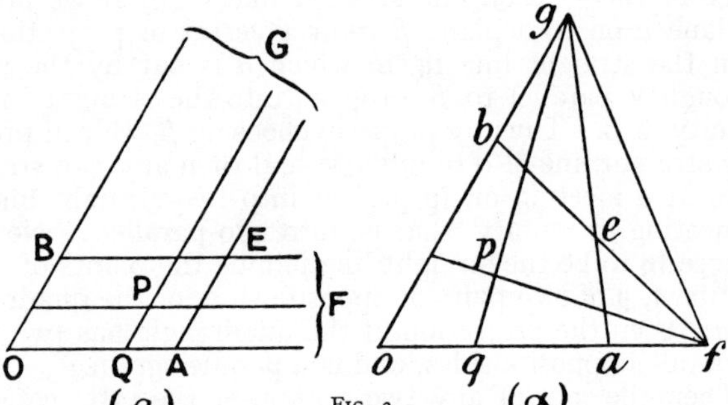

FIG. 3.

the parallelogram. F is the point at infinity in which the parallels OA, BE meet, and G is the point at infinity in which the parallels OB, AE meet. Let P(**x**, **y**) be any point in β, and PQ its ordinate, drawn parallel to OB to meet OA in Q; then QP produced also passes through G at infinity, and it is the projection of the straight line **qp**, joining **p** to the diagonal point **g** of the quadrangle, produced to meet **oa** in **q**.

Now the coordinates of P are **x**, **y**, the numerical measures of OQ and QP; and since OA is the unit of abscissae, the numerical measure of OQ is the ratio $\frac{OQ}{OA}$. Since F is at infinity, the ratio $\frac{QF}{AF} = \mathbf{1}$, and we may divide the coordinate by this without altering its value.

$$\therefore \quad x = \frac{OQ}{OA} = \frac{OQ}{OA} \cdot \frac{AF}{QF}$$

= the cross-ratio of the two pairs of points OF, QA, which is written {OF, QA},

= {**of**, **qa**}, since a cross-ratio is unaltered by projection,

= **g**{**of**, **qa**}, that is, the cross-ratio of the pencil subtended at **g** by **of**, **qa**, so that the Cartesian coordinate of the projection of **p** is equal to the cross-ratio of this pencil, one of whose rays is **gp**, and the others are the sides and diagonal of the quadrangle of reference which pass through **g**.

Similarly, $y = f\{og, pb\}$.

We can therefore take the cross-ratios of these two pencils as the *projective coordinates* of the point **p**. They are unaltered by projection, and become the same as the Cartesian coordinates for the particular projection in which **fg** passes to infinity.

Then the points	o ;	a ;	b ;	e ;	f ;	g ;
have the coordinates	0, 0 ;	1, 0 ;	0, 1 ;	1, 1 ;	∞, 0 ;	0, ∞.

We suppose that any four of the given points are taken to be **o**, **a**, **b**, **e** ; then **f** is determined as the intersection of **oa**, **be**, and **g** as the intersection of **ob**, **ae**. The system of projective coordinates is determined, and to them can be applied the results obtained before for Cartesian coordinates. The final conclusion is therefore :

If a set of points is given in a plane, then by ruler constructions those points and those points only can be obtained whose projective coordinates are rational functions of those of the given points, where any four of the given points, no three of which are in a straight line, are taken to determine the frame of reference, that is, to have the coordinates (**0**, **0**) ; (**1**, **0**) ; (**0**, **1**) ; (**1**, **1**).

Trivial Cases.

We have to assume that the data furnish four points, no three of which are in a straight line ; if there are not so many, the constructions fail. When all that is given is a single point, we can do nothing ; when two points are given, we can join them ; when we have three points, we can draw the sides (which may all coincide) of the triangle of which they are the vertices. More generally, when all the given points except one lie on the same straight line, we can draw this line and also the rays of the pencil subtended at the last point by the others. In all these cases we are not able to construct any new points.

Domains.

If the system consists of exactly four points, no three of which are on a straight line, they furnish the quadrangle of reference; the set of given coordinates is $(\mathbf{0}, \mathbf{1}, \boldsymbol{\infty})$, and the coordinates which can be obtained are the set of all rational numbers, positive, negative, or zero. All the numbers of this set are rational functions of $\mathbf{1}$ alone, for $\mathbf{0}$ is the difference $\mathbf{1}-\mathbf{1}$, and $\boldsymbol{\infty}$ is the reciprocal of $\mathbf{0}$. We say that the *domain* of the coordinates in this case is the set of all rational numbers, which we denote by $[\mathbf{1}]$.

Now let a fifth point $(\mathbf{a}, \mathbf{b})$ be added to the data; then if $\mathbf{a}$ and $\mathbf{b}$ are both rational, nothing is added to the domain of the coordinates; for if the point had not been given, it could have been constructed. But if $\mathbf{a}$, say, is an irrational number, we are now able to construct a much larger set of points; the domain is extended from all rational numbers to all rational functions of $\mathbf{a}$. This domain is denoted by $[\mathbf{1}, \mathbf{a}]$; it includes the former domain $[\mathbf{1}]$. If $\mathbf{b}$ is an independent irrational number, and not a rational function of $\mathbf{a}$, it also extends the domain, which becomes $[\mathbf{1}, \mathbf{a}, \mathbf{b}]$; but if either $\mathbf{a}$ or $\mathbf{b}$ is a rational function of the other, it can be omitted, for its inclusion adds nothing. In the same way, if there is any greater number of given points, whose coordinates are $\mathbf{a}, \mathbf{b}, \mathbf{c}, \ldots$, the corresponding domain is $[\mathbf{1}, \mathbf{a}, \mathbf{b}, \mathbf{c}, \ldots]$, in which any rational coordinates can be omitted, and also any which can be expressed as rational functions of the remaining irrational coordinates; and $\mathbf{a}, \mathbf{b}, \mathbf{c}, \ldots$ may be replaced by another set of numbers $\mathbf{a}', \mathbf{b}', \mathbf{c}', \ldots$, provided that each of these two sets can be expressed as a set of rational functions of the other set of numbers.

Suppose, for example, that we are given a square with an equilateral triangle described on one of its sides. Referred to this side and an adjacent side of the square as axes, the five given points may be taken to be

$$\mathrm{O}(\mathbf{0}, \mathbf{0}), \quad \mathrm{A}(\mathbf{1}, \mathbf{0}), \quad \mathrm{B}(\mathbf{0}, \mathbf{1}), \quad \mathrm{E}(\mathbf{1}, \mathbf{1}), \quad \mathrm{C}(\tfrac{1}{2}, \tfrac{1}{2}\sqrt{\mathbf{3}}),$$

and the domain is most simply expressed as $[\mathbf{1}, \sqrt{\mathbf{3}}]$. We can now construct equilateral triangles on all the other sides of the square and on its diagonals, for the coordinates of all the other points required involve no surd other than $\sqrt{\mathbf{3}}$. For example, if ABF is the equilateral triangle on AB on the side remote from O, the coordinates of F are each $\frac{1}{2}(\mathbf{1}+\sqrt{\mathbf{3}})$, and F can be constructed by drawing CD

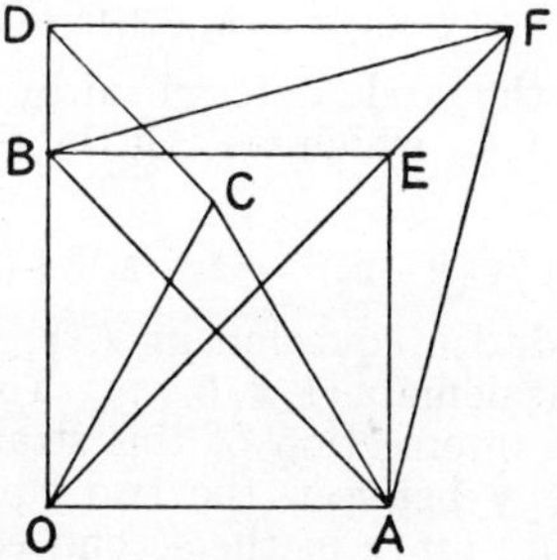

FIG. 4.

parallel to AB to meet OB in D, and DF parallel to OA to meet OE in F. But we cannot construct a regular pentagon on any straight line in the figure, for the coordinates of one or more of its corners would involve $\sqrt{\mathbf{5}}$, which is not a rational function of $\sqrt{\mathbf{3}}$, and therefore does not belong to the domain of the coordinates of points which can be constructed from the five given points with ruler only.

II. Ruler and Compasses.

When we also use compasses, a new point may be determined not only as the intersection of a pair of straight lines, but also as one of the intersections of a circle and a straight line, or of two circles. The case of two circles is much the same as that of a circle and a straight line, for the second circle may be replaced by the common chord or radical axis as below, p. 22.

When a circle is drawn, any two diameters cut the circumference in the corners of a rectangle, so that we can use rectangular Cartesian coordinates; and since we can cut off equal lengths from the two axes, we can use the same scale for abscissae and ordinates Now the equation of the straight line joining two points is linear, with coefficients rational in the coordinates $1, a, b, \ldots$ of the points already constructed, say

$$Lx + my + 1 = 0.$$

The circle determined by its centre $A_1(a_1, b_1)$ and a point $A_2(a_2, b_2)$ on its circumference, which we call the circle $A_1(A_2)$, has for its equation

$$(x - a_1)^2 + (y - b_1)^2 = (a_2 - a_1)^2 + (b_2 - b_1)^2 ;$$

more generally, the circle determined by its centre A_1 and radius equal to A_2A_3, which we call the circle $A_1(A_2A_3)$, has for its equation

$$(x - a_1)^2 + (y - b_1)^2 = (a_2 - a_3)^2 + (b_2 - b_3)^2,$$

which is a quadratic equation in x, y, whose coefficients also belong to the domain $[1, a, b, \ldots]$. To find the abscissae of the points of intersection of this straight line and this circle, eliminate y between the two equations and solve for x. Substitute for y in the second equation the value $-\dfrac{Lx+1}{m}$ obtained from the first. We find

$$(x - a_1)^2 + \left\{-\frac{Lx+1}{m} - b_1\right\}^2 = (a_2 - a_3)^2 + (b_2 - b_3)^2,$$

or say

$$px^2 + 2qx + r = 0,$$

which is a quadratic equation for x, whose coefficients p, q, r belong to the same domain as before. The expression which we obtain for x is $\dfrac{1}{p}(-q \pm \sqrt{q^2 - pr})$, which does not in general belong to $[1, a, b, \ldots]$, but involves the square root of the rational expression $q^2 - pr$; and y involves the same square root.

In the case of two circles we can take the equations to be

$$x^2 + y^2 - 2a_1x - 2b_1y + c_1 = 0,$$
$$x^2 + y^2 - 2a_2x - 2b_2y + c_2 = 0,$$

with coefficients belonging to $[1, a, b, \ldots]$. The coordinates of the points of intersection satisfy both these equations, and satisfy the equation found by subtracting,

$$2(a_2 - a_1)x + 2(b_2 - b_1)y + c_1 - c_2 = 0,$$

which is linear, and therefore represents the straight line

joining the two real or imaginary points of intersection, that is, the common chord of the two circles. The intersections are therefore the same as those of either circle with this straight line, and since the coefficients of the linear equation also belong to $[\mathbf{1}, \mathbf{a}, \mathbf{b}, \ldots]$, so that it is of the type $lx + my + 1 = 0$ considered above, this case is included in the last.

Thus the geometrical operation of describing a circle with given centre and radius to cut a given straight line in two points, is equivalent to the analytical operation of taking a square root, together with some rational operations. Conversely, we can find a geometrical operation equivalent to taking the square root of any number $\mathbf{p}$,

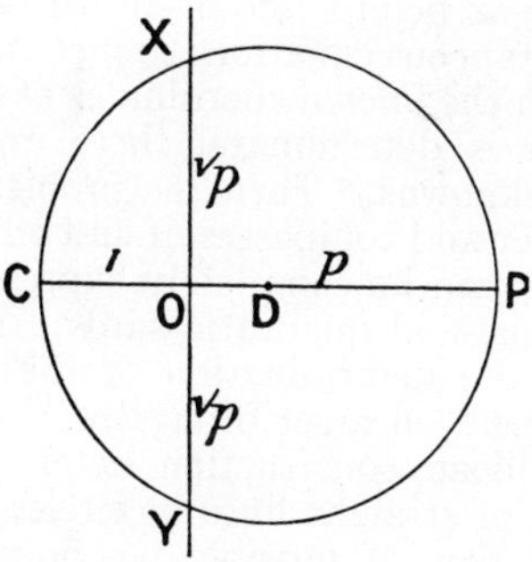

FIG. 5.

provided that $\mathbf{p}$ is a coordinate of a point which can be constructed.

Let P be the point $(\mathbf{p}, \mathbf{0})$ and let C be $(-\mathbf{1}, \mathbf{0})$. Bisect CP in D; then the circle D(P) meets the axis of $\mathbf{y}$ in two points X, Y, such that

$$YO \cdot OX = CO \cdot OP = 1 \cdot p;$$

$$\therefore\ OX = \sqrt{p}, \qquad OY = -\sqrt{p}.$$

Quadratic Surds.

Thus the analytical steps equivalent to the various steps of a ruler and compass construction are rational operations together with the new operation of taking a square root; and these may be combined and repeated in any way a finite number of times and carried out upon the coordinates

(**1**, **a**, **b**, ...) of the given points. The result of such an analytical process, if it does not happen to be rational, is called a *quadratic surd*; the simplest type is the square root of a rational expression, but a general quadratic surd may consist of several terms, and the quantity under each radical need not be rational, but may be another quadratic surd. But the whole expression must involve no roots but square roots, and others, such as fourth roots, which can be expressed by means of square roots only; and it must not involve an infinite series of operations.

Now if we want to know whether some definite problem can be solved by means of ruler and compasses, we can express the conditions laid upon the elements to be constructed, which, as before (see p. 8), we take to be points, as a set of simultaneous equations connecting their unknown coordinates with the known coordinates of the given points. If the problem is determinate, there are just as many equations as unknowns. Then the problem can be solved by means of ruler and compasses, if and only if the solution of the equations can be completely expressed by means of rational functions and quadratic surds. It is a necessary condition that the determination of the coordinates can be made to depend on a set of *algebraic* equations. Each step of a Euclidean construction determines a point as the intersection of straight lines or circles, either uniquely or as one of a pair of possible positions. Since a construction has only a finite number of steps, when all the alternatives are considered the final coordinates have one of a finite number of sets of possible values, and their determination can be made to depend on a set of equations having only a finite number of solutions, that is, of finite degree and therefore algebraic, of the form

$$\mathsf{F}(\mathbf{x}, \mathbf{y}, \dots, \mathbf{a}, \mathbf{b}, \dots) = \mathbf{0},$$

where **x**, **y**, ... are the unknown, and **a**, **b**, ... the known coordinates, and F is an algebraic expression.

For the sake of simplicity, let us now take the case in which all the given coordinates are rational, so that the domain is [**1**]. In the general case, the whole argument follows the same course, and leads to corresponding results, if for the set of rational numbers we substitute the domain [**1**, **a**, **b**, ...]. By the known processes of the theory of equations, $F=0$ can first be rationalized, so that we can take F to be a polynomial with rational coefficients; next, we can eliminate all but one of the unknown coordinates, and suppose each unknown given by a separate equation. Thirdly, $F(x)$ may be reducible, that is, it may fall into two or more factors with rational coefficients, of any degrees in x from 1 up to $n-1$ if n is the degree of F itself; in this case processes are known for finding the factors and reducing the equation. We therefore need only consider an algebraic, irreducible equation in a single unknown, with rational coefficients.

Now consider a quadratic surd x; suppose that its denominator has been rationalized (p. 27), and that it is written before us. It consists of a collection of rational numbers, of symbols denoting rational operations, and of radicals which are symbols denoting the extraction of square roots. In reading this collection, the first time we come to a radical let us put the expression underneath it equal to y_1^2, so that we can write y_1 instead of the whole square root, and pass on till we come to another square root, for which we write y_2, and so on. Since each square root may be multiplied by a rational factor, x is now expressed in the form

$$x=c_1 y_1+c_2 y_2+\ldots+c_k y_k+c', \qquad \text{(I)}$$

where each c is rational and each y^2 is either rational or a quadratic surd simpler than x; for even in the extreme case when $k=1$, there is one radical fewer in y_1^2 than in x.

Now treat y_1^2, y_2^2, ... y_k^2 in the same way as we have treated x; then there are k equations of the type

$$y_k^2=c_{k,1}y_{k,1}+c_{k,2}y_{k,2}+\ldots+c_{k,l}y_{k,l}+c_k', \qquad \text{(II)}$$

where each c is rational and each $y_{k,l}^2$ is a quadratic surd simpler than y_k.

Continue this process with $y_{1,1}^2$, ... $y_{k,l}^2$, until there are no radicals left. We thus introduce as many new unknowns y as there are different square roots in x, and the last set of equations corresponding to (II) are of the type

$$y^2_{k,l,m,\ldots}=c'_{k,l,m,\ldots} \qquad \text{(III)}$$

For example, let

$$x=\sqrt{5a+\sqrt{6b}}+\sqrt{8b+\sqrt{7a}}.$$

We put
$$x = y_1 + y_2, \qquad \text{(I)}$$
$$y_1^2 = 5a + y_{11}, \qquad y_2^2 = 8b + y_{21}, \qquad \text{(II)}$$
$$y_{11}^2 = 6b, \qquad y_{21}^2 = 7a. \qquad \text{(III)}$$

Or again, let $x = \sqrt{1+\sqrt{3}} + \sqrt{3}$.

Then
$$x = y_1 + y_2,$$
$$y_1^2 = 1 + y_2, \qquad y_2^2 = 3.$$

Now take the equations in reverse order. Each of the last set of y s is connected with rational terms by one of the equations (III); that is, it is determined by a quadratic equation with rational coefficients; we can solve these equations and find this set of y's. Each of the set of y's which came just before is determined by an equation of type (II), whose coefficients are either rational, or rational functions of the y's which have just been found. At every step a set of y's are each found from a quadratic equation whose coefficients are rational in the y's already found, till finally we find $y_1, \ldots y_k$, and then x itself is given by a linear equation connecting it with these. Thus we can find x by solving a series of equations of degrees 1 and 2 only; hence *If an equation can be solved by quadratic surds, its solution can be made to depend on that of a series of equations of first and second degrees.* In the first place, all linear and quadratic equations can be so solved, but no irreducible cubic; for we shall now show, more generally, that the degree of an irreducible equation which can be solved by quadratic surds must be a power of 2; this is a necessary condition, but it is not sufficient.

Conjugate Surds.

If the signs of any particular set of these surds y are changed wherever they occur in x, the result is another quadratic surd x_1 which is said to be *conjugate* to x. In this way we obtain a set of 2^n conjugate surds, of which x is one. For example, if

$$x = \sqrt{2} + \sqrt{3} + \sqrt{2} \cdot \sqrt{3},$$

the other three conjugate surds are

$$-\sqrt{2} + \sqrt{3} - \sqrt{2} \cdot \sqrt{3}, \qquad \sqrt{2} - \sqrt{3} - \sqrt{2} \cdot \sqrt{3},$$
$$-\sqrt{2} - \sqrt{3} + \sqrt{2} \cdot \sqrt{3}.$$

But if we write $\mathbf{x} = \sqrt{2} + \sqrt{3} + \sqrt{6}$, we must not consider all three square roots as independent $\mathbf{y}$'s; for to change the sign of $\sqrt{2}$, without at the same time changing the sign of $\sqrt{6}$, does not give a conjugate surd. In this case there exists a relation $\sqrt{2} \cdot \sqrt{3} = \sqrt{6}$, which does not hold when the sign of $\sqrt{2}$ only is changed. We avoid the case in which the $\mathbf{y}$'s are dependent in this way; for a fuller discussion of independent surds, see Enriques, p. 140 ff. The equations (II), (III) may be used to eliminate all squares of $\mathbf{y}$'s from any relations that exist; if they do not then become identities, they are reduced to a form linear in each $\mathbf{y}$, and can be used first of all to eliminate some of the $\mathbf{y}$'s altogether. We suppose this carried out; the remaining $\mathbf{y}$'s are then linearly independent, that is, there exists no rational relation between them which is linear in each separately. It is a consequence of this, that any relation that exists remains true if the sign of any one $\mathbf{y}$ is changed throughout.

We shall now prove that if a given rational equation $\mathbf{f}(X) = \mathbf{0}$ is satisfied by $X = \mathbf{x}$, then it is also satisfied by $X = \mathbf{x}_1$, where $\mathbf{x}_1$ is any of the surds conjugate to $\mathbf{x}$.

Rationalizing Factor.

Any symmetric function of a complete set of conjugate surds is rational. For if it were irrational, it would depend really, and not only apparently, on some one of the radicals, and would be altered in value by changing the sign of that radical. But to change the sign of one radical changes each of the surds into a conjugate surd, and permutes the set of conjugates among themselves, without changing the set as a whole, and therefore does not alter a symmetric function of the set; these symmetric functions therefore do not really depend on the sign of any radical, but are rational. In particular, the continued product of the whole set is rational, and any surd has as a *rationalizing factor* the product of all the other surds conjugate to it. We can rationalize the denominator of the fraction considered above (p. 25), which is the quotient of two rational integral functions of the $\mathbf{y}$'s, by multiplying numerator and denominator by the product of all the surds conjugate to the denominator. This rationalizing factor is by construction an integral function of the different surds $\mathbf{y}$, of which the denominator is a function, and of their conjugates, which are other surds of the same nature.

Consider the result of substituting $\mathbf{x}$ for X in the function $\mathbf{f}(X)$. In the first place, we substitute for X the expression (I) linear in each $\mathbf{y}$, and carry out on this the operations indicated by $\mathbf{f}$. Now, since $X=\mathbf{x}$ is a root of $\mathbf{f}(X)=\mathbf{0}$, this last expression vanishes identically, as a direct consequence of the defining equations (I), (II), which, combined in a certain way, lead to $\mathbf{f}(\mathbf{x})=\mathbf{0}$.

Now instead of $\mathbf{x}$ let us substitute $\mathbf{x}_1$ in $\mathbf{f}$, where $\mathbf{x}_1$ only differs from $\mathbf{x}$ in the sign of a single surd $\mathbf{y}_1$. If $\mathbf{y}_1$ occurs anywhere under a radical this has the effect of changing some of the other surds $\mathbf{y}$ into their conjugates $\mathbf{y}'$ wherever they occur, but these new surds $\mathbf{y}'$ are defined by equations of exactly the same forms (II), (III) as before, so that $\mathbf{f}(\mathbf{x}_1)$ vanishes as a direct consequence of these, combined in the same way as before.

For example, consider the surd

$$\mathbf{x}=\sqrt{3}+\sqrt{1+\sqrt{3}}.$$

In order to find the rational equation satisfied by this, we first transpose the term $\sqrt{3}$ and square,

$$(\mathbf{x}-\sqrt{3})^2=1+\sqrt{3},$$

then rearrange in the form

$$\mathbf{x}^2+2=\sqrt{3}(2\mathbf{x}+1)$$

and square again. Therefore $\mathbf{x}$ is a root of the equation

$$\mathbf{f}(X)\equiv X^4-8X^2-12X+1=0.$$

Substitute and expand :

$$\begin{aligned}\mathbf{f}(\mathbf{x})=3^2&+4\,.\,3\sqrt{3}\sqrt{1+\sqrt{3}}+6\,.\,3(1+\sqrt{3})\\&+4\sqrt{3}(1+\sqrt{3})\sqrt{1+\sqrt{3}}+(1+\sqrt{3})^2\\&-8\{3+2\sqrt{3}\sqrt{1+\sqrt{3}}+1+\sqrt{3}\}-12\{\sqrt{3}+\sqrt{1+\sqrt{3}}\}+1\\=(9&+18+1+3-24-8+1)+\sqrt{3}(18+2-8-12)\\&+\sqrt{1+\sqrt{3}}(12-12)+\sqrt{3}\sqrt{1+\sqrt{3}}(12+4-16),\end{aligned}$$

each term of which vanishes separately. If in this we change the sign of $\sqrt{3}$, we must replace the second independent surd $\sqrt{1+\sqrt{3}}$ by the conjugate $\sqrt{1-\sqrt{3}}$; we have four terms

$$f(x_1) = (0) - \sqrt{3}(0) + \sqrt{1-\sqrt{3}}(0) - \sqrt{3}\sqrt{1-\sqrt{3}}(0),$$

each of which vanishes as before. But the quartic equation is not satisfied by $-\sqrt{3}+\sqrt{1+\sqrt{3}}$, which is not one of the surds conjugate to x.

In the general case, it follows that $f(X) = 0$ is also satisfied by $X = x_2$, where x_2 is a surd conjugate to x_1 formed by changing the sign of any other independent surd y_2 in x as well as the sign of y_1; and so on. If the signs of any number of the surds are changed wherever they occur in x, the result is a conjugate surd x_p, which also satisfies $f(X) = 0$.

Thus, *if a rational equation has a quadratic surd for one root, all the conjugate surds are roots.* But these 2^n conjugate surds may not all have different values, so that the number of different roots of $f(X) = 0$ given in this way is N say, where $N \leqq 2^n$. For example, if

$$x = \sqrt{5+\sqrt{3}} + \sqrt{5-\sqrt{3}},$$

its value is not altered by changing the sign of $\sqrt{3}$ in both terms.

Degree a Power of 2.

We next prove that N is a factor of 2^n, and is therefore itself a power of 2.

Let $x_q'(q = 1, 2, \ldots N)$ be the set of all the N different values of all the 2^n conjugate surds $x_p(p = 1, 2, \ldots 2^n)$.

Consider the two products

$$F(X) = \Pi(X - x_p), \quad F'(X) = \Pi(X - x_q')$$

of degrees 2^n, N respectively. A change of sign of any particular square root wherever it occurs in the x, interchanges the x in pairs, and therefore merely alters the order of the factors in F, and does not affect the value of the coefficient of any power of X in F; these coefficients therefore, being independent of the sign of any of the square roots that occur, are rational: F is a rational function of X with rational coefficients. Again, the change of sign of a square root changes the x' into another set of N of the surds x, all different, and therefore the same as the X' in another order; the change merely alters the order of the factors in F' and does not affect its coefficients, so that F' also is a rational function of X with rational coefficients. Moreover F' is irreducible; for if it fell into two rational factors F_1' and F_2', each would be the product of some but not all of the differences $X - X'$, and $F_1' = 0$ would have some but not all of the X as roots; but we have seen that if any rational equation has one of the X as roots, it has them all, so that this is a contradiction, and it follows that F' is irreducible. But F' consists of N from among the factors of F, so that F has F' as a factor, and the remaining factor F'' must also be rational. F'' may be a mere constant; in this case F and F' are of the same degree, $N = 2^n$, and the X are all different. But if not, F'' is of degree $2^n - N$; it is the product of some of the factors of F, and $F'' = 0$ has at least one of the X as a root; hence it has all the N different surds X' as roots, and, by the same argument as before, F'' has F' as a factor. Since both F'' and F' are rational, the remaining factor F''' of F'' is rational, and its degree is $2^n - 2N$. The same argument applies to F'''; it is either a constant or it contains F' as a factor, and then we can treat its other factor in the same way. We go on, at each step removing the factor F', and the process cannot stop as long as the remaining factor is of positive degree. Since the degree of F is finite, there can only be a finite number of steps, and we come at last to a factor of degree 0, that is, a constant. Hence F is the product of a power of F' and this constant, and its degree 2^n is a multiple of the degree N of F. Therefore N is a factor of 2^n, and is itself a power of 2, say $N = 2^k$.

We have now proved that any equation $f(X) = 0$ with rational coefficients which has a quadratic surd X as a root has as roots all the 2^k different values X' of the set of conjugate surds of which X is one; and f has all the differences $X - X'$ as factors, and therefore their product F', so that f has the rational factor F'. Therefore either f is reducible or it is a constant multiple of F', and its degree is 2^k. Therefore:

If an irreducible algebraic equation with rational coefficients can be solved by quadratic surds, its degree is a power of 2. Note that *all* its roots are conjugate surds.

Duplication and Trisection.

These theorems have been stated with regard to algebraic equations whose coefficients are rational, that is, belong to the domain $[\mathbf{1}]$. They can be extended to apply to equations whose coefficients belong to any given domain $[\mathbf{1}, \mathbf{a}, \mathbf{b}, \ldots]$.

It follows that the ancient problems of the duplication of a cube and the trisection of an angle cannot be solved by means of ruler and compasses. The problem of finding the side $\mathbf{x}$ of a cube whose volume is twice that of a cube of given side $\mathbf{a}$ is equivalent to solving the cubic equation

$$\mathbf{x}^3 = 2\mathbf{a}^3,$$

whose coefficients are rational in $[\mathbf{1}, \mathbf{a}]$, but whose real root $\mathbf{x} = \mathbf{a} \cdot \sqrt[3]{2}$ involves a surd which is not quadratic. And the trisection of an arbitrary angle θ carries with it the determination of $\cos \frac{1}{3}\theta$ $(= \mathbf{y}$ say) when $\cos \theta$ $(= \mathbf{b}$ say) is given; but since $\cos \theta = \mathbf{4} \cos^3 \frac{1}{3}\theta - \mathbf{3} \cos \frac{1}{3}\theta$, this is equivalent to solving the cubic equation

$$\mathbf{4y}^3 - \mathbf{3y} - \mathbf{b} = \mathbf{0},$$

which is irreducible except for special values of the given cosine $\mathbf{b}$. These two famous problems are therefore beyond the range of ruler and compass constructions.

III. Regular Polygons.

A good example of this theory is to determine what regular polygons can be constructed with ruler and compasses. If $\mathbf{n}$ the number of sides is not prime, say $\mathbf{n} = \mathbf{a}_1\mathbf{a}_2$, then the $\mathbf{n}$-gon also furnishes a regular $\mathbf{a}_1$-gon by taking $\mathbf{a}_1$ of the vertices, each separated from the next of the $\mathbf{a}_1$ by $\mathbf{a}_2 - \mathbf{1}$ consecutive vertices of the $\mathbf{n}$-gon. For example, the 1st, 4th, 7th and 10th vertices of a regular dodecagon form a square. So if we could construct the $\mathbf{n}$-gon, we could also from it construct the $\mathbf{a}_1$-gon; and if the latter is known to be impossible, it follows that the former must be impossible also, and it is useless to consider it. In the first place, therefore, we consider prime values of $\mathbf{n}$. The construction can be made to depend on finding the irrational roots of the equation $\mathbf{x}^n - \mathbf{1} = \mathbf{0}$. This is reducible, but removing the factor $\mathbf{x} - \mathbf{1}$, we have

$$\mathbf{x}^{n-1} + \mathbf{x}^{n-2} + \ldots + \mathbf{x} + \mathbf{1} = \mathbf{0},$$

which can be shewn to be irreducible if n is prime (Enriques, p. 152). Then, if the polygon can be constructed with ruler and compasses, this equation can be solved by quadratic surds, and its degree $n-1$ must be a power of 2; hence

$$n=2^k+1.$$

Further, since n is prime, k must also be a power of 2; for if k had an odd factor l other than 1, let $k=lm$; then

$$n=(2^m)^l+1,$$

which is not prime, but has 2^m+1 as a factor. Hence k has no odd factor, and we may write

$$k=2^p, \quad n=2^{2^p}+1.$$

The only exception to the condition $k=2^p$ is the trivial case $k=0$, $n=2$.

It has also been proved, conversely, that if n is a prime of this form, the equation can be solved and the polygon constructed. For example, if $p=1$, $n=5$, one root is

$$x=\tfrac{1}{4}\{-1-\sqrt{5}+\sqrt{-10+2\sqrt{5}}\},$$

and the corresponding geometrical construction may be carried out as in the fourth book of Euclid.

Putting	$p=$	0,	1,	2,	3,	4,
we have	$n=$	3,	5,	17,	257,	65537.

The next values $p=5, 6, 7$ do not give prime values for n, and little is known of the higher numbers of the series. Hence $3, 5, 17, 257$ are the only prime values of n less than $10{,}000$ for which the regular n-gon can be constructed with ruler and compasses.

Next, if n is composite, let

$$n=a_1^{k_1}a_2^{k_2}\ldots,$$

where $a_1, a_2 \ldots$ are the different prime factors of n. Then, if the n-gon can be constructed, so can the a_1-gon; and since a_1 is prime, it is of the form $2^{2^p}+1$, unless $a_1=2$.

Now all the roots of $\dfrac{x^n-1}{x-1}=0$ are quadratic surds; but the left-hand side has the factor $\dfrac{x^{ak}-1}{x-1}$ (with any suffix to k), and this again has the factor $\dfrac{x^{a^k}-1}{x^{a^{k-1}}-1}$; hence the rational equation of degree $a^{k-1}(a-1)$,

$$x^{a^{k-1}(a-1)}+x^{a^{k-1}(a-2)}\ldots+x^{a^{k-1}}+1=0,$$

can be solved by quadratic surds. Now this equation can be shown to be irreducible when a is prime, so its degree $a^{k-1}(a-1)$ is a power of 2, and therefore both a^{k-1} and $a-1$ are powers of 2. Now if $a=2$, both these conditions hold for all values of k; but if $a=2^{2^p}+1$, then $a-1$ is a power of 2, but a is not, and a^{k-1} is not, unless $k=1$, when $a^{k-1}=a^0=1=2^0$. Thus we must have either $a=2$ or $k=1$, and n can have no repeated factor other than 2; we may write

$$n=2^k(2^{2^{p_1}}+1)(2^{2^{p_2}}+1)\ldots,$$

where p_1, p_2 ... are different integers, and all the numbers $2^{2^p}+1$ are prime.

Conversely, if a_1, a_2 are two numbers prime to one another, and the regular a_1-gon and a_2-gon can be constructed, so can the regular a_1a_2-gon. For we can find two positive integers N_1, N_2 such that $N_1a_2-N_2a_1=\pm 1$. Then the difference between the angles subtended at the centres of the polygons by N_1 sides of the first and N_2 sides of the second is

$$N_1\frac{2\pi}{a_1}-N_2\frac{2\pi}{a_2}=(N_1a_2-N_2a_1)\frac{2\pi}{a_1a_2}=\pm\frac{2\pi}{a_1a_2},$$

which is the angle subtended at the centre of a regular a_1a_2-gon by one of its sides; so that we can construct this angle, and therefore the whole polygon. This can be extended to any number of factors prime to one another. Also any angle can be bisected any number of times, so that the factor 2^k can be provided for. Therefore if n has the form given above, the regular n-gon can certainly be constructed.

Thus with ruler and compasses we can construct regular polygons of

2, 3, 4, 5, 6, 8, 10, 12, 15, 16, 17, 20, ... sides,
but not of 7, 9, 11, 13, 14, 18, 19, ... sides.

Regular 17-gon.

Here is a construction for the regular polygon of **17** sides (Richmond).

Take perpendicular radii OA, OC of a circle centre O; bisect OC and bisect again, making $OD = \frac{1}{4}OC$; join DA. Bisect $\angle ODA$ and bisect again, making $\angle ODE = \frac{1}{4}\angle ODA$. Draw a perpendicular at D to DE, and bisect the right

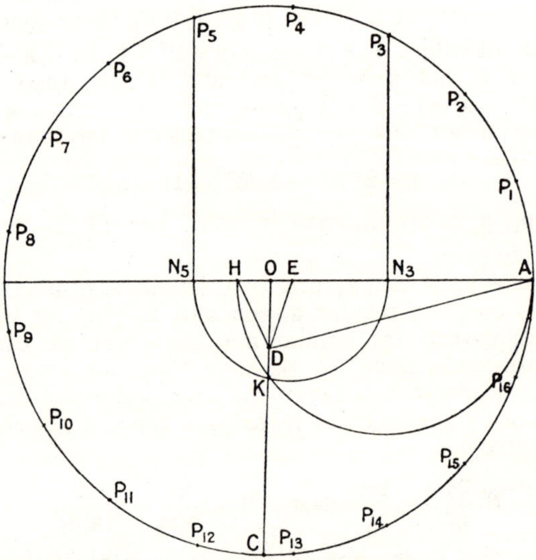

FIG. 6.

angle, making $\angle EDH = \frac{1}{4}\pi$, its arms meeting OA in E, H. Describe the circle on AH as diameter to meet OC in K, and describe the circle E(K) to meet OA in N_3, N_5. Draw ordinates through these points perpendicular to OA to meet the circle O(A) in P_3, P_5; then these are the third and fifth vertices of a regular **17**-gon, of which A is the last.

CHAPTER III.

RULER CONSTRUCTIONS.

In this chapter we consider the actual carrying out of some of the ruler constructions which we have found to be possible when a set of points are given in a plane. First of all, an important distinction must be made according as the positions of the points have or have not any special relations to one another. If we project all the points and lines that we obtain, upon any other plane, the result is a figure in this second plane, which can also be constructed with a ruler if we are given in the second plane the projections of the points which are given in the first plane. The two constructions correspond step by step, and every incidence of a point with a line in either plane corresponds to the incidence of their projections in the other plane.

Now if the data have no special relations, the only properties possessed by the figure which we first constructed are due to the various incidences secured by the steps of the construction; for these steps consist either of drawing a new straight line which passes through two points already obtained, or of obtaining a new point which lies on two straight lines already drawn. Since all these incidences occur in the second figure also, it follows that all the properties of each figure hold of the other; they are said to be *descriptive* or *projective properties*. They are unchanged by any projection or by any series of projections, and so belong to a whole set of figures, any one of which can be obtained from any other by one or more projections. All the figures of the set are exactly alike as regards projective properties.

If a figure possesses any properties which are not of this nature, they are called *non-projective*, or, more usually, *metrical*. These cannot be due to the steps of the construction, when this is carried out with ruler only, and must be due to the data, which are therefore a set of points whose positions are not general, but which enjoy certain special relations which do not hold between their projections, and so do not lead to the same properties of the second figure. But to a metrical property of the first figure there must correspond *some* property of the second, and the latter must be a projective property, since it holds of all the projections of the first figure; it must therefore hold of all the projections of these projections, among which the first figure itself is included. So there exists a projective property, arising out of the metrical property, which holds of all the projective set of figures, including the first figure; and when applied to this original figure, it must give us exactly the same information about it as the metrical property from which we started. Thus *any metrical property can be stated in projective language*, and then it is true of all the projective set to which the original figure belongs; stated in metrical language, which is usually simpler, it is true of the original figure only.

Conversely, from a figure having a projective property we can obtain a figure having a corresponding metrical property, by projecting so that certain elements assume special positions. The most familiar case is that of projecting a straight line to infinity, when two straight lines intersecting at a point on that particular straight line become two parallel straight lines. Here "intersecting at a point on that particular straight line" and "parallel" are respectively the projective and metrical ways of stating the same property.

We can at once apply this way of looking at things to the projection used on p. 18. The first figure there has one metrical property, namely that the straight line FG is at infinity, which is not true of its projection **fg**; we can avoid mentioning this if for "parallel straight lines" we substitute "straight lines meeting on FG", and in particular, if for "straight lines parallel to the axes" we substitute "straight lines through F or G"; to these there correspond, in the second figure, "straight lines meeting on **fg**, or in **f** or **g** " respectively. Then a ruler construction for carrying out any rational operation upon the projective coordinates of points in the second plane can be deduced from those given in chapter II for Cartesian coordinates by merely making these changes in the wording. This gives a theoretical method of solution for any linear problem whatever.

The simplest metrical properties are concerned with length and angular measure, each regarded as a ratio. Among the simplest metrical statements that can be made are: that a given segment of a straight line is bisected at a given point, that two straight lines are parallel, that two straight lines are at right angles. The idea of area is metrical, and so is that of similarity; the usual definition of a circle is entirely metrical. The simplest projective property is incidence, as explained above, and next come the cross-ratios of ranges of four points on a straight line or of pencils of four straight lines through a point.

I. Projective Properties.

Desargues' Theorem.

An example of a projective property concerned with incidence only is given by *Desargues' theorem: If two triangles are in perspective, the pairs of corresponding sides meet in collinear points.* For a full discussion, see Mathews, chap. V.

Two triangles ABC, A′B′C′ are said to be in perspective if the three straight lines AA′, BB′, CC′ joining corresponding vertices all pass through one point O, called the centre of perspective or of homology. The theorem states that if the pairs of corresponding sides BC, B′C′; CA, C′A′; AB, A′B′ meet in X, Y, Z respectively, then the three points XYZ lie on a straight line called the axis of perspective or of homology.

This is true whether the two triangles ABC, A′B′C′ are in the same plane or in different planes; in the second case, the axis XYZ is the straight line of intersection of the planes of the two triangles.

We can apply Desargues' theorem to draw a straight line through a given point to pass through the point of intersection of two given straight lines, when for any reason it is not possible to lay the edge of a ruler against this point of intersection (see chapter VI). In fig. 7, let AA′, BB′ be the given straight lines and O their inaccessible point of intersection. If C is the given point, we first take any two points A, B, one on each of the given straight lines, and then construct a triangle A′B′C′ which is in perspective with ABC, with O as centre of perspective.

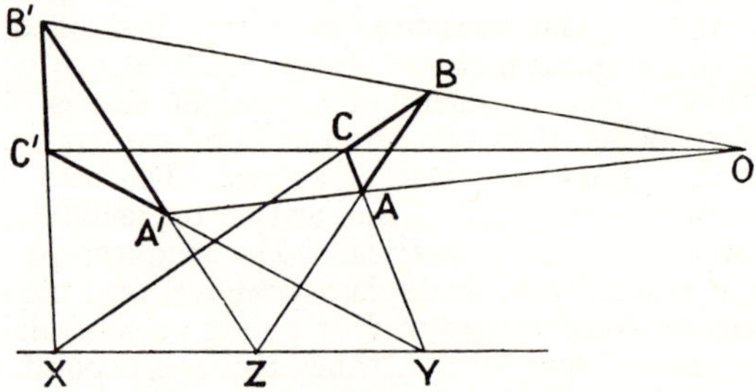

FIG. 7.

Then CC′ is the required straight line. Take any convenient axis of perspective and let it cut the sides of the triangle ABC in XYZ. Draw any straight line through Z to meet the given straight lines in A′, B′ respectively. Join YA′, XB′ to meet in C′; then CC′ passes through O, the point of intersection of AA′, BB′. In particular, if AA′, BB′ are parallel, so that O is at infinity, we can thus draw a third parallel through any given point C. Several other constructions for the general problem are given on p. 102, ex. 1.

Cross-ratio.

Most projective properties are connected with cross-ratio. The fundamental fact that a cross-ratio is unaltered by projection has already been assumed in several places. It is easily proved by reference to the angles at the vertex of projection.

Let there be given a range of four points A, B, C, D on a straight line, and let them be projected from any point O on to another transversal. Draw OP perpendicular to AB. Then, since

$$AC \,.\, OP = 2\triangle AOC = OA \,.\, OC \sin AOC,$$

the cross-ratio of the range is

$$\{AB, CD\} = \frac{AC . DB}{CB . AD} = \frac{AC . OP . DB . OP}{CB . OP . AD . OP}$$

$$= \frac{OA . OC \sin AOC . OD . OB \sin DOB}{OC . OB \sin COB . OA . OD \sin AOD}$$

$$= \frac{\sin AOC . \sin DOB}{\sin COB . \sin AOD},$$

and therefore depends only on the angles at O, and not

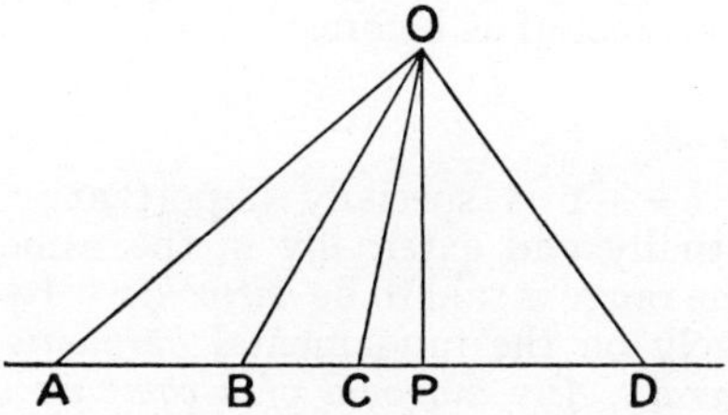

FIG. 8.

on the particular transversal AB, but is the same as the cross-ratio of the range cut out on any other transversal.

The first problem which presents itself is the construction of a range of given cross-ratio: more definitely, given three points OAF in a straight line, to find a point X in the

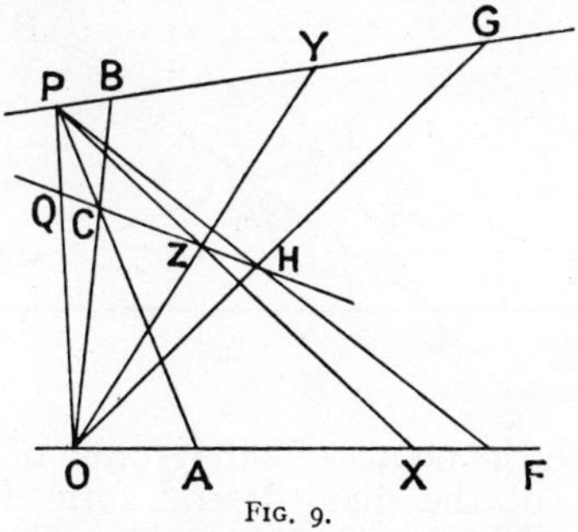

FIG. 9.

same straight line such that $\{OF, XA\} = \frac{OX . AF}{XF . OA} = \lambda$, where λ is a given number. The values $\lambda = \mathbf{0}, \mathbf{1}, \boldsymbol{\infty}$ are obtained by making X coincide with O, A, F respectively. If the assigned value λ is given as the cross-ratio {PG, YB} on another straight line, we can carry out the construction as follows.

Let PA, OB meet in C, and PF, OG in H. Let CH meet OP in Q and OY in Z. Then PZ meets OA in X; for

$$\{OF, XA\} = \{QH, ZC\} = \{PG, YB\},$$

the vertices of projection being P and O respectively.

If the given and required ranges are on the same straight line, we can first project the former on to any other straight line, and then proceed as before.

Harmonic Range.

The case $\lambda = -1$ is specially important; then OF is divided internally and externally in the same ratio at A and P, and the range is said to be *harmonic.* Its importance depends largely on the fundamental *harmonic property of the quadrilateral: Any diagonal of a complete quadrilateral is divided harmonically by the other two diagonals.* This property can also be taken as the definition of a harmonic range (see Mathews, chap. VI).

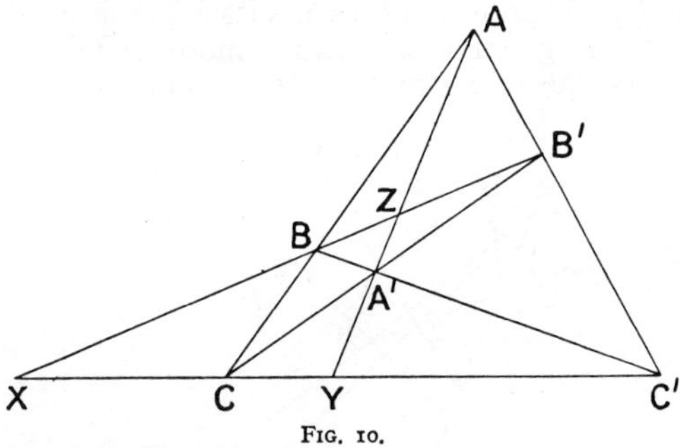

FIG. 10.

Let X, Y, Z be the points of intersection of the diagonals AA′, BB′, CC′ of the quadrilateral formed by the four straight lines ABC, AB′C′, A′BC′, A′B′C. Then the theorem states that each of the ranges

$$\{AA', YZ\}, \quad \{BB', ZX\}, \quad \{CC', XY\}$$

is harmonic. This is proved by projecting the range lying on one diagonal from each end in turn of a second diagonal on to the third.

Involution.

The harmonic property of the quadrilateral is a particular case of the more general *involution property of the quadrangle* (Mathews, p. 83) : *The pairs of opposite sides of a complete quadrangle cut any transversal in involution.* And this is in turn a particular case of a similar theorem on conics through four points (see p. 63). An *involution* is a set of pairs of points on a straight line, the two points of each pair being said to be *conjugate* to one another, such that the cross-ratio of any four of the points is equal to that of their four conjugates.

Let P, P′ ; Q, Q′ ; R, R′ be the intersections, with any transversal PQ, of the pairs of opposite sides AB, A′B′ ; AB′, A′B ; AA′, BB′ of a quadrangle ABA′B′. Then for example,

$$\{PQ, RR'\} = A\{BB', ZR'\} = A'\{Q'P', RR'\} = \{P'Q', R'R\},$$

and, similarly, the involution property can be verified for any four points of the range. If we now make the transversal coincide with CC′, we can identify P and P′ with C ;

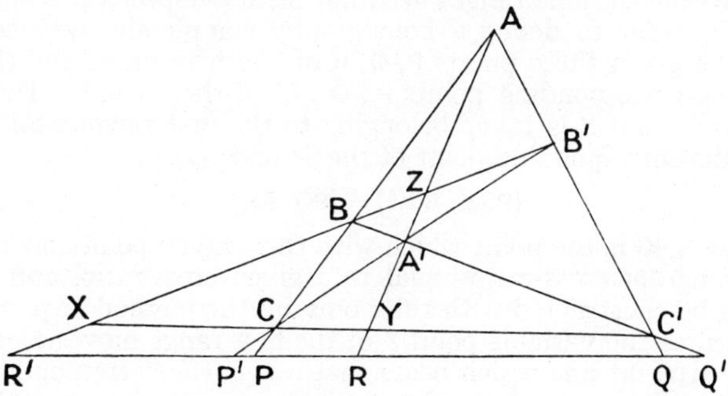

FIG. 11.

Q and Q′ with C′ ; R with Y ; R′ with X. The involution property just proved becomes {CC′, YX} = {CC′, XY}, which expresses that the range is harmonic.

If two pairs PP′, QQ′ of the involution are given, the quadrangle property enables us to construct the conjugate R′ of any given point R of PQ. Take any point A of the plane, not lying on the straight line PQR, and join AR; take any other point A′ on AR. Let AP, A′Q′ meet in B, and let AQ, A′P′ meet in B′; join BB′ and let it meet PQ in R′, which is the required conjugate to R. The point R′ so obtained is quite independent of the choice of the auxiliary points A, A′, and is completely determined by means of the equation $\{PQ, RR'\} = \{P'Q', R'R\}$ as the conjugate of R in the involution determined by the two given pairs of points PP′, QQ′.

Homography.

An involution is a particular case of a *homography*, defined as follows. Let there be a correspondence established between pairs of points P, P′; Q, Q′; R, R′; ... belonging respectively to two different ranges PQR..., P′Q′R′..., on the same or on different straight lines. Then if the cross-ratio of any four points of the first range is equal to that of the four corresponding points of the second range, the two ranges are said to be *homographic*, and the correspondence between the points is a homography. Two ranges in perspective give the simplest case of homography, but two homographic ranges need not be in perspective.

In order to define a homography completely, we need to be given three points P, Q, R of the first range and the three corresponding points P′, Q′, R′ of the second. Then if any point X is taken belonging to the first range, and X′ is the corresponding point of the second,

$$\{P'Q', R'X'\} = \{PQ, RX\};$$

that is, X′ is the point which with three given points makes a range of cross-ratio equal to a given cross-ratio, and it can be constructed with ruler only by the method of p. 39. Then as the variable point X of the first range moves along the straight line which bears that range, the corresponding point X′ moves along the straight line which bears the second range. If these two straight lines coincide, then X, X′ are in general different points of this line; when X is at P, then X′ is at P′; when X moves along and comes to the position P′, then X′ will in general take up some position P″ different from P′ and from P. But it may happen that for every position of P, this point P″ coincides with P; that is to say, to every point P′ of the straight line there corresponds the same point P, whether we regard P′ as belonging to the first range or to the second. Then the points of the straight line fall into conjugate pairs, and we need not specify which point of the pair belongs to either range; the cross-ratio of any four points is equal to that of their four conjugates, and the homography is an involution.

Common Points.

For example, let the two ranges (X), (X′) on the same straight line PQ be each in perspective, with centres of perspective C, C′ respectively, with the same range (Y)

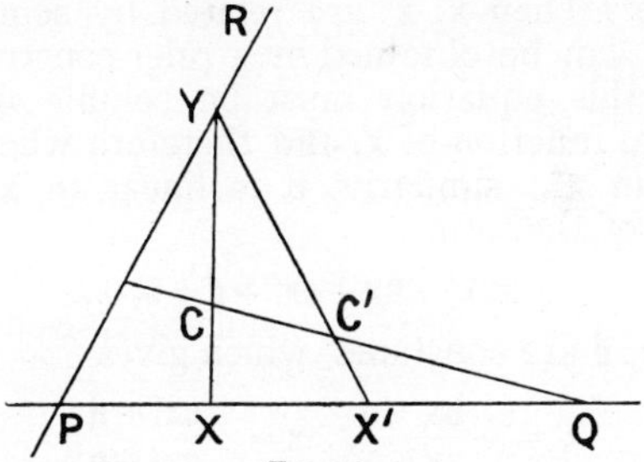

FIG. 12.

on the straight line PR. The three ranges are all homographic, and the common points of the first two are P the point of intersection of the two given straight lines PQ, PR, and Q the point of intersection of CC′, PQ.

As another example, the points of the same axis, whose Cartesian coordinates are x and λx, where λ is any constant, form homographic ranges. Then if p is the coordinate of P, that of P′ is λp and that of P″ is $\lambda^2 p$. If this homography is an involution, we must have $\lambda^2 p = p$ for all values of p, or $\lambda = \pm 1$. If $\lambda = +1$, the two ranges are identical; if $\lambda = -1$, each range is the reflexion of the other in the origin. In the homography, the origin corresponds to itself, and so does the point at infinity; in any homography on one straight line, or in any involution, there are always two points X_1, X_2, real or imaginary, which coincide with their corresponding or conjugate points X_1', X_2', but the construction requires the solution of a quadratic equation, and cannot be carried out with ruler only, except in special cases.

The two double points of an involution have the important property of dividing harmonically any pair of conjugates PP'; for if X_1, X_2 are their own conjugates, $\{X_1X_2, PP'\} = \{X_1X_2, P'P\}$, which is the characteristic property of a harmonic range. An involution can be defined as the set of pairs of points PP' which divide a given pair X_1X_2 harmonically.

We can prove analytically that any homography on one straight line has two common points. Suppose that the positions of X and X' are determined by their Cartesian coordinates x, x' measured from any fixed origin in the straight line. Then x, x' are related by some equation, and since x' can be obtained by a ruler construction when x is given, this equation must be soluble in the form x' = a rational function of x, and therefore when simplified it is linear in x'; similarly, it is linear in x, and must therefore be of the form

$$axx' + bx + cx' + d = 0, \qquad \text{(I)}$$

where a, b, c, d are constants, which gives

$$x' = -\frac{bx + d}{ax + c}, \quad x = -\frac{cx' + d}{ax' + b}.$$

If p is the coordinate of P, etc., where PQR, $P'Q'R'$ are the given points which determine the homography, we have $\{PQ, RX\} = \{P'Q', R'X'\}$, or

$$\frac{(r - p)(q - x)}{(q - r)(x - p)} = \frac{(r' - p')(q' - x')}{(q' - r')(x' - p')},$$

which can be rearranged in the form (I). The coefficients a, b, c, d are rational functions of the six given coordinates $p, \dots r'$, and these coefficients, or any rational functions of them, can be constructed with ruler only from the six given points.

Now the common points X_1, X_2 are found by putting $x = x'$ in equation (I), and their coordinates x_1, x_2 are therefore the roots of the quadratic equation

$$ax^2 + (b + c)x + d = 0, \qquad \text{(II)}$$

so that there are always two common points, real, coincident or imaginary, and finite or infinite.

In general, x_1, x_2 involve the square root $\sqrt{(b+c)^2-4ad}$, and so cannot be constructed with ruler only, except in particular cases, as for example when one root is known to be rational. If $a=0$, and also $b+c=0$, both roots are infinite. In this case the relation (I) between x and x' reduces to

$$x'=x+\frac{d}{b},$$

so that the second range is equal to the first displaced through a constant distance d/b; it is clear geometrically that there are no finite common points.

If the homography is an involution, the value of x' corresponding to any assigned value p of x is the same as the value of x when the value p is assigned to x'; in other words, the bilinear relation (I) is symmetrical in x, x', and the condition for this is $b=c$. Then the relation is

$$axx'+b(x+x')+d=0,$$

and the coordinates x_1, x_2 of the double points are the roots of the quadratic

$$ax^2+2bx+d=0.$$

The point O midway between the double points is called the *centre* of the involution; its coordinate is

$$x_0=\frac{x_1+x_2}{2}=-\frac{b}{a}.$$

We can change the origin to O by writing

$$x=\xi-\frac{b}{a}, \quad x'=\xi'-\frac{b}{a},$$

and the relation between the new coordinates ξ, ξ' reduces to

$$\xi\xi'=\frac{b^2}{a^2}-\frac{d}{a}=\left(\frac{x_1+x_2}{2}\right)^2-x_1x_2=\left(\frac{x_1-x_2}{2}\right)^2$$
$$=e^2, \text{ say,}$$

where e is the distance of either double point from the centre.

The centre is defined by a metrical property, and is the conjugate of the point at infinity. It can be constructed with ruler only, provided we can draw parallels (p. 47). The involution property expressed in metrical language is that the product of the distances of any pair of conjugates from the centre is constant. If the involution is projected on to any other straight line, conjugate pairs project into conjugate pairs, and double points into double points, but the first centre does not project into the second centre, except in the special case when the point at infinity on the first line projects into the point at infinity on the second line. The constant e may be real or imaginary; if it is real, so are the double points, and conjugate points are on the same side of the centre; if e is pure imaginary, there are no real double points, and conjugate points are on opposite sides of the centre.

II. Metrical Properties.

(i) Lengths.

A figure can have metrical properties only when some of its elements have special positions. The harmonic property of the quadrilateral (p. 40) is still the basis of discussion. First let one of the diagonal points, Z of fig. 10 say, lie at infinity; then the straight lines AA′, BB′ are parallel, and Z may be considered to be defined as the

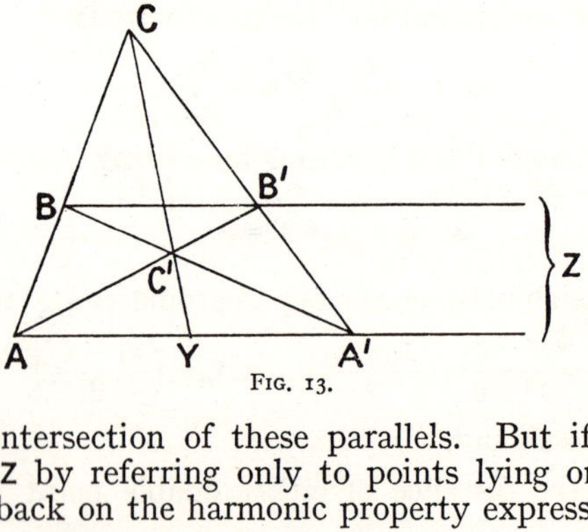

Fig. 13.

point of intersection of these parallels. But if we wish to define Z by referring only to points lying on AA′, we must fall back on the harmonic property expressed by the equation

$$\{AA', YZ\} = \frac{AY \,.\, ZA'}{YA' \,.\, AZ} = -1.$$

Since Z is at infinity, in this case the ratio $\frac{ZA'}{ZA}=\mathbf{1}$, and therefore

$$AY=YA'.$$

The point Z at infinity may be considered as the fourth harmonic of the three points A, A' and the midpoint Y of the bisected segment AYA'. Thus a pair of parallels, and a bisected segment on one of them, are equivalent data. If AA', BB' are two given parallels, we can bisect a given segment AA' on one of them: take any point B on the parallel BB', and any point C on AB; then to complete fig. 13, join AC to meet BB' in B'; join AB', $A'B$ to meet in C'; join CC' to meet AA' in Y, which is the required midpoint of AA'. Conversely, if we are given a bisected segment AYA', we can draw a parallel to AA' through any given point B: take any point C on AB; then to complete the figure, join $A'B$, YC to meet in C'; join AC', $A'C$ to meet in B'; join BB', which is the required parallel to AA' through the given point B. In the first case, we are given the point Z at infinity by means of the second straight line BB' passing through it, and then we can construct the fourth point Y of the harmonic range AA', YZ; in the second case, we are given Z by means of the other three points $AA'Y$ of a harmonic range to which it belongs, and then we can draw BB' to pass through Z. In either case, the point Z is determined, in the sense that we can draw the straight line joining it to any given point, that is, we can draw a parallel to AA' through any given point. This is exactly what is meant by saying that a finite point, such as B' in the last construction, is determined as the intersection of the two straight lines AC', $A'C$: it is so determined that we can use it in the next step of that construction, which is to join it to B.

Infinity.

It is important to notice that if AA' is produced to any *finite* point Z, the segment AZ is greater than the segment $A'Z$, because their difference AA' is greater than zero *and* their ratio $AZ/A'Z$ is greater than unity. For finite segments, the two properties of zero difference and unit ratio are inseparable, and each implies that the segments are equal. The converse is what is expressed by saying that the whole is greater than the part; this is not so much an axiom as the *definition of finiteness.* As Z recedes along AA′, the difference AZ − A′Z remains finite and constant, being equal to AA′; but the ratio AZ/A′Z is always greater than unity and tends to unity as its limit. When Z takes up its position at infinity, this ratio passes to its limit and is equal to unity, the difference remaining finite; but both the segments AZ and A′Z are infinite. They have become equal, which for infinite segments means that their ratio is infinite and not necessarily that their difference has any special value. The two properties of unit ratio and zero difference have become separable, and this fact is the *definition of infiniteness*: if from a whole a part is removed, and the whole is equal (that is, in a unit ratio) to the remaining part, then the whole and this part are both infinite.

Addition of Segments.

When the point at infinity on AA′ is determined, so that we can draw parallels to AA′, then besides bisecting any segment, we can transfer it to any other part of the straight line. Let it be required to lay off a segment equal to AA′ from any point B of AA′ or AA′ produced.

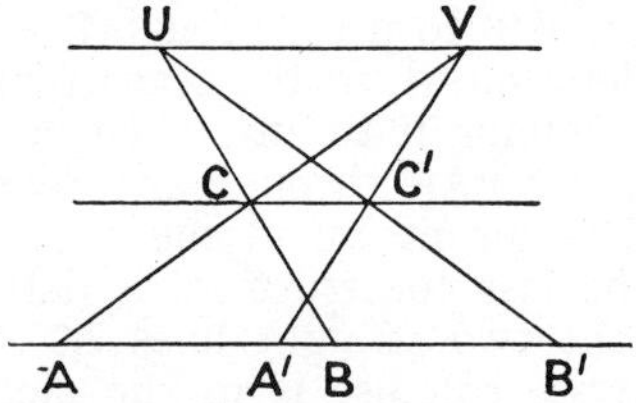

FIG. 14.

To AA′ draw any two parallels CC′, UV; take any point V on one of these, and join VA, VA′ to cut the other parallel in C, C′; join BC to meet UV in U, and join UC′ to meet AB in B′. Then B′ is the required point, for by similar triangles,

$$\frac{BB'}{CC'}=\frac{BU}{CU}=\frac{AV}{CV}=\frac{AA'}{CC'},$$

so that

$$BB'=AA'.$$

If we had wished to lay off the segment on the other side of B, we should have joined BC′ instead of BC.

Thus a bisected segment on a given straight line enables us to construct the sum or difference of any two given segments of that line, or, by repetitions, any linear function with integral coefficients of any number of segments of the same straight line.

For example, given a bisected segment ABA′ and another point C of the straight line AB, let it be required to lay off from a given point X of AB a segment XY of length $3\mathfrak{b} - \mathfrak{c}$, where $AB = \mathfrak{b}$, $AC = \mathfrak{c}$.

We might first draw two parallels to AB by means of the bisected segment ABA′; then transfer the segment CA′ to the position FA, and the segment FB to the position XY. The whole construction may be arranged more compactly as follows.

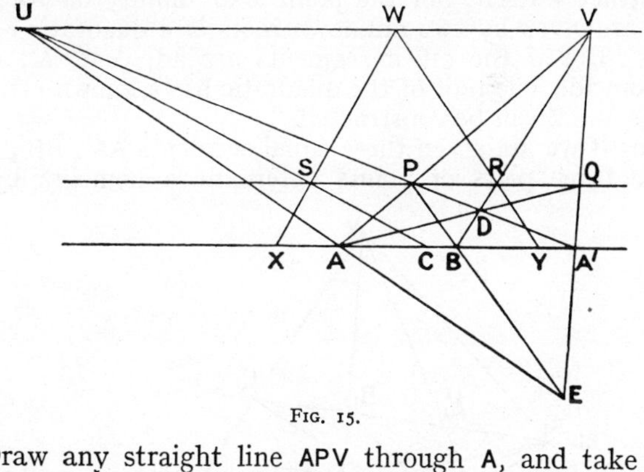

FIG. 15.

Draw any straight line APV through A, and take any two points P, V on it; join PA′, VB to meet in D, and join DA, VA′ to meet in Q. Join PB to meet VA′ in E, and join EA to meet PA′ in U. Then PQ, UV are each parallel to AB. Let VB meet PQ in R. Join UC to meet PQ in S, join XS to meet UV in W, join WR to meet AB in Y. Then XY is the required segment; the proof is left to the reader.

If we are only given two segments AA′, BB′ of the straight line, which are equal but not adjacent, we cannot draw parallels. Suppose, for example, that the four points occur in the order named above, so that the given segments do not overlap. Then taken in pairs AA′, BB′, they determine an involution whose centre is Y the midpoint of AB′ or of A′B, and whose double points are real; taken in pairs AB, A′B′, they determine an involution with Y as

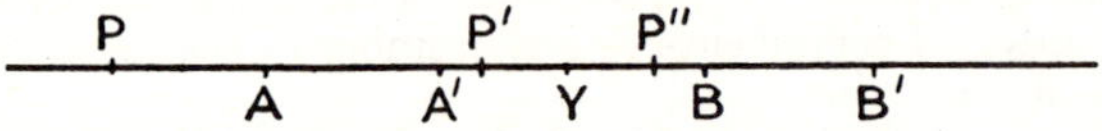

FIG. 16.

centre and imaginary double points; taken in pairs AB′, A′B, they determine an involution with its centre and one double point at infinity, and with Y as the other double point. If any point P is conjugate to P′ in the first of these involutions and to P″ in the second, then P′, P″ are conjugate in the third, and are equidistant from Y. Thus we can construct the reflexion of any point in Y, but we cannot construct Y itself, nor the point Z at infinity on AA′, for Y, Z are given by two unknown roots of a quadratic equation. But if the given segments are adjacent, A′, B, Y all coincide, one root of the quadratic has a known rational value and Z can be constructed.

But if we are given three equal segments AA′, BB′, CC′, these three pairs of points determine a homography in

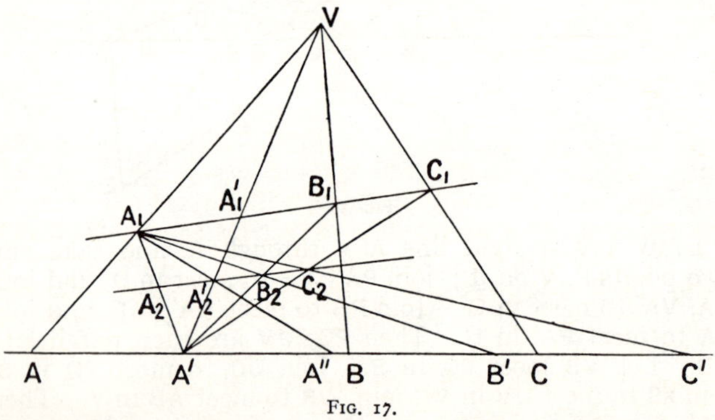

FIG. 17.

which the distance between corresponding points is con stant (p. 45). We can therefore construct the point A″, which corresponds to A′ when A′ is regarded as a point of the first range ABC; we make $\{A'B'C'A''\} = \{ABCA'\}$, by the construction of p. 39. Then $A'A'' = AA'$, and we have a bisected segment $AA'A''$, and can therefore draw parallels. The actual construction of A'' is shown in fig. 17. The range $ABCA'$ is first projected from any vertex V into $A_1B_1C_1A_1'$ on any straight line; then from A' as vertex into $A_2B_2C_2A_2'$, and then from A_1 into $A'B'C'A''$.

Parallelogram Given.

But none of these data help us to draw parallels to any straight line in a direction different to that in which we are given two parallels or a rationally divided segment. In other words, we are only given one point on the straight line at infinity, namely the point in which it meets the given parallels, and so we cannot consider it as fully determined, nor can we obtain the point in which it meets any other straight line, which is the point at infinity on the latter, and so we cannot draw parallels in a second direction. But if we are given two pairs of parallels in different directions, then we have two points on the straight line at infinity, which is therefore completely determined, and we can obtain its intersection with any other straight line, and so draw parallels in any direction. Since two pairs of parallels in different directions are the sides of a parallelogram, we may say that *if a parallelogram is given, parallels can be drawn to any straight line with ruler only*. This is the theorem which was assumed in chapter II, p. 14. In fact, if we have, or can obtain, a bisected segment BYB', and if we can also draw through these three points three straight lines BX, YZ, $B'X'$ all parallel to a second direction, these three equidistant parallels meet any straight line EF in a bisected segment XZX', by means of which we can draw parallels to EF. When a parallelogram is given, each diagonal is bisected by the other; we can draw parallels to one diagonal through the ends of the other, and these cut off from an arbitrary straight line a segment which is bisected by the parallel diagonal.

In the figure, AYA', BYB' are the diagonals of the given parallelogram $ABA'B'$, bisecting one another in Y. Using the bisected segment AYA', we draw through B a straight line BX parallel to AA' by the construction of p. 47; the sides AB, $A'B$ of the given parallelogram are used as sides of the complete quadrilateral required, and AA′ is one diagonal. We therefore draw any straight line YC_1C through Y to meet AB, A′B in C, C_1; join A′C, AC_1 to meet in B_1, and join BB_1, which is the required parallel. In constructing the other parallel through B′, we can use the

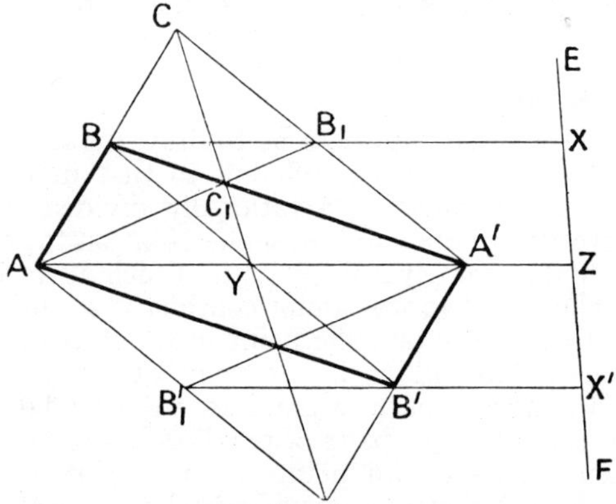

FIG. 18.

quadrilateral which is the reflexion of the first in the point Y, which includes the straight lines AB′, A′B′, CY produced, and the third parallel $B'B_1'$. Then these three parallels meet any straight line EF in points XZX′, which furnish a bisected segment, by means of which we can draw parallels to EF.

Rotations.

Then, as we saw in chapter II, we can perform any rational operations upon the Cartesian coordinates of the given points, including exchange of abscissa and ordinate; but these are not absolute lengths, but ratios of segments of the axes to the unit segments of the same axes; we cannot assume that these units are equal, or that the scale is the same for ordinates as for abscissae; we can transfer coordinates from one axis to the other, but we cannot transfer lengths. If we are able to choose equal lengths as units along the two axes, this is due to a separate, additional metrical relation between the data. When it is given, we can at once draw a rhombus, which is the special form assumed by the unit parallelogram, and its diagonals are a pair of straight lines at right angles. We can now transfer a length from a straight line in any direction to its reflexion in either diagonal of the unit rhombus; for the distance between the points whose coordinates are ($\mathbf{p}$, $\mathbf{q}$) and ($\mathbf{r}$, $\mathbf{s}$) is equal to the distance between ($\mathbf{q}$, $\mathbf{p}$) and ($\mathbf{s}$, $\mathbf{r}$). But unless the angle α between the axes has some special value, we cannot do more. For if the distance $\mathbf{d}$ between the two points ($\mathbf{p}$, $\mathbf{q}$) and ($\mathbf{r}$, $\mathbf{s}$) could be transferred to an axis, $\mathbf{d}$ would belong to the domain [1, $\mathbf{a}$, $\mathbf{b}$, ...] determined by the coordinates of the given points, to which domain $\mathbf{p}$, $\mathbf{q}$, $\mathbf{r}$, $\mathbf{s}$ also belong. But the formula for the distance between two points in oblique coordinates is

$$\mathbf{d}^2 = (\mathbf{r}-\mathbf{p})^2 + (\mathbf{s}-\mathbf{q})^2 + 2(\mathbf{r}-\mathbf{p})(\mathbf{s}-\mathbf{q}) \cos \alpha,$$

which gives a value for cos α belonging to the same domain, except in the cases $\mathbf{r}=\mathbf{p}$ or $\mathbf{s}=\mathbf{q}$, when the segment $\mathbf{d}$ is already parallel to one or other axis. But in general, the cosine of the angle between the axes is not rational, nor a rational function of the given coordinates; if this is so, we have a special case.

For example, if OA = OB and $\cos \alpha = \frac{1}{2}$, $\alpha = \frac{1}{3}\pi$, we have an equilateral triangle OAB, which is half the unit rhombus, and by drawing parallels, we can construct a regular hexagon ABCDEF, which gives three pairs of perpendiculars,

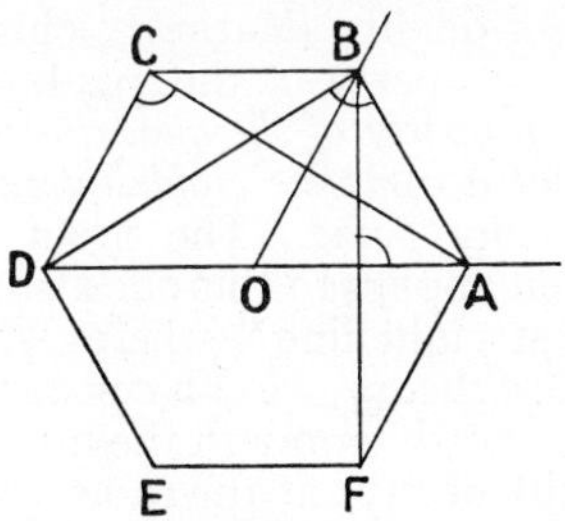

FIG. 19.

AB, BD; AC, CD; AD, BF. We can now reflect any segment in any of the diagonals of the hexagon, and two such reflexions give a rotation through an angle $\frac{2}{3}\pi$; we can erect an equilateral triangle on each side of any given segment, and by joining the vertices obtain a perpendicular, so that we can draw right angles and trisect them, but not bisect them.

If the units are not equal, but the axes are rectangular, then the unit parallelogram is a rectangle, and its diagonals give a new pair of axes, not at right angles, along which we have equal segments; so that the case of rectangular axes with unequal units is equivalent to the case of oblique axes with equal units. But if we are given both these metrical data at once, the unit parallelogram is a square, and its diagonals give a second pair of rectangular axes with equal units. We can reflect any segment in the sides and in the diagonals of the unit square, and therefore turn it through a right angle; for the straight line joining the points whose coordinates are ($\mathbf{p}$, $\mathbf{q}$) and ($\mathbf{r}$, $\mathbf{s}$) is now equal and perpendicular to that joining ($\mathbf{q}$, $\mathbf{r}$) to ($\mathbf{s}$, $\mathbf{p}$). We can erect a square on any given segment, and draw its diagonals; we can draw right angles and bisect them, but not trisect them.

(ii) Angles.

The chief interest of metrical properties of angles is based on the right angle, and the theory of right angles can be worked out in projective language, in terms of pencils in involution. The involution property involves cross-ratios only, and therefore belongs equally to a range of points on a straight line and to a pencil of rays through a point, by the principle of duality (Mathews, chap. I). Reciprocal to the involution property of the quadrangle (p. 41) there is the *involution property of the quadrilateral: The pairs of opposite vertices of a complete quadrilateral subtend a pencil in involution at any point.* The proof of p. 41 for the reciprocal theorem applies to this one also, if we interchange "point" and "straight line" wherever they occur; and just as in the other theory, we can construct with ruler only the conjugate of any given ray in the involution determined by two given pairs of rays at the same vertex.

Orthogonal Involution.

The application of the theory of involution to right angles is based on the theorem: *Pairs of straight lines at right angles through any point form a pencil in involution,* called the *orthogonal involution* at the point. For if OA′ is perpendicular to OA, etc., the pencil O{A′B′C′D′} is the same as O{ABCD} turned through a right angle; the angles between corresponding pairs of rays are equal, and the two pencils have the same cross-ratio. But this involution is not determined unless two different pairs of straight lines at right angles are given, or can be drawn, at the same vertex O; then we can construct the perpendicular to any other straight line through O, as its conjugate in the involution determined by the two given pairs.

The accompanying figure gives the actual ruler construction of the perpendicular to a given straight line **r** through a point O at which two right angles are given.

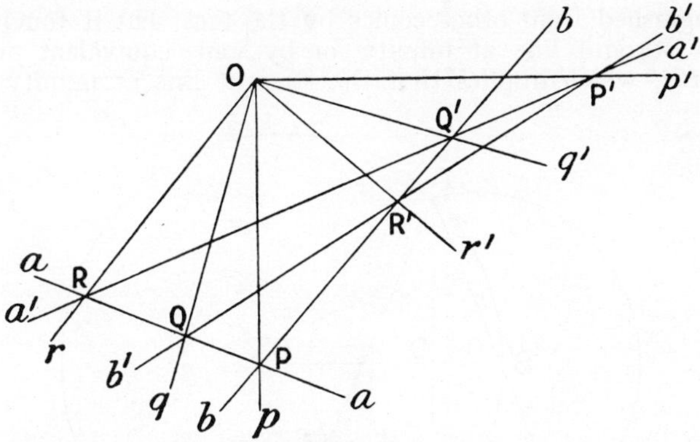

FIG. 20.

Small letters denote straight lines; **p**, **p**′; **q**, **q**′ are the two given pairs of straight lines at right angles, and **r** the straight line to which a perpendicular **r**′ is required. Take any straight line **a** of the plane, not passing through the common vertex O, and let it meet **p**, **q**, **r** in P, Q, R; take any straight line **a**′ through R and let it meet **p**′, **q**′ in P′, Q′. Draw the straight lines **b** joining PQ′ and **b**′ joining P′Q; let **bb**′ meet in R′. Then the straight line **r**′ joining OR′ is the required perpendicular to **r**. For **p**, **p**′; **q**, **q**′; **r**, **r**′ are the pairs of straight lines joining O to the pairs of opposite vertices P, P′; Q, Q′; R, R′ of the complete quadrilateral **aba**′**b**′, and are therefore in involution; and since two pairs of rays are at right angles, this is the orthogonal involution at O, and the two rays of every other conjugate pair, including **r**, **r**′, are also at right angles to each other. The figure is lettered so as to be reciprocal to fig. 11. The steps of the two constructions correspond exactly (see p. 42).

Focus of a Parabola.

As another example, consider the problem: to find the focus of a parabola touching four given straight lines. Since the focus is uniquely determined, this is possible by a ruler construction. The form of the question implies that the metrical data are given. For a parabola is distinguished from other conics by the fact that it touches the straight line at infinity, or by some equivalent property, which implies that the straight line at infinity is

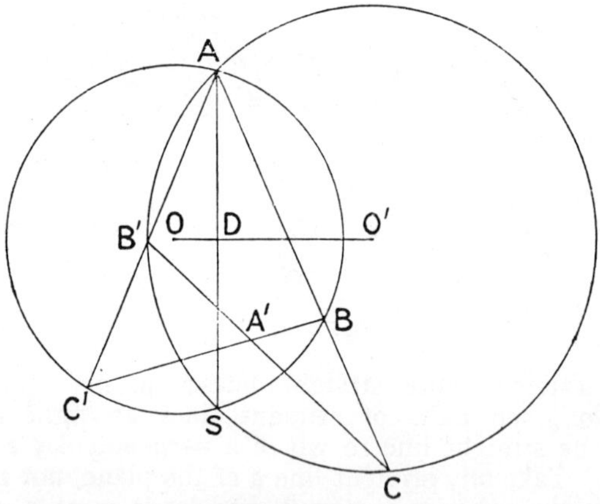

FIG. 21.

determined. And any definition of a focus involves right angles directly or indirectly. The usual focus and directrix definition of a parabola is as the locus of a point P such that its distance from a given point S is equal to its perpendicular distance PN from a given straight line NX. This assumes that we have the means of comparing a segment in an arbitrary direction PS with one in a fixed direction PN, so that we must have some instrument beyond a ruler, equivalent to an Einheitsdreher (p. 71); and we suppose that this has been used first of all to provide the metrical data for drawing parallels and perpendiculars.

In order to construct the focus S from the four given tangents AB, AB′, A′B, A′B′, we can use the known property that S lies on the circumcircle of each of the four triangles ABC′, etc., formed by three out of the four tangents, and it therefore lies on the common chord of each pair of these circles. We can construct these common chords without actually describing the circles, because each pair of circles has one common point already given. Find the centres O, O′ of the circles ABC′, AB′C in Euclid's way, by bisecting the sides at right angles. Draw AD perpendicular to OO′, and produce it to S, making DS = AD; then AD is the common chord, and S is the required focus.

Summary of the Theory of Ruler Constructions.

When no metrical data are given, we can carry out all rational operations upon the coordinates of the given points, these coordinates being projective, that is, expressed as cross-ratios.

If we are given that two straight lines are parallel, or that one segment is bisected, or one equivalent metrical fact, we can replace the corresponding coordinate by a Cartesian coordinate, which is a ratio of lengths instead of a cross-ratio. We can draw parallels in one direction, and carry out rational operations upon lengths of segments of any one of these parallels, but we cannot transfer lengths from one parallel to another. A single point at infinity is determined, but not the whole straight line at infinity. If we are given a separate metrical fact with regard to each of two different directions, we can operate upon a Cartesian system of two coordinates; we can draw parallels in any direction, and transfer a length from one parallel to another, but not from one direction to another. The straight line at infinity is completely determined.

If we are also given one right angle, or two equal lengths in different directions, we can choose a pair of oblique axes with the same unit of length, and we can reflect any length in the arms of the right angle. We cannot draw right angles with their arms in other directions.

If two pairs of parallels and two right angles are given, we can draw any parallels and perpendiculars. We can reflect a length in any axis, but we cannot transfer it into an arbitrary direction; the straight line at infinity is determined. This gives the greatest powers that can be obtained with ruler only. Since the sides of a square and its diagonals give two pairs of parallels and two right angles, *a square gives all the metrical data for ruler constructions*.

CHAPTER IV.

RULER AND COMPASS CONSTRUCTIONS.

I. Ruler and Compasses.

WHEN compasses are used as well as ruler, we can obtain at once all the metrical data required for that part of the construction in which the ruler is used. When a circle is drawn with a given centre, any diameter gives a bisected segment, so we can draw parallels to any straight line; and any two diameters give the corners of a rectangle, so we can draw right angles; and all the radii are equal, so we can have the same unit of measurement in all directions. We can also, by the use of ruler and compasses, bisect an arbitrary angle, but not trisect it, and carry out all the other constructions of the first six books of Euclid.

As we have seen, the steps of a ruler and compass construction which require compasses are each equivalent to the solution of a quadratic equation.

(i) The solutions of the equation

$$\mathbf{x}^2 = \mathbf{a}, \quad \text{or} \quad \mathbf{x} = \pm\sqrt{\mathbf{a}},$$

can be constructed as the mean proportionals between $\mathbf{a}$ and $\mathbf{1}$ by means of an ordinate in a circle on $\mathbf{1} + \mathbf{a}$ as diameter.

Take a unit length AO and produce it to Q, making $OQ = a$; draw a circle on AQ as diameter, and let it meet the perpendicular at O to AQ in X_1, X_2.

Then $$OX_1 = \sqrt{a}, \quad OX_2 = -\sqrt{a}.$$

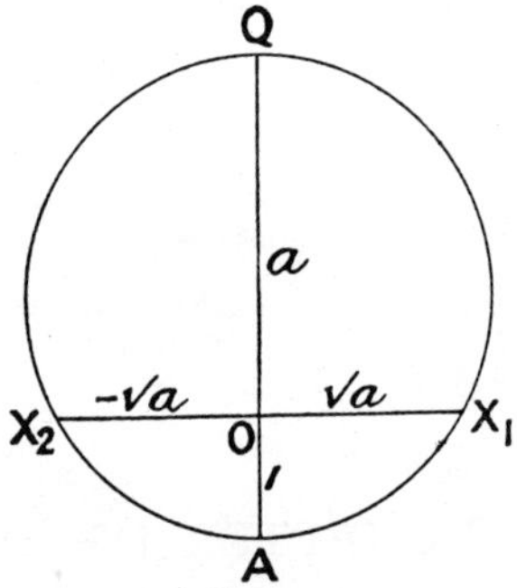

FIG. 22.

(ii) The solutions of the equation

$$x^2 - 2px + q = 0, \quad \text{or} \quad x = p \pm \sqrt{p^2 - q},$$

can be constructed by laying off a length $\sqrt{p^2 - q}$ in both senses from the end of a segment of length p. If OX_1,

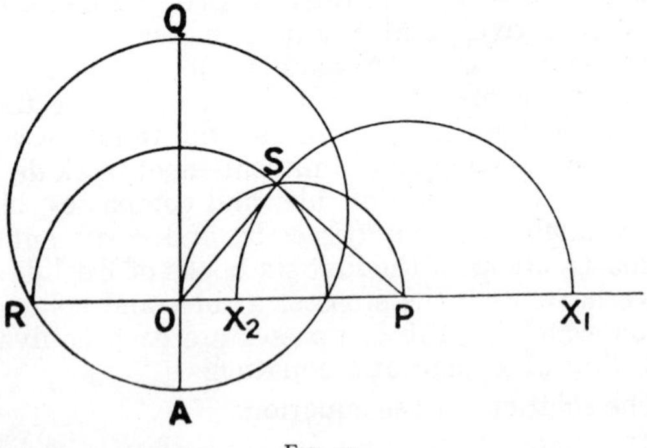

FIG. 23.

OX_2 are the roots, and $OP = p$, all measured along the same axis, P is the midpoint of X_2X_1, and $PX_1 = -PX_2 = \sqrt{p^2 - q}$.

We therefore take a unit length AO and produce it to

Q, making $OQ = q$; draw a circle on AQ as diameter, and let it meet the perpendicular at O to AQ in R; then $RO = \sqrt{q}$. Produce RO to P, making $OP = p$, and draw a circle on OP as diameter to meet the circle O(R) in S; then OSP is a right-angled triangle, with hypotenuse $OP = p$ and one side $OS = OR = \sqrt{q}$; therefore the other side $PS = \sqrt{p^2 - q}$. Describe the circle P(S) to meet OP in X_1, X_2; then the distances of X_1, X_2 from O are the roots of the given equation.

The construction fails if the circle O(R) does not meet the semicircle OSP in real points; this can only be when the latter lies entirely within the former, and $p < \sqrt{q}$; then the surd is imaginary.

V. Staudt has given a prettier and less obvious construction.

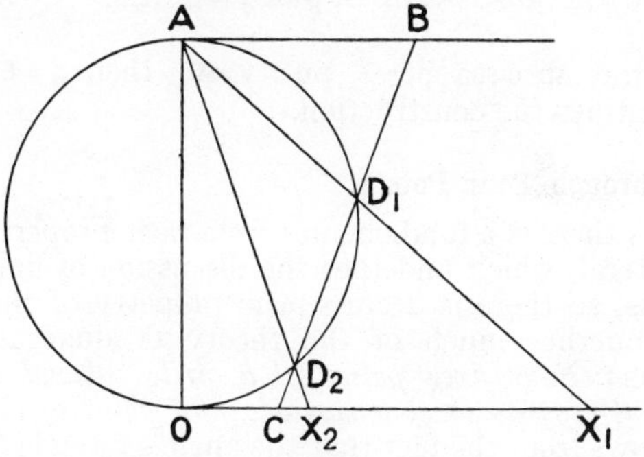

FIG. 24.

Take a circle of unit radius, and from two parallel tangents OC, AB cut off lengths $OC = \frac{q}{2p}$, $AB = \frac{2}{p}$, measured from the points of contact O, A. Join BC to cut the circle in D_1, D_2; join AD_1, AD_2 to cut OC in X_1, X_2; then OX_1, OX_2 are the roots of the given equation.

This can be verified analytically. Take OC, OA as axes of x and y; then the equation of AB is $\qquad y = 2$,
and of the circle OD_2D_1A, $\qquad x^2 + y^2 - 2y = 0$.

If X_1 has the coordinates $(x_1, 0)$, where x_1, x_2 are the roots of the given quadratic, the equation of AX_1 is

$$\frac{x}{x_1} = \frac{2 - y}{2};$$

this meets the circle again in D_1, whose coordinates we find to be $\frac{4x_1}{x_1^2+4}, \frac{2x_1^2}{x_1^2+4}$, and similarly for D_2.

The equation of D_1D_2 is therefore

$$\begin{vmatrix} x, & y, & 1 \\ 4x_1, & 2x_1^2, & x_1^2+4 \\ 4x_2, & 2x_2^2, & x_2^2+4 \end{vmatrix} = 0.$$

But since x_1, x_2 are the roots of $x^2-2px+q=0$, we may write $2p$ for x_1+x_2 and q for x_1x_2; by use of these relations, the equation of D_1D_2 may be reduced to the form

$$4(px-y)+q(y-2)=0.$$

To find the abscissa of B, put $y=2$; then $x=AB=\frac{2}{p}$; to find the abscissa of C, put $y=0$; then $x=OC=\frac{q}{2p}$; which justifies the construction.

Conics through Four Points.

Just as there is a fundamental cross-ratio property of the quadrilateral, which underlies the discussion of linear constructions, so there is a cross-ratio property of the circle, which underlies much of the theory of quadratic constructions: *Four fixed points of a circle subtend a pencil of constant cross-ratio at a variable fifth point of the circle.* This follows from the fact that the angle subtended by any two of the fixed points at the variable point is constant as the latter moves round the circumference. Hence the angles between the rays of the pencil are constant, and since the cross-ratio can be expressed in terms of the sines of these angles, the cross-ratio of the pencil is constant also. Thus we can give a meaning to the cross-ratio of four points upon a circle, defining it as the cross-ratio of the pencil subtended by the four points at any fifth point of the same circle. If four points lie upon a straight line, they subtend a pencil of constant cross-ratio at any vertex whatever; if they do not lie upon a straight line, the cross-ratio can have any value if the vertex can be any point of the plane; but if the four points all lie on a circle, and the vertex is taken to lie on the same circle, we get a definite, constant value for the cross-ratio.

This is a projective property, and is therefore true of any conic; for a conic can always be projected into a circle. Thus we have the *projective definition of a conic*, as the locus of a point at which four fixed points subtend a pencil of fixed cross-ratio.

From this there follows a theorem which is a generalization of the involution property of the quadrangle. In the first place: If a quadrangle is inscribed in a circle, the two intersections of any straight line with the circumference are a pair of conjugate points of the involution in which the pairs of opposite sides cut the chord. For let PP′ be

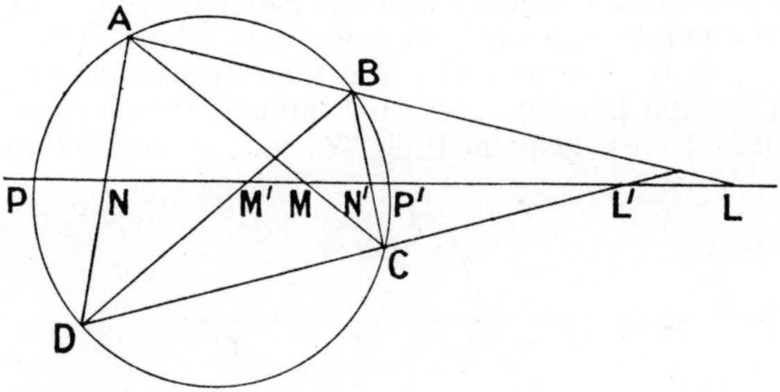

FIG. 25.

the chord and ABCD the quadrangle, with intersections as in fig. 25. Then

$$\{PLMP'\} = A\{PBCP'\} = D\{PBCP'\} = \{PM'L'P'\} = \{P'L'M'P\},$$

so that PP′ belong to the involution determined by the pairs LL′, MM′. This is a projective property, and is therefore true of the range cut out on a transversal by the sides of a quadrangle and any conic through the four vertices; hence all such conics cut the transversal in pairs of conjugate points of the same involution determined by the pairs of opposite sides. But these pairs of sides are themselves particular, degenerate cases of conics through the four points, so the theorem can be stated in the simple and general form: *Conics through four fixed points cut any transversal in involution.*

Hence if five points ABCDE on a conic are given, the other point of intersection X of any straight line EX through one of the five points is determined as the conjugate of E in the involution cut out on EX by the sides of the quadrangle ABCD. It can therefore be constructed with ruler only. Thus we have a ruler construction for any number of points on the conic through five given points, for we can find its intersection with any number of straight lines through any of the given points. Thus with ruler only, and without calculation, we can plot the curve with as much accuracy as time, patience and other limitations permit, and this is the nearest approach to actually drawing the curve that we can ever get with Euclidean instruments.

But if we wish to find the two intersections of a conic with an arbitrary straight line not passing through any of the five points, the problem cannot be solved with ruler only. Let it be required to find the intersections XY of a given straight line with the conic through five given points ABCDE. Take any point P on XY; join AP and determine

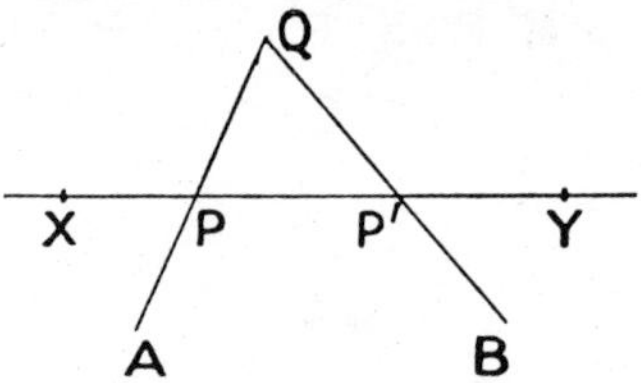

FIG. 26.

its other intersection Q with the conic; join BQ to meet XY in P′. Then if P lies on the conic, it coincides with Q and also with P′; and conversely, if P coincides with P′, it lies on the conic. If not, take any four positions P_1, P_2, P_3, P_4 of P, and let $\{P\}$ stand for their cross-ratio, with a similar notation for Q, P′. Then, since $ABQ_1 \dots Q_4$ lie on the same conic,

$$\{P\} = A\{Q\} = B\{Q\} = \{P'\},$$

so that P, P′ describe homographic ranges, and the points required are their common points X, Y, which demand the quadratic construction given below. But in the case in which XY passes through one of the five given points, E say, then one of the common points is E, and the other only demands a ruler construction, as we have just seen.

To the cross-ratio property of points on a conic there corresponds the cross-ratio property of its tangents: *Four fixed tangents to a conic cut out a range of constant cross-ratio on a variable fifth tangent to the conic.* And there is a construction for tangents to the conic touching five fixed straight lines, which corresponds step by step to any construction for points on the conic through five fixed points.

Common Points of a Homography.

In order to complete this discussion, we must go back to the theory of homography, and give the construction for the common points of two homographic ranges. Let there be given two homographic ranges PQR... , P′Q′R′... on the same straight line PQ, whose common points X, Y are required. Take any circle, and any point V upon it; with V as vertex project the ranges into **pqr**... , **p′q′r′**... on

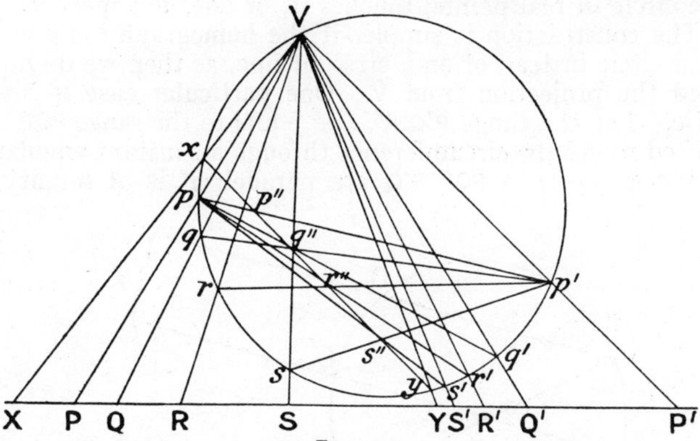

FIG. 27.

the circle. Now if three pairs of corresponding points P, P′; Q, Q′; R, R′ are given on the straight line, we know that the point S′ corresponding to any point S is determined; and it may be found thus.

Join **pq′**, **p′q** to meet in **q″**; join **pr′**, **p′r** to meet in **r″**; join **q″r″** to meet **pp′** in **p″**, to meet **p′s** in **s″**, and the circle in **x**, **y**. Join **ps″** to meet the circle in **s′**, and join V**s′** to meet PQ in S′. Then S′ is the point corresponding to S; for

$$\{PQRS\} = V\{pqrs\} = p'\{pqrs\} = \{p''q''r''s''\}$$
$$= p\{p'q'r's'\} = V\{p'q'r's'\} = \{P'Q'R'S'\}.$$

We have not only the two given homographic ranges on PQ, but also the two ranges **pqr**... , **p′q′r′**... on the circle, homographic with the former and with each other; and the common points on the straight line project into common points on the circle. But the common points on the circle are **x**, **y**, where **q″r″** meets the circumference, and therefore the common points of the given homography are X, Y, the projections from V of **x**, **y**.

One common point X is at infinity if Vx is parallel to PQ; in this case any problem that requires a finite common point of the homography has only one solution, and demands only a linear construction. Both X and Y are at infinity if **q″r″** touches the circle where it is met again by the parallel to PQ through V. In general, the common points are real, coincident, or imaginary according as **q″r″** cuts the circle in real points, touches it, or does not meet it.

The construction is simpler if the homography is given on a circle instead of on a straight line, as then we do not need the projection from V. One particular case is historic. Let the range P′Q′R′... be equal to the range PQR... shifted round the circumference through a constant angular distance α. Then PQ′, P′Q are parallel, **q″** is at infinity,

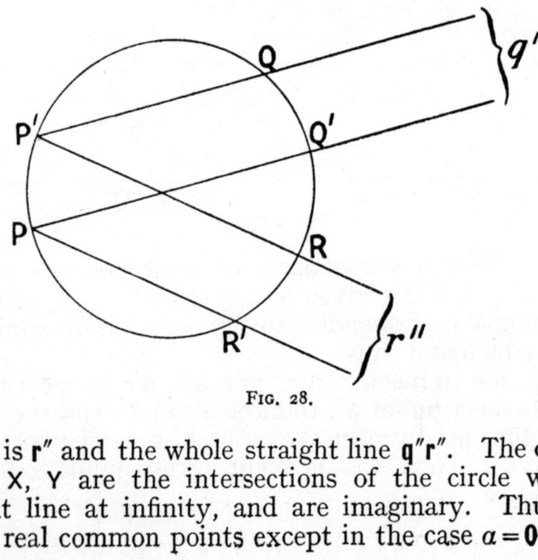

FIG. 28.

and so is **r″** and the whole straight line **q″r″**. The common points X, Y are the intersections of the circle with the straight line at infinity, and are imaginary. Thus there are no real common points except in the case $\alpha = 0$ or $2n\pi$, when the two ranges coincide, and every point is the same as its corresponding point.

Porisms.

Now suppose, for example, that we wish to draw a triangle whose corners lie on one circle, and whose sides touch a concentric circle. Take any point P on the outer circle, and draw successive tangents PQ, QR, RP′ to the inner circle to meet the outer circle again in Q, R, P′. Then

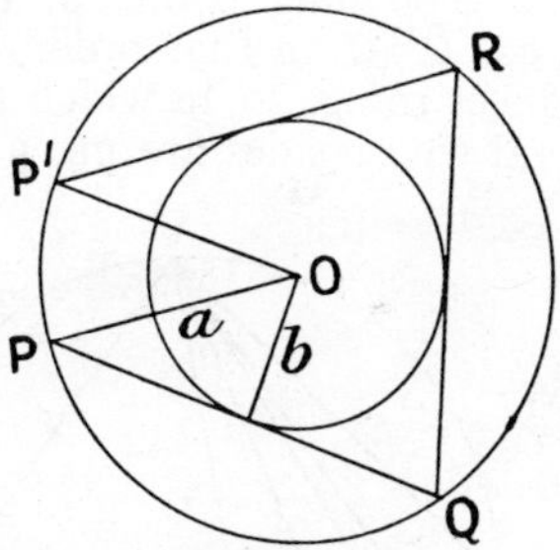

FIG. 29.

if P, P′ coincide, PQR is the required triangle, and the problem is solved; if not, P and P′ describe homographic ranges on the circle, and the solution is given by the common points. But if O is the centre of the circles, and **a**, **b** the outer and inner radii, $\angle POP' = 6\cos^{-1}(b/a) = \text{constant}$; the homography is of the kind just mentioned, and no real positions of P, P′ can coincide, unless we have the condition

$$6\cos^{-1}\frac{b}{a} = 2n\pi, \quad \text{or} \quad a = 2b,$$

and in that case, P, P′ always coincide, and there are an infinite number of triangles of the kind required, one with a corner at any point P of the outer circle. The problem is *poristic*, that is to say, although the number of independent conditions is equal to the number of variables, yet these conditions are in general inconsistent, and there is no solution, except in the case when a special relation holds among the constants, when the conditions are no longer independent, and there are an infinite number of solutions. The same is true when the triangle is replaced by a polygon of any given number of sides; it can also be extended to any pair of circles, and, by projection, to any pair of conics: *The problem of circum-inscribing a polygon to two given conics is poristic.*

Double Points of an Involution.

In the case of an involution, the construction for the double points can be simplified, for one pair of conjugate points of the involution, taken in reverse order, gives a second pair of corresponding points of the homography. In fig. 27 we can use Q′, Q, in that order, in place of R, R′, and the figure reduces to fig. 30, in which **q″r″** is a diagonal of the quadrilateral whose sides are **pq**, **p′q′**, **pq′**, **p′q**.

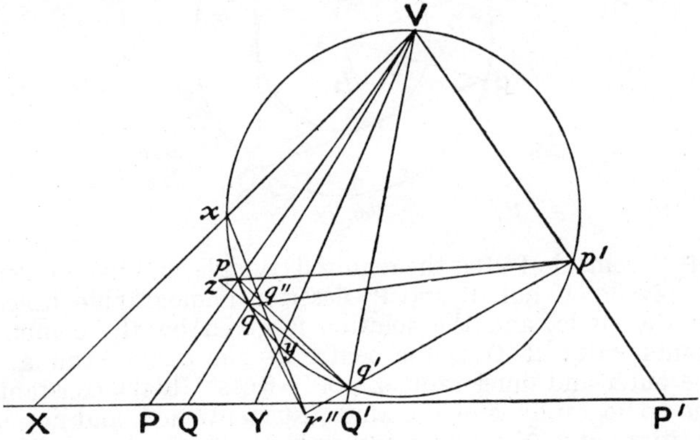

FIG. 30.

This figure enables us to see when two given pairs of points PP′, QQ′ determine an involution whose double points are real. This is so if **q″r″** meets the circle in real points. Let **pp′**, **qq′** meet in **z**; then since **q″r″** divides harmonically the two chords **zpp′**, **zqq′** through **z**, therefore **q″r″** is the polar of **z**, and meets the circle in real points if **z** is outside the circle and in imaginary points if **z** is inside the circle.

Now **z** is inside the circle, provided **q**, **q′** lie one on each of the two arcs into which **p**, **p′** divide the circumference. In this case, on whatever arc V may lie, one and only one of the straight lines Vq, Vq' lies between Vp and Vp', so that one and only one of Q, Q' lies between P and P'. Hence *the double points are imaginary if the given pairs overlap*; and are real if one pair lies wholly between the other pair (as when the points are in the order $PQQ'P'$), or when each pair is outside the other pair (as $PP'QQ'$).

The graphical solution of any quadratic equation can be made to depend on finding the double points of an involution. The general equation

$$x^2 - 2px + q = 0$$

can be regarded as the equation for the coordinates of the double points of an involution, in which the coordinates x, x' of any pair of conjugate points are connected by the relation

$$xx' - p(x + x') + q = 0.$$

We can at once obtain two pairs of conjugates; for example,

$$x = 0,\ x' = \frac{q}{p}; \quad x = 1,\ x' = \frac{q - p}{p - 1}.$$

If p, q are given, we can find these four points by a ruler construction, and thence find the double points of the involution by the construction given above, and so solve the quadratic equation.

One Fixed Conic.

Now that construction only requires one circle to be drawn, which is an arbitrary circle; and only uses projective properties of that one circle, namely the cross-ratio property and the property of meeting any straight line in two points. Hence the circle can be replaced by any conic actually drawn; the rest of the construction requires a ruler only. From this it follows that if a single conic of any species is drawn once for all, we can find the double points of any involution of which two pairs of conjugates are given, and hence we can construct the roots of any quadratic equation with given coefficients. But this is exactly the power which the use of compasses adds to the use of a ruler; which proves that *a single fixed conic can completely replace the use of compasses.* The case in which the fixed conic is a circle is discussed in chapter VII, where some actual constructions are given.

II. Other Elementary Instruments.

Geometrical construction of to-day is a very different thing from what it was in the time of Euclid. A modern schoolboy collects far more apparatus, when he sits down to write out the theorem of Pythagoras, than its discoverer ever possessed; but both for accuracy and for general utility, the two Euclidean instruments still hold the field. Moreover, most other modern instruments depend for their manufacture on ruler and compass constructions, and in all their commoner uses they do not enable us to draw any figure that could not have been drawn, if enough time and trouble were taken, by Euclid's methods.

In this section we shall refer to a few of the instruments that are commonly used along with ruler and compasses. It would carry us too far to describe the devices that extend the range of geometric constructions to the solution of cubic, higher and transcendental problems, such as trisectors and integraphs. But there are three instruments that call for some notice, as providing short cuts in Euclidean constructions, without adding to the range; these are *dividers*, *parallel ruler*, and *set-square*. We shall determine the extent of their powers by means of the analytical ideas of chapter II.

(i) Dividers.

When we speak of a pair of compasses as one of Euclid's instruments, we have to remember that the ordinary modern article is a complex instrument, combining the powers to do two quite different things. First we may make the two legs stand upon two given points A, B, whose distance apart is equal to the radius of the required circle; then we pick up the compasses and set them down again with one leg standing upon a given point C, which is to be the centre of the circle, and then we describe the circumference with the other leg. Now when Euclid marked his circles, probably on the sand, he required to be given the centre C and a point A on the circumference of the required circle before he could describe it. It was not sufficient to have the radius AB given in any other position; one of the points, B say, had to coincide with C.

It is true that Euc. I. 2 removes this restriction, so that, as far as regards the possibility of a construction, there is no difference between the two forms of compasses; but the figure of Euc. I. 2 contains three straight lines and four circles, and though a simpler figure of five circles may be used instead (see p. 132), there is still a great difference between the two forms of instrument as regards the simplicity of a construction.

Mascheroni, who wrote at the end of the eighteenth century, found that the makers of scientific instruments of his day, in graduating the reading circles of telescopes where great accuracy was required, preferred to use compasses as far as possible, as being more accurate than rulers. He regarded it as a special feature of a pair of compasses that it retained accurately a radius once taken up with it, until he chose to alter it (" compas fidèle "). Where the same radius occurred more than once in a construction, he kept a separate pair of compasses for it, laying them aside until they were wanted again; for there was less chance of error from the compasses slipping than there was from having to take up the radius afresh. We shall use the terms *Euclidean* and *modern* compasses where it is necessary to make the distinction; there are quite a large number of constructions in which the modern use has no advantage over the ancient. Modern compasses combine the power of describing circles, which belongs to Euclidean compasses, with the power of carrying distances, which belongs to a separate instrument called *dividers*, whose elementary operation is exactly that described in Euc. I. 2: " from the greater of two straight lines to cut off a part equal to the less."

All that can be done with ruler and dividers can be done, possibly with greater labour, with ruler and Euclidean compasses. The converse is not true; ruler and dividers can do more than ruler alone, but not so much as ruler and compasses.

Einheitsdreher.

Dividers give all the metrical data for drawing parallels and perpendiculars; for we can draw two straight lines intersecting at O say, and lay off along them equal lengths AO, OA′, BO, OB′, and so obtain the corners of a rectangle

ABA′B′. This enables us to draw parallels, and we have only to repeat the construction to obtain a second right angle with its arms in other directions, and this enables us to draw perpendiculars. Since we can draw parallels, we can make similar figures, and dividers have no wider range of construction than an instrument, such as a ruler or strip of paper with two fixed marks on it, that will transfer from one part of the plane to another one definite segment, say the unit of length. The range is also just as wide if all we can do is to rotate the unit of length about one end which remains fixed, that is, to cut off a unit length from any straight line through a fixed origin; an instrument which can do this is called by Hilbert *Einheitsdreher* (unit rotator). For with an Einheitsdreher we can construct a rhombus as above, and then with a ruler we can draw parallels. Then let it be required to cut off, from a given segment PQ, a part PX equal to a given segment

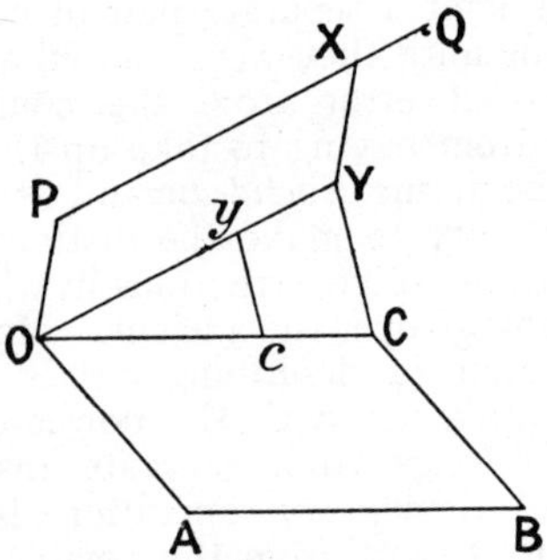

Fig. 31.

AB. Let O be the origin; draw OC, OY parallel to AB, PQ respectively, and cut off from these unit lengths Oc, Oy. Complete the parallelogram OABC; draw CY parallel to cy, and YX parallel to OP to meet PQ in X. Then PX is the required segment, for, by parallels,

$$PX = OY = OC = AB.$$

Hilbert in his *Grundlagen* gives a proof that the range of constructions possible with this instrument corresponds to the set of quadratic surds which are essentially real, so as to remain real however the quantities on which they depend vary within their domain ; the full proof is too hard to give here. For example, we can bisect any angle ; this is a quadratic problem which always has two real solutions, the internal and external bisectors, however the given angle may vary. Cut off equal lengths OA, OA′, OB from the vertex O along the arms of the angle ; complete the rhombuses OACB, OA′C′B ; then OC, OC′ are the

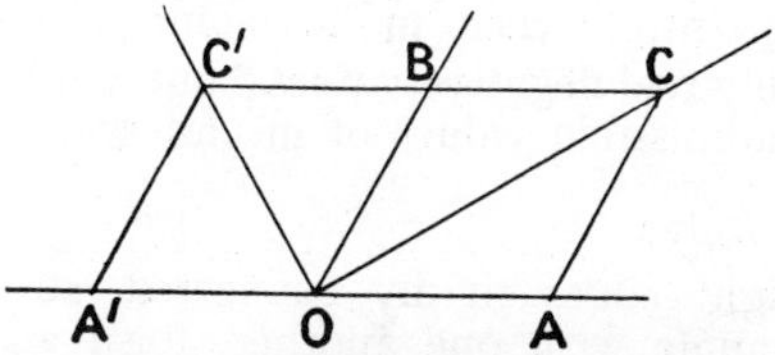

FIG. 32.

two bisectors. Again, if the centre and radius of a circle are given, we can draw straight lines through the centre, and mark off on each its two intersections with the circumference, at a distance from the given centre equal to the given radius ; we can obtain the intersections of the circle with any diameter, for these are always real ; but we cannot obtain its intersections with an arbitrary straight line which does not pass through the centre, for these intersections become imaginary for some positions of the given straight line.

We can see more exactly what can be done with an Einheitsdreher if we turn to analysis. Take the centre as origin ; we can construct the two intersections of the circle $x^2+y^2=1$ with any straight line through the centre $y=mx$ that we can draw, that is, such that m belongs to the domain. The coordinates of these points of intersection are

$$\left(\pm\frac{1}{\sqrt{1+m^2}}, \quad \pm\frac{m}{\sqrt{1+m^2}}\right),$$

and since we can carry out rational operations on these coordinates, the effect is to add $\sqrt{1+m^2}$ to the domain, which is therefore formed from the coordinates of the given points by rational operations, together with this new operation of passing from m to $\sqrt{1+m^2}$, where m is any quantity belonging to the domain. More generally, if m, n are any two quantities of the domain, we can construct in turn $\frac{\mathsf{m}}{\mathsf{n}}$, $\sqrt{1+\left(\frac{\mathsf{m}}{\mathsf{n}}\right)^2}$, and $\mathsf{n}\sqrt{1+\left(\frac{\mathsf{m}}{\mathsf{n}}\right)^2}$ or $\sqrt{\mathsf{m}^2+\mathsf{n}^2}$; so that the new operation is equivalent to taking the square root of the sum of squares. This operation is less general than that of taking the square root of an arbitrary quantity of the domain, for $\sqrt{\mathsf{m}}$ is an imaginary quantity when m has some of its admissible values, namely real negative values; but $\sqrt{\mathsf{m}^2+\mathsf{n}^2}$ remains real for all admissible values of m and n.

(ii) **Parallel Ruler.**

Two straight edges rigidly connected, so as to make a constant angle with one another, form an instrument various types of which are common. If the angle is a right angle, we have a T-square; $\frac{1}{6}\pi$, $\frac{1}{4}\pi$, $\frac{1}{3}\pi$ are usual in set-squares; these will be discussed below. If the angle is zero, we have a parallel ruler.

First of all, this can be used as an ordinary ruler. Next, it gives the metrical data, for if we lay it down anywhere and rule along both edges, and then do it again with the edges in a different direction, we get a rhombus, of which each altitude is equal to the breadth of the ruler, which we shall take as the unit of length, and the diagonals of the rhombus give a right angle. The positions of the two edges are completely determined as soon as they are made to pass through two given points of the plane; we must distinguish between two uses of the parallel ruler, according as the same edge is made to pass through both of the given points, or a different edge through each.

First Use Equivalent to Dividers.

In the first use, one edge is made to coincide with a given straight line, joining the given points AB; and we rule a parallel straight line at unit distance from AB; in fact we can rule two, one on each side of it. Since there are these two positions, this elementary operation with the parallel ruler is equivalent to solving a quadratic equation, but as before, it is one whose roots are always real. The first use is exactly equivalent to the use of dividers; for we can cut off a unit distance from any straight line OA, by first drawing the perpendicular OB and then drawing a parallel to OB at unit distance from it, to meet OA in the point required; and conversely, dividers enable us to draw a parallel to a given straight line OB at unit distance from it, by first cutting off a unit length from the perpendicular OA. With dividers, we can find the intersections of the unit circle with any diameter, and then draw the tangents at these points, perpendicular to the radii; with a parallel ruler we can draw the tangents to the unit circle parallel to any given radius, and then determine their points of contact as the feet of the perpendiculars through the centre.

Second Use Equivalent to Compasses.

In the second use of the parallel ruler, one edge is made to pass through one given point A, and the other through B. This is only possible provided $AB \geqq 1$. Then by ruling along the second edge we obtain, by an elementary operation, the solution of the problem: through a given point B to draw a straight line BC at unit distance from another given point A, that is to say, to draw from B a tangent to the circle centre A of unit radius. This has two solutions in general, and is again equivalent to solving a quadratic equation; but it is one whose roots become imaginary when B lies within the unit circle, that is, when the parameter $m = AB$ takes real values less than unity, which are permissible values for the distance between two arbitrary points of the plane. Now if C is the point of contact of the tangent, which is the foot of the perpendicular from A to BC, we have $BC = \sqrt{m^2 - 1}$, so that we can construct surds of this form. But then we can construct any quadratic surd, for if a is any given quantity of the domain, and $m = \frac{a+1}{a-1}$, we can obtain in turn m, $\sqrt{m^2-1}$, and $\frac{a-1}{2}\sqrt{m^2-1}$ or $\sqrt{a}$. Thus a parallel ruler, in its second use, can wholly replace compasses.

For example, let it be required to find the intersections of a given straight line PQ with a circle whose centre O and radius are given, but which is not to be drawn. We assume that the radius is equal to the breadth of the ruler; if this were not so, we could first construct a figure similar to the required figure but with a unit circle, and then alter the scale to the desired size.

If X, Y are the required intersections, the tangents to the circle at X, Y meet in R the pole of PQ. These tangents can be drawn at once if R is first obtained. Now R lies on TT′, the polar of any point P of PQ, and also on OS, the

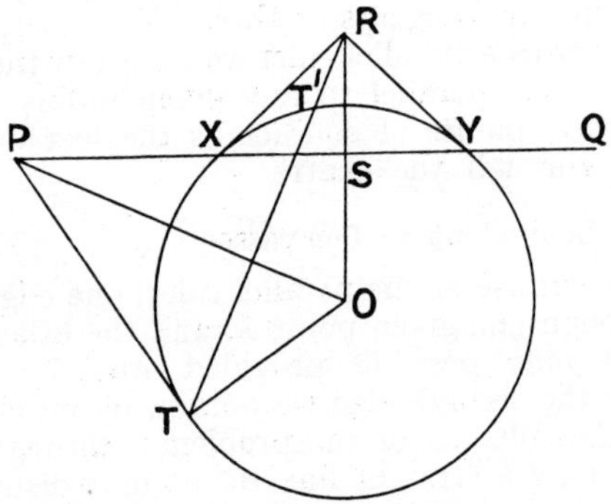

FIG. 33.

perpendicular from O to PQ. We therefore take any point P on PQ, and draw PT through P at unit distance from O ; draw OT perpendicular to PT ; then TR perpendicular to OP and OSR perpendicular to PQ to meet in R ; draw RX, RY at unit distance from O to meet PQ in X, Y, which are the required points.

In this second use of the parallel ruler, we are only using a single point of the first edge, and the essential parts of the instrument are one straight edge and a point at a fixed distance from it, which can be made to coincide with any given point of the plane, while at the same time the straight edge is made to pass through a second fixed point which is far enough away from the first.

(iii) **Set-square.**

If the ruler has two straight edges which are not parallel, but inclined at a constant angle α, its position is not determined unless we make its edges pass through three fixed points, two on one edge and one on the other ; the elementary operation is to draw through a given point A a straight line making the angle α with the straight line joining two other given points BC. We can construct a parallelogram and draw parallels; in particular, we can draw a rhombus with BC as diagonal and an angle 2α at B; the other diagonal is then at right angles to BC, and we have all the metrical data. But so far we have added nothing to the range of ruler constructions, except that $\tan \alpha$, if irrational, must be added to the set of independent coordinates, of which the rational functions can be constructed.

But if we assume the power to make a marked point of one edge coincide with a fixed point of the plane, while the second edge passes through another fixed point, then since the marked point is at a fixed distance from the second edge, this use of the set-square is exactly equivalent to the second use of a parallel ruler, and so can replace the use of compasses. A different use of the set-square, which can also replace the use of compasses, is when we assume the power to make each edge pass through one of two given points A, B, and also the vertex, or point of intersection of the edges, lie on a given straight line **c**. The two positions of the vertex are the two intersections of **c** with the arc of a circle standing on AB capable of containing an angle α; and since we can determine the intersections of any straight line with this fixed circle, we can (p. 69) carry out all constructions that are possible with ruler and compasses.

CHAPTER V.

STANDARD METHODS.

THE only quite general methods which apply to all ruler and compass problems are those which follow the steps of the analytical discussion in each case; these are seldom the best. But there are certain standard methods, each of which applies to a more or less well-defined group of cases, though the groups overlap considerably. The idea underlying many of these is the *separation of properties*. If the problem is definite, we have to construct certain elements which have certain relations to the data, and each of these required elements has to satisfy just as many independent conditions as are necessary to determine it: two for a point or a straight line, three for a circle. If these conditions have more than one solution, there may be further requirements, which exclude some, but not all, of the elements which satisfy the other conditions. Now it is often possible to separate the conditions into two distinct statements, each expressing a single requirement or group of requirements, and to consider each statement by itself. Then in general, each statement determines an infinite set of elements, and the solution of the problem is given by the common elements of the two sets.

(i) Method of Loci.

The simplest example is the method of loci, when a point is to have two properties, each of which is satisfied by all points of a certain locus; then the point or points of intersection of these two loci are the solutions of the problem. In applying this, the art is so to separate the properties that the two loci are both easy to discover and possible to draw: they must for our purpose consist of straight lines and circles. The method fails to give a ruler and compass construction if one of the loci is a curve other than a circle. Easy examples of this method are given by the usual constructions for the circumcentre and incentre of a triangle, and many similar problems.

Ex 1. The circumcentre S of a triangle ABC is defined as a point equidistant from the three corners A, B, C. This can be separated into the two statements, SA = SB and SA = SC. The locus of points having the first of these two properties is the perpendicular bisector of AB; the locus of points having the second is the perpendicular bisector of AC; the single point S common to these two loci is the solution of the problem.

Ex 2. To construct the radical axis of two given circles. This is defined as the locus of points from which equal tangents can be drawn to the two circles; it is known to be a straight line perpendicular to the straight line joining the centres A, B. It can therefore be drawn if a single point X on it is obtained, by drawing the perpendicular from X to AB; or we may obtain two such points X, Y and join them.

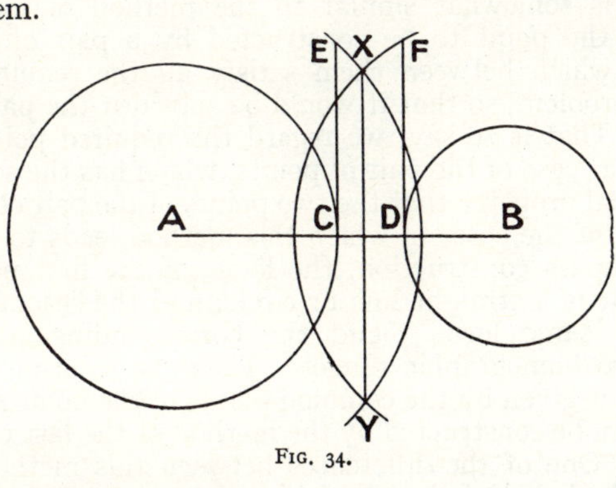

FIG. 34.

If X is the point from which the tangent to each circle is of a given length t, then X may be constructed by the method of loci. The point X has two properties: (i) the tangent to the first circle is of length t, (ii) the tangent to the second circle is of length t. The locus corresponding to the statement (i) is a circle concentric with the first given circle; in order to describe it, we first find one point E on its circumference, by cutting off from any tangent a length CE = t measured from the point of contact C, and then describe the circle A(E). Similarly the locus corresponding to (ii) is another circle B(F). We can take either of the points of intersection of these two loci to be X and the other to be Y; the radical axis of the two given circles is the straight line XY. The whole construction can be arranged as follows.

Let the line of centres AB meet the circumferences in C, D. Draw CE, DF at right angles to AB and cut off CE, DF, each equal to any convenient length t. Describe the circles A(E), B(F) to meet in X, Y; join XY, which is the radical axis required. We must take t large enough for the circles A(E), B(F) to meet in real points. This is always possible except when the given circles are concentric, and then the radical axis is wholly at infinity. For a shorter construction of the radical axis see pp. 108, 117.

(ii) **Method of Trial and Error.**

This is somewhat similar to the method of loci; we replace the point to be constructed by a pair of related points, which between them satisfy all the requirements of the problem, so that it would be solved if the pair coincided. That is to say, we regard the required point as a particular case of the pair of points, which has the separate additional property that the two points of the pair coincide. In most of the cases in which this method leads to a ruler and compass construction, the locus of the first point of each pair is a straight line or circle, and the second point has the same locus; and the corresponding positions form two homographic ranges. Then the solution of the problem is given by the common points of the homography, which can be constructed by the method of the last chapter (p. 65). One of the differences between this method and the method of loci is that there we have to find a point of the first locus which coincides with some point, no matter which, of the second locus; while here we have to find a point of the first range which coincides with a definite corresponding point of the second range, namely the point corresponding to the first by means of the construction adopted.

Ex. 1. We have already had an example of this method on p. 64, in determining the intersections of an arbitrary straight line with the conic through five given points.

Ex. 2. To draw a triangle XYZ with its sides parallel to fixed directions and its corners lying upon the sides of a given triangle ABC.

Start from any point X of BC; draw in succession XY, YZ, ZX′, parallel to the proper fixed directions, to meet CA in Y, AB in Z, and BC in X′ respectively. Then the problem is solved if X′ coincides with X. This will not

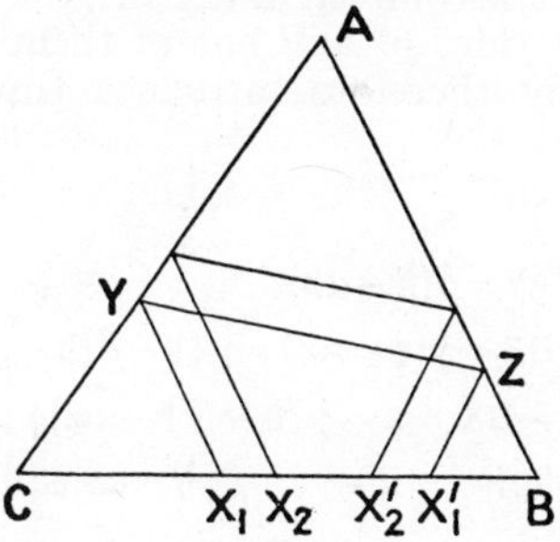

FIG. 35.

happen in general, but if X moves on BC, and describes a certain range, X′ describes a homographic range also on BC; for the cross-ratio of any four positions of X is equal to those of the corresponding positions of Y, of Z and of X′. The common points of the homography X, X′ give the solution of the problem.

If the lines are arranged as in fig. 35, when X moves from X_1 to X_2 towards X′, both Y and Z move towards A, and X′ moves toward X; thus X, X′ come together at some point between B and C, so that there is at least one real finite common point of the homography. It follows that the other common point is also real, but it lies at infinity; in general, the problem has one and only one solution. With a different figure, there may not be a common point between B and C, but in the general case, as X goes to infinity along BC produced, Y, Z and X′ all go to infinity also, and X, X′ both ultimately coincide with the point at infinity on BC, which is the second common point of the homography. Certain special cases are an exception to this. If the first fixed direction is parallel to BC, then Y always coincides with C, and Z, X′ are fixed points; similarly X′ is fixed if YZ is parallel to CA, or ZX′ to AB. These cases we exclude, and also the similar cases in which XY, YZ or ZX′ is parallel to CA, AB or BC respectively, when X′ is at infinity always.

These facts about the common points may also be proved analytically. Let $\mathbf{a}$, $\mathbf{b}$, $\mathbf{c}$ be the sides of the triangle ABC, and let $CX=\mathbf{x}$, $CX'=\mathbf{x}'$. Since the triangles XCY, YAZ, ZBX′ have their sides in fixed directions, they are of constant shapes, and the sides of any one of them are in constant ratios. We may therefore introduce three constants α, β, γ, such that

$$\begin{aligned}
CY &= \alpha CX = \alpha\mathbf{x},\\
AZ &= \beta AY = \beta(\mathbf{b}-\alpha\mathbf{x}),\\
BX' &= \gamma BZ = \gamma(\mathbf{c}-AZ) = \gamma\{\mathbf{c}-\beta(\mathbf{b}-\alpha\mathbf{x})\},\\
\mathbf{x}' = CX' &= \mathbf{a}-BX' = \mathbf{a}-\gamma\{\mathbf{c}-\beta(\mathbf{b}-\alpha\mathbf{x})\};\\
\therefore\ \mathbf{x}'+\alpha\beta\gamma\mathbf{x} &= \mathbf{a}-\gamma\mathbf{c}+\beta\gamma\mathbf{b} = \text{constant}.
\end{aligned}$$

The excluded cases correspond to the values $\mathbf{0}$ and ∞ of α, β or γ. To find the common points, put $\mathbf{x}=\mathbf{x}'$; this gives one finite solution

$$\mathbf{x}=\frac{\mathbf{a}-\gamma\mathbf{c}+\beta\gamma\mathbf{b}}{1+\alpha\beta\gamma};$$

if we regard this as a degenerate quadratic equation, the other root is infinite.

In the particular case $\alpha\beta\gamma=\mathbf{1}$, the equation of the homography is symmetrical in $\mathbf{x}$, $\mathbf{x}'$, and we have an involution; starting from X′ and repeating the construction, we get back to X. This condition is satisfied, for example, if XY, YZ, ZX′ are parallel to the sides of the pedal triangle PQR formed by joining the feet of the perpendiculars from A, B, C on to the opposite sides of the triangle. Then the pedal triangle itself is the solution of the problem.

In this case each of the triangles XYC, AYZ, X′BZ is similar to ABC, and we have

$$\alpha=\frac{\mathbf{a}}{\mathbf{b}},\quad \beta=\frac{\mathbf{b}}{\mathbf{c}},\quad \gamma=\frac{\mathbf{c}}{\mathbf{a}},\quad \alpha\beta\gamma=\mathbf{1}.$$

Also XY, YZ are equally inclined to CA, and similarly at Z, X′, so that XYZX′Y′Z′X is the path of a ray of light

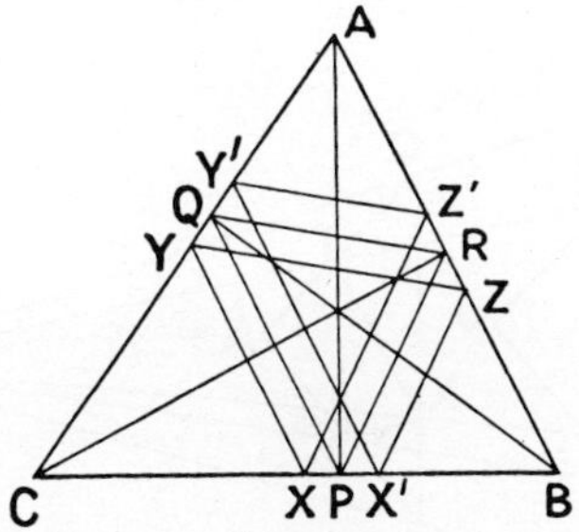

FIG. 36.

incident along Z′X and reflected at the sides in turn ; if such a ray after three reflexions is parallel to its original direction, then after six reflexions it repeats its path.

Another exceptional case is when $\alpha\beta\gamma=-\mathbf{1}$; then both common points are at infinity, and

$$\mathsf{XX}'=\mathbf{x}'-\mathbf{x}=\mathbf{a}-\gamma\mathbf{c}+\beta\gamma\mathbf{b}=\text{constant},$$

so that X, X′ can never coincide unless this constant vanishes, and then they always coincide. When this special relation $\alpha\beta\gamma=-\mathbf{1}$ holds, the problem is poristic (p. 67), and there is no solution unless the further relation $\mathbf{a}-\gamma\mathbf{c}+\beta\gamma\mathbf{b}=\mathbf{0}$ is satisfied, and then there are an infinite number of solutions. But if both these conditions hold, the three given directions must be all the same.

(iii) Method of Similar Figures.

This same problem can also be treated by the method of similar figures, in which the properties separated from the rest are the size and position of the required figure. We construct a figure similar to the required one in some other part of the plane, and then copy it on the right scale and in the right position. The similar figure is sometimes constructed by reversing the data and desiderata of the problem. This *method of the inverse problem* can often be used together with other methods.

Ex. 1. In the example we have just been discussing, let any three straight lines in the given directions form a triangle LMN. Through the corners of this draw parallels to the sides of the given triangle ABC, to form a third

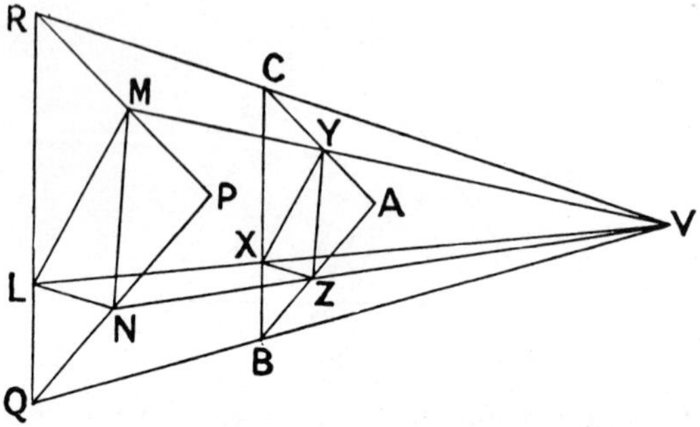

FIG. 37.

triangle PQR. Then PQRLMN is a figure similar to the required figure ABCXYZ, and X divides BC in the same ratio as L divides QR; if QB, RC meet in V, then VL meets BC in the required point X. The two figures are not only similar, but similarly situated, with V as centre of similarity; and Y, Z can be determined either by drawing parallels to LM, LN through X, or by joining VM, VN to meet CA, AB respectively.

Ex. 2. To draw a triangle OXY similar to a given triangle LMN, having one vertex at a given point O and the others on two given straight lines AB, BC.

We can complete the construction by drawing a similar figure, if we can find the point P which has the same relation to the triangle LMN that B has to OXY; for then PM, PN correspond to BA, BC. Now P lies upon the arc of a circle standing on MN capable of containing an angle equal to $\angle$ABC; and we can also determine the straight line LP by means of its second intersection Q with the same circle. For Q corresponds to D the point of intersection of OB with the circle XBY and we have

$$\angle MNQ = \angle XYD = \angle XBD \text{ or } ABO,$$

which is a given angle. Hence we can draw NQ and determine Q, draw LQ and determine P, and then construct OXBY similar to LMPN.

Now let AB, BC be adjacent sides of a parallelogram ABCE, and let O be its centre; and let MN be a side and

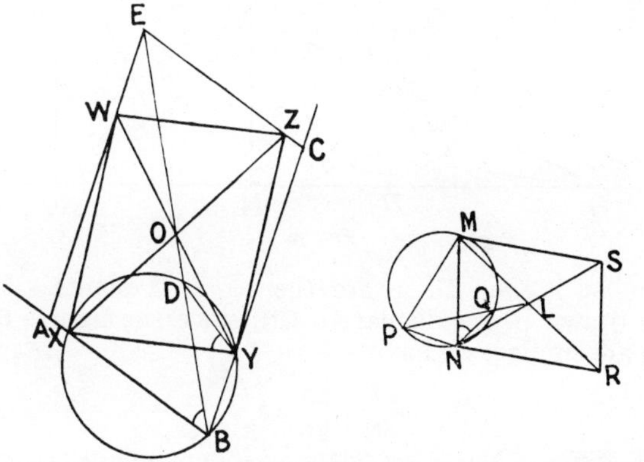

FIG. 38.

L the centre of another given parallelogram MNRS. Then, if XO, YO are produced to meet CE, EA in Z, W respectively, XYZW is a parallelogram similar to MNRS, and we have a solution of the problem: In a given parallelogram to inscribe a parallelogram similar to another given parallelogram.

Ex. 3. **Eccentric Circle.**—To find the intersections of a given straight line with a conic whose focus, directrix and eccentricity are given. In this case the method of loci would apply at once if we had an instrument for drawing the conic, for the two properties of lying on the straight line and lying on the conic have for their obvious loci the straight line and the conic respectively. With ruler and compasses we cannot draw the second locus, but we can fall back on the method of similar figures.

Let the given straight line meet the given directrix QN in Q and the conic in X, X′; let S be the focus and e the eccentricity. Join **SQ**. Take any point **p** on **QX**, draw **pn** perpendicular to **QN**, and describe the circle centre **p**, radius **e . pn** ; this is the *eccentric circle* of **p** ; let it meet **QS** in **s, s′**. Draw **SX, SX′** parallel to **sp, s′p** to meet

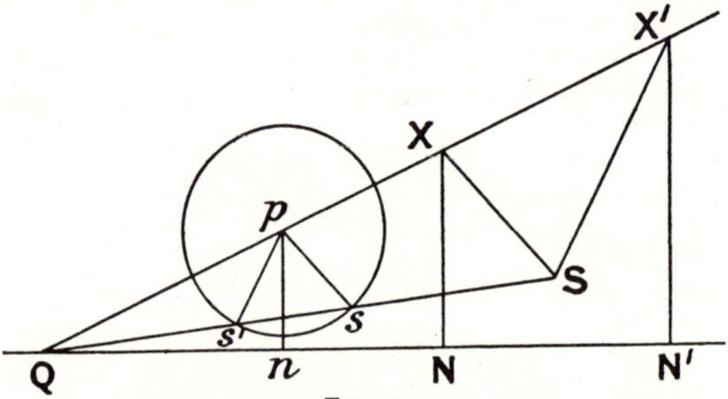

Fig. 39.

QX in **XX′** ; then these are the required points. For if **XN** is drawn perpendicular to **QN**, since the figures **SXQN**, **spQn** are similar, we have

$$\frac{SX}{XN} = \frac{sp}{pn} = e,$$

so that **X** lies on the conic, and similarly **X′**. The solutions are real, provided **SQ** meets the eccentric circle of **p** in real points. If **SQ** touches this circle, the given straight line touches the conic, and the construction gives the point of contact.

(iv) Methods of Perspective, Translation, Rotation and Reflexion.

These are all particular cases of the method of similar figures, in which abstraction is made of some but not all of the properties which determine the size and position of the required figure. Examples 1 and 3 of pp. 84, 85 have already illustrated the method of perspective. It is often useful to apply these methods to part only of the figure, and so transform it into something simpler.

Ex. 1. To find the locus of a point such that the difference between its perpendicular distances **x, y** from two given straight lines **a, b** is a constant length **d**.

Translate **b** parallel to itself through a distance equal to **d**; it will take up one of two positions **b**′, **b**″ according to the sense of the translation. Then the required points are equidistant either from **a**, **b**′ or from **a**, **b**″, and therefore lie on one or other of the four bisectors of the angles between these two pairs of straight lines. If **x**, **y** are always taken to be positive, the figure shows the parts of these bisectors on which $x - y = d$; on the other parts, either $x + y = d$ or $y - x = d$.

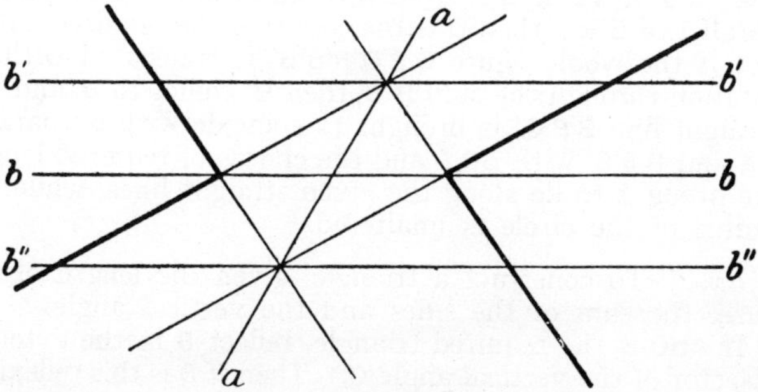

FIG. 40.

Ex. 2. To describe a circle of given radius **a** to cut off intercepts PQ, RS of given lengths from two given straight lines AB, BC.

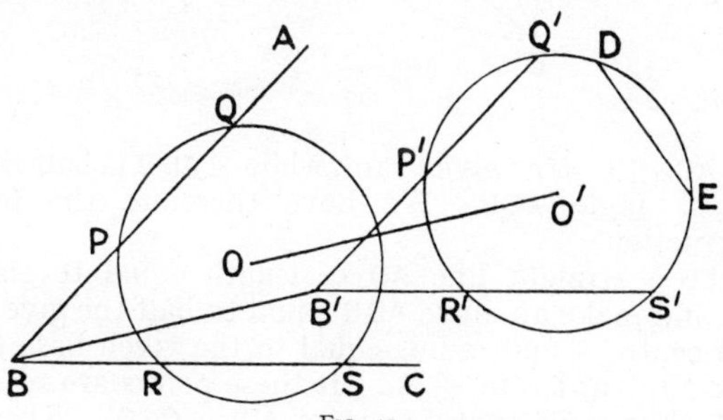

FIG. 41.

Take any circle O′(**a**) of the same radius; place it in a chord DE of length PQ; rotate DE about O through an angle equal to that between DE and AB. We thus have a chord P′Q′ equal and parallel to the required intercept PQ. Similarly place in the same circle a chord R′S′ equal and parallel to the required intercept RS on BC, and let P′Q′, R′S′ meet in B′. Join B′B, and draw O′O equal and parallel to B′B; then O is the centre of the required circle. For if the whole figure O′P′Q′R′S′B′ is translated without rotation through the step B′B, then B′ comes to B, and the straight line B′P′Q′ is brought to coincide with its parallel BA and B′R′S′ with BC; and the chords of required length are brought to lie along the given straight lines, while the radius of the circle is unaltered.

Ex. 3. To construct a triangle, given the length of the base, the sum of the sides and the vertical angle.

If ABC is the required triangle, reflect B in the external bisector of the vertical angle C. Then if D is this reflexion, ACD is a straight line, BCD is an isosceles triangle, and

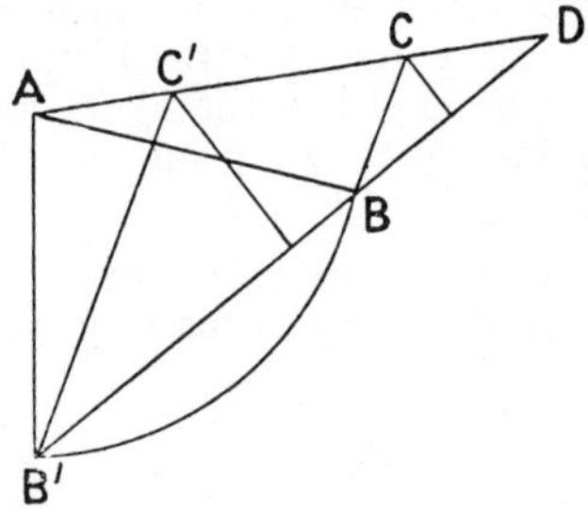

Fig. 42.

AD = AC + CB = the given sum, while ∠ ADB is half the given vertical angle ACB. We have therefore the following construction.

Take a straight line AD of length equal to the given sum, and make an angle ADB equal to half the given angle. With centre A and radius equal to the given base describe a circle to cut DB in B, B′. If these points are real, bisect DB, DB′ at right angles to meet AD in C, C′. Then either ABC or AB′C′ is the required triangle.

The method of reflexion naturally finds its classic examples in the theory of geometrical optics.

Ex. 4. **Optical Image.**—To find the path of a ray of light from one given point to another by one reflexion at a given plane surface, and to show that it is the shortest of all such broken lines joining the two points.

We have to find the point X at which the ray AXB between the given points A, B is reflected at the given plane XC. The laws of optical reflexion are that AX, XB lie in a plane through the normal at X to the reflecting

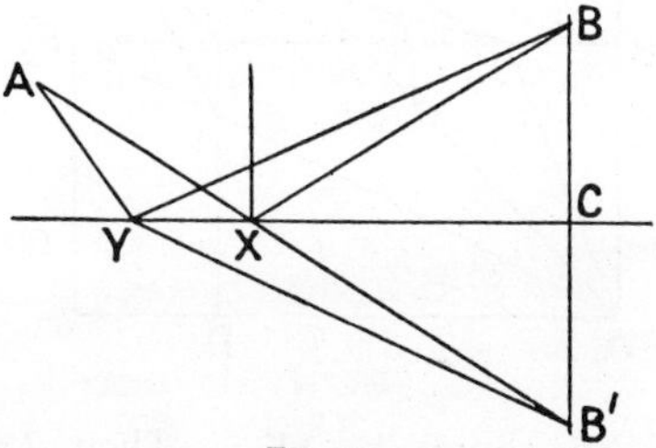

FIG. 43.

surface, and make equal angles with the normal on opposite sides. Hence if B′ is the geometrical reflexion of B in the plane XC, found by drawing BC perpendicular to the plane and producing it an equal length, then AXBB′ lie in a plane and AXB′ is a straight line. X can therefore be found as the intersection of AB′ with the given plane.

Also, since B, B′ are equidistant from any point Y of the plane, the total length of any other broken line AYB from A to B is the same as AY + YB′, which is greater than AB′ the third side of the triangle AYB′, and therefore greater than the length of the actual path of the ray.

The following problem involves much the same idea.

Ex. 5. A spider in one corner of a room wants to reach a fly on the opposite corner of the ceiling by crawling across the floor and up one wall; what is his shortest path?

Let ABCD be the floor, A the spider, E the fly, E being vertically above C. There are two types of path, according as the spider leaves the floor at a point X on the longer side, BC say, or at a point Y on the shorter side CD. While on the floor his path must be a straight line, and also while on the wall. We have to determine, first for what point

X of BC the path AXE is a minimum, then for what point Y of CD the path AYE is a minimum, and lastly, which of these two minima is the lesser. Now let the wall BCE be rotated about BC till it comes into the same plane as the floor. Then E comes to E_1 in DC produced, where

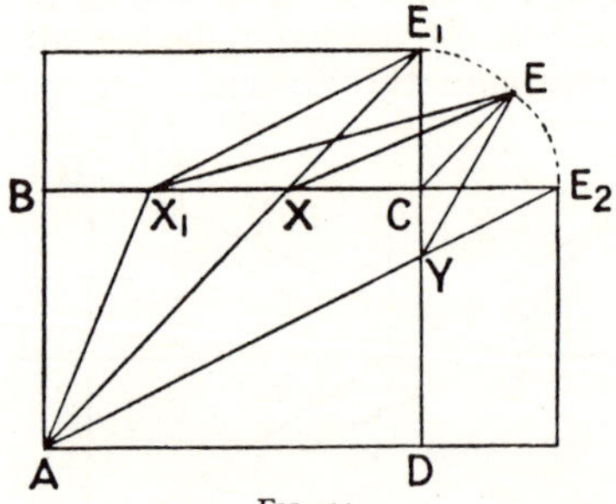

FIG. 44.

$CE_1 = CE$, and XE comes to XE_1. Then $AX + XE_1 \geqq AE_1$, and the minimum length for paths of this type is AE_1, when AXE_1 is a straight line; the point X is constructed as the intersection of AE_1 and BC, and X divides BC in the ratio AB : CE. Similarly, for paths that leave the floor at a point Y on BC, the minimum is of length AE_2, where E_2 is the point in BC produced, such that $CE_2 = CE$. The length of the required shortest path is therefore the lesser of AE_1, AE_2. Now

$$AE_1^2 = AD^2 + (AB + CE)^2 = AD^2 + AB^2 + CE^2 + 2AB \cdot CE,$$
$$AE_2^2 = (AD + CE)^2 + AB^2 = AD^2 + AB^2 + CE^2 + 2AD \cdot CE,$$

and the first is the lesser, provided $AB < AD$. The shortest path therefore leaves the floor at a point in the longer side, which divides it in the ratio of the breadth to the height of the room.

(v) Method of Projection.

The method of similar figures and its varieties are all particular cases of the method of projection, in which we separate the projective from the metrical properties of the required figure. By a suitable projection we transform the problem into one in which the same set of projective properties are combined with a different metrical set, which makes the case easier to solve. Then we reverse the projection, and pass either from the final solution of the transformed problem to the solution of the original problem, or else from each step in the construction of the second figure to the corresponding step in the construction of the required figure.

From one point of view the method of projection is an example of the separation of properties; but from another it illustrates the more important *principle of transformation*. That is to say, we do not directly solve the proposed problem, but another in which each element, whether given or required, stands in some definite relation to the corresponding element in the given problem and its solution. If the modified problem cannot be solved, or is no easier than the original problem, then the transformation adopted is not suitable. When we have solved the transformed problem, by a reversal of the transformation we pass back to the original figure, and so obtain a construction for the problem in the form in which it was given. Many examples of this have already been met with in chapter III.

Ex. I. **Auxiliary Circle.**—To draw the tangent at a given point of an ellipse whose principal axes are given.

Project the ellipse into a circle. We may take the vertex of projection at infinity in the direction of the normal to the plane of the ellipse, and project upon a plane through the major axis inclined at an angle $\cos^{-1}\frac{CB}{CA}$ to the plane of the ellipse, where CA, CB are the given semi-major and -minor axes. The given point of the ellipse

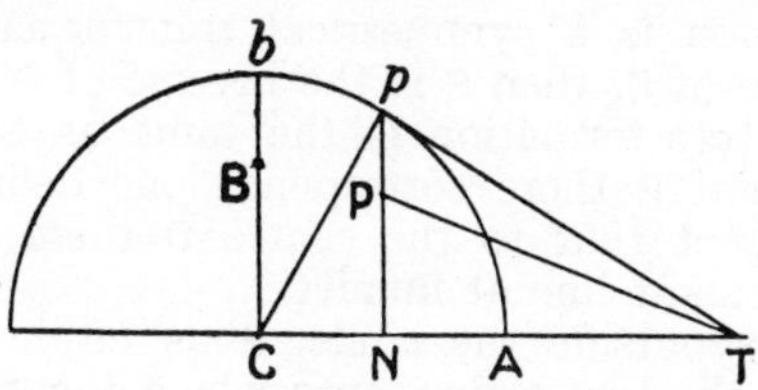

FIG. 45.

projects into a point of the circle, and the required tangent into the tangent to the circle at this point. If the plane of projection is rotated about the major axis till it coincides with the plane of the ellipse, then the circle comes to coincide with the major auxiliary circle ; corresponding points of the two figures lie on straight lines perpendicular to CA, at distances from it which are in the ratio of the axes. Points on CA correspond to themselves, and the corresponding tangents meet on CA, so that we have the following ruler construction for the problem :

Let CA, CB be the given semi-axes and P the given point. Draw PN perpendicular to CA, and produce NP to p, making $\frac{pN}{PN}=\frac{CA}{CB}$. Join Cp, and draw pT perpendicular to Cp to meet CA in T. Join PT, which is the required tangent to the ellipse at P.

(vi) Method of Inversion.

This is another important example of the method of transformation, which under certain circumstances changes circles into straight lines, and so may transform a problem into a much simpler one. Inversion replaces every point by its inverse with regard to a fixed circle, where two points P, P′ are defined to be the *inverses* of one another with regard to the circle O(k) if OPP′ is a straight line and $OP \cdot OP' = k^2$. Here O and k are called the *centre* and *radius of inversion* ; when the radius is arbitrary, we may speak of inversion with regard to the point O. If the radius k is pure imaginary, then $OP \cdot OP'$ is negative, and P, P′ are on opposite sides of the centre O.

We assume without proof the following properties of inversion, which are used in this and later chapters.

(1.) Inversion is a symmetrical transformation : if P′ is the inverse of P, then P is the inverse of P′, so that to reverse the transformation is the same as to repeat it. To every point P there corresponds one definite inverse point P′, except that to the centre O there corresponds the whole straight line at infinity.

(2.) If P is outside the circle, P′ is inside ; inversion interchanges the two regions into which the circle divides the plane. Every point on the circumference of the circle, which is the common boundary of the two regions, inverts into itself.

(3.) Any circle inverts into a circle, except that a circle through O inverts into a straight line; the centre of the original circle inverts into the point inverse to O with regard to the inverse circle.

(4.) Any straight line inverts into a circle through O, except that a straight line through O inverts into itself.

(5.) A circle orthogonal to the circle of inversion inverts into itself.

(6.) Two curves cut at the same angle as their inverses; and, in particular, contact is unaltered by inversion.

Ex. 1. To find the locus of the centre of a circle orthogonal to two given circles.

Let the given circles OAP, OBP cut in O, P. Invert with regard to O as centre and OP as radius; then P inverts into itself, and the given circles become straight lines PA′, PB′ intersecting at P. A circle centre X, orthogonal

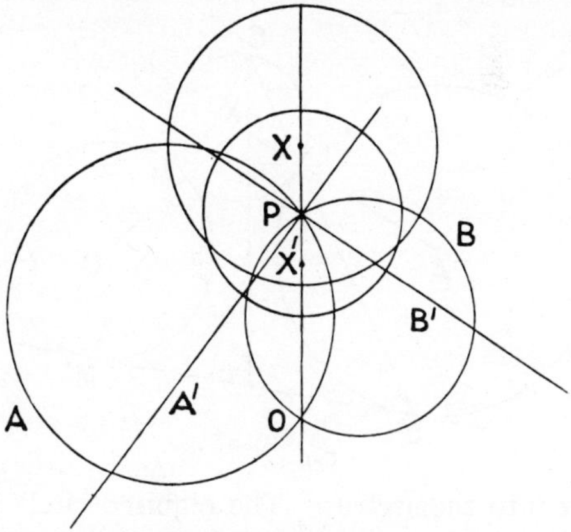

FIG. 46.

to both the given circles, inverts into a circle to which PA′, PB′ are both normals, that is, radii, so that the centre is P. Now X′, the inverse of X, is the point inverse to O with regard to this last circle, whose centre is P; hence X′ lies on OP; and since X, X′ are inverse with regard to O as centre, therefore X lies upon the same straight line OPX′. The required locus is therefore OP, the radical axis of the two given circles. It is easy to find other proofs of this. The proof by inversion is valid whether the point O and the inverse figure are real or imaginary.

Since inversion with regard to a common orthogonal circle changes each of the given circles into itself, the locus found is that of the centre of an inversion which leaves two given circles unaltered.

We can invert three circles into themselves by choosing their radical centre as centre of inversion; but we cannot invert four or more circles into themselves unless their radical axes all meet in a point, so that they have a common orthogonal circle.

Ex. 2. To describe a circle to touch two given circles and to pass through a given point on their radical axis.

Invert with regard to the given point C, and choose the radius of inversion equal to the tangent from C to either of the given circles α, β; it follows from (5) that

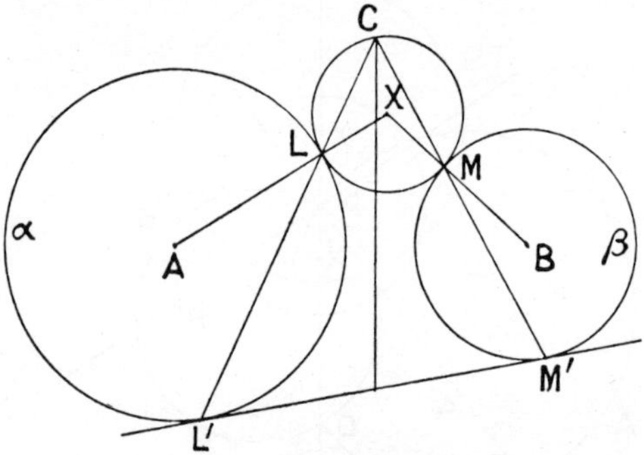

FIG. 47.

they invert into themselves. The required circle becomes a common tangent to α, β. We must therefore construct the common tangents, and then invert them with regard to C. There are in general four real common tangents and four solutions of the problem, if α, β are external to each other. But there is an exceptional case when the given point C lies upon one of the common tangents; for then, by (4), this straight line inverts into itself, that is, into a straight line and not into a circle, so that there are only three solutions to the problem as stated. If α, β touch externally, there are three real common tangents; but one, the tangent at the point of contact, is the radical axis and passes through C, so that it does not invert into a circle, and there are only two solutions. There are in general two solutions also if α, β cut in real points, but only one if C lies on one of the two real common tangents; and if one of α, β touches the other internally, or lies wholly within it, there is no solution of the problem. The actual construction is as follows.

Draw a common tangent, by Euclid's or any other method, to touch the given circles α, β in L′, M′. If A, B are the centres and C the given point on the radical axis, join CL′, CM′ to meet α in L and β in M respectively. Join AL, BM to meet in X; then X is the centre of the required circle, XC is its radius, and it touches α at L and β at M. The proof is left to the reader.

If we regard the point C as a circle of zero radius, we have a particular case of the famous *Problem of Apollonius*:

Ex. 3. To describe a circle to touch three given circles.

If the three circles are all external to one another, there may be as many as eight solutions, for each of the given circles may be either inside or outside the touching circle. If the second circle is inside the first, and the third inside the second, as, for example, when they are concentric, there is no real solution; in other cases there may be any intermediate number; one real solution is lost when one of the touching circles becomes a straight line or reduces to a point, and two are lost when a pair of touching circles first coincide and then become imaginary.

If two of the given circles cut in real points, we may invert the whole figure with regard to one of these points; then these two circles become straight lines, and the problem becomes the simpler one: to describe a circle to touch two given straight lines and also a given circle. We shall first solve this, and then show how to pass to the less simple case when none of the given circles cut in real points.

Let LM, LN be the two given straight lines and P(**p**) the given circle. If Y(**y**) is the required circle, it touches P(**p**) either externally or internally. Assume first that contact is external; then suppose **y** increased by **p**: we have a circle centre Y, which passes through P and touches two straight lines L′M′, L′N′ parallel to LM, LN, each at a distance p from its parallel on the side remote from Y. Hence Y is a solution of the problem: to find a point equidistant from two given straight lines L′M′, L′N′ and from a given point P. The first requirement tells us that Y lies on one or other of the bisectors of ∠M′L′N′, and

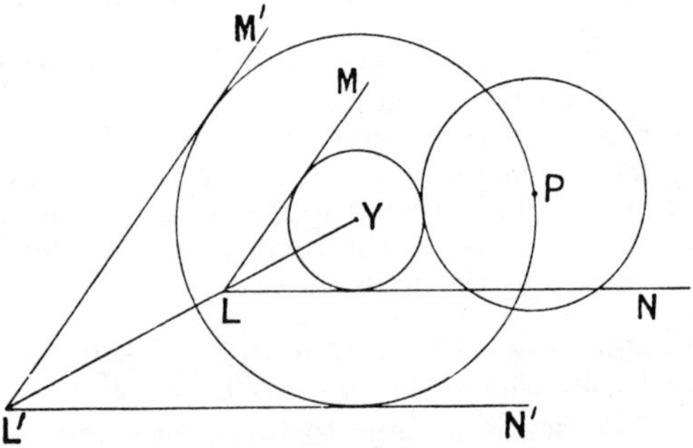

Fig. 48.

the second requirement tells us that it lies on the two parabolas with focus P and directrices L′M′, L′N′ respectively. Since Y is not only equidistant from L′M′, L′N′, but also from LM, LN, it lies on the bisector LL′ of ∠M′L′N′, and not on the other bisector. Corresponding to this assumption there may be two real positions of Y, the intersections of LL′ with a parabola whose focus and directrix are given. The determination of these is a particular case of p. 85, ex. 3. If we make a different assumption as to the position of Y, or assume that Y(y) touches P(p) internally, the parallels may have to be drawn on the other sides of LM or LN or both; so we have four cases to consider, in each of which we have to look for the intersection of one of the bisectors of ∠MLN with a parabola focus P. This leads to eight real positions of Y in the case in which P(p) cuts each of LM, LN in two real points on opposite sides of L. We can thus complete the construction of all the positions of Y that are real. The inverses of these, in the inversion used at the beginning to find LM, LN, are the positions X of the centres of the circles required by the problem of Apollonius.

Now return to the general case, in which none of the given circles are assumed to meet in real points. Let $A(\mathbf{a})$, $B(\mathbf{b})$, $C(\mathbf{c})$ be the given circles, and $X(\mathbf{x})$ the required circle. Now two circles touch if the distance between the centres is equal to either the sum of the radii, or to the difference taken positively. The conditions of tangency are therefore

$$XA = \pm \mathbf{a} \pm \mathbf{x}, \quad XB = \pm \mathbf{b} \pm \mathbf{x}, \quad XC = \pm \mathbf{c} \pm \mathbf{x},$$

with any arrangement of signs that makes the right-hand side positive in each equation. These equations still hold, if we add any quantity $\mathbf{d}$ to $\mathbf{x}$, and subtract the same quantity from $\mathbf{a}$, $\mathbf{b}$ and $\mathbf{c}$ if the sign is the same as that of $\mathbf{x}$ in the corresponding equation, or add it if the sign is the opposite. We are now solving a different problem, the three given circles having the same centres as before, but modified radii, and the solution is a circle whose centre X is the same as in the solution of the original problem. This process may require us to consider a circle of negative radius. This must be supposed to coincide with the circle of corresponding positive radius, but where one would touch any circle internally, the other must touch the corresponding circle externally, and vice versa. In any case in which there is a real position of X, it is possible to choose $\mathbf{d}$ so that two of the modified circles meet in real points, so that X can be found by the above method. Then the corresponding radius of the required circle is $\pm XA \pm \mathbf{a}$. For if a circle exists touching the three given circles, of the three contacts there must be two which are either both external or both internal, and in the two corresponding equations the two signs are either both like or both unlike. By adding a positive or negative quantity to $\mathbf{x}$, we can therefore increase or decrease both of these two radii by any quantity, the same for both circles. Now if these two circles are external to one another, then by increasing the radii by a suitable quantity, they can be modified so as to meet in real points; while if one is internal to the other, the same can be done by decreasing the radii by a suitable quantity, the smaller radius becoming negative, and the inner circle passing outside the touching circle. If the two circles are neither external nor internal one to the other, then they meet in real points without modification.

The figure shows the case

$$XA = x + a, \quad XB = x + b, \quad XC = x - c.$$

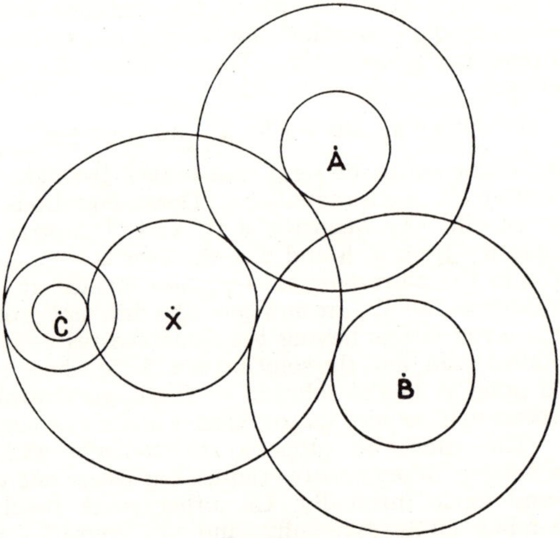

FIG. 49.

The radii of the modified circles are

$$a + d, \quad b + d, \quad c - d, \quad x - d,$$

of which the third is negative, and the corresponding circle has passed from inside to outside the touching circle.

(vii) Method of Reciprocation.

The method of reciprocation is a transformation which makes points correspond to straight lines and straight lines to points. In its simplest form, a point is made to correspond to its polar, and a straight line to its pole, with regard to a fixed circle of reciprocation. When the radius is arbitrary we may speak of reciprocation with regard to the centre of this circle. To points on a fixed straight line there correspond straight lines through a fixed point, and it may be shown that cross-ratio is unaltered by reciprocation. To any curve, regarded as the locus of a point, there corresponds the envelope of a straight line, that is, another curve. To the points on the circle of reciprocation there correspond the tangents at these points, which envelope the same circle; this curve therefore reciprocates into itself.

The connection between reciprocation and inversion is given by the theorem: *the reciprocal of any curve is the inverse of the pedal of the same curve with regard to the centre of reciprocation.*

If SY is the perpendicular from S the centre of reciprocation upon any straight line p, then the pole P of p with

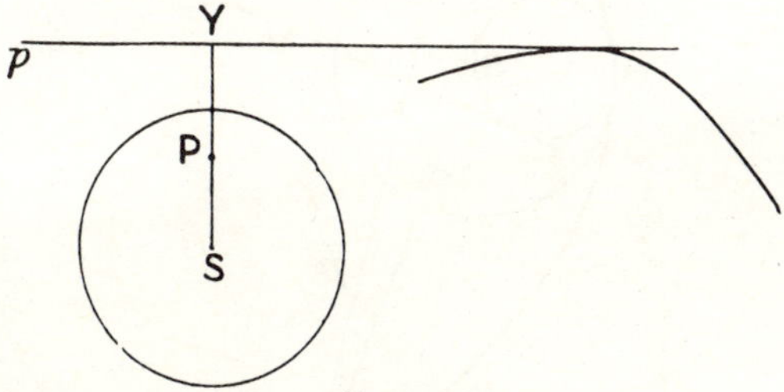

Fig. 50.

regard to the circle of reciprocation is the same as the inverse of Y. If p moves so as to envelope a curve, the locus of Y is called the *pedal* of the curve with regard to S, and the locus of P is the reciprocal of the same curve; hence the reciprocal is the inverse of the pedal.

Ex. 1. The reciprocal of a conic with regard to the focus is a circle.

The pedal of a conic with regard to the focus S is the auxiliary circle; hence, if we reciprocate with regard to S, the conic becomes the inverse of the auxiliary circle, which is another circle. The centre of the conic is also the centre of the auxiliary circle, which inverts into the point T inverse to S with regard to the circle reciprocal to the conic, p. 93 (3). Hence conics with a common focus can be simultaneously reciprocated into circles; and confocal conics, which are also concentric, reciprocate into a family of circles for each of which S, T are inverse points, that is, a coaxal family with S, T as limiting points.

If the conic is a parabola, the pedal is the tangent at the vertex A, and the reciprocal is a circle through S.

This gives another construction for the intersections X, Y of a given straight line **p** with a parabola whose focus S and vertex A are given (p. 85). Find the pole P of **p**

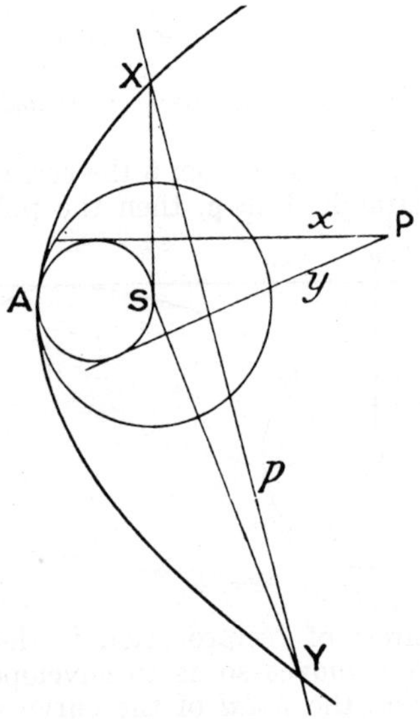

FIG. 51.

with regard to the circle S(A), and draw the tangents **x**, **y** from P to the circle on SA as diameter. Then the poles X, Y of **x**, **y** with regard to S(A) are the required points of intersection of **p** with the parabola.

CHAPTER VI.

COMPARISON OF METHODS.

HITHERTO we have spoken of the possibility of various constructions, and in the examples we have merely given one or more methods which lead to what is wanted, without making any comparison between them. Now whether one method is to be preferred to another depends upon many things; such as the context in which the problem arises, the extent and accuracy of our apparatus, and our skill in using it; or again, it may depend upon the ideas involved, rather than upon the operations; upon whether the construction can be easily put into words; upon the beauty and symmetry of the solution; or we may prefer a construction which combines several of these advantages in a moderate degree.

Thus it is very simple, from the point of view of language, to say in the course of a construction, "draw a figure similar to a certain figure," or "invert the whole figure with regard to a certain circle," but these may involve drawing a few dozen straight lines and circles. Euclid's proof of II. 9 and 10 are simple and attractive; but if we are dealing with the whole series II. 1-11, we shall probably treat them all by the same method of dissecting squares and rectangles. Simplicity is much the same as economy, and we have to choose what we shall economize: space in the diagram or words in the description, thought or labour or chance of error, or the indefinable quality of tiresomeness. The rest of this book is devoted to constructions which, as far as possible, avoid first, too large figures; next, certain ill-defined intersections; thirdly, more operations than need be; and lastly, in chapters VII and VIII, the use respectively of compasses and ruler.

I. Constructions in Limited Space.

It may happen that points which we might wish to use fall outside the paper, or too far off for the ruler to reach; or in surveying operations on a large scale, important points may fall in lakes or precipices or other inaccessible places. One "simple" way out of the difficulty is given by the method of inversion. Invert with regard to any circle which satisfies the two conditions: (i) every point inside it is accessible; (ii) every point or line given or required is outside it; then, by p. 92 (2), every point of the inverse figure is inside, and therefore accessible. If we have a construction for the problem which only fails because it requires inaccessible points, we can thus invert and construct the whole of the inverse figure, and finally, by inverting again, construct as much of what is required as comes on to the paper. In this way we have a proof that it is always possible to do what we want by means of lines and points which are not too far off, and so we are encouraged to look for more practical ways of doing it. Very often we can use the method of similar figures; that is, we copy the figure on a small scale, so that much more of it becomes accessible; and after completing the construction on this model, we enlarge as much of it as possible back to the former scale. The art lies in choosing the centre of similitude so as to make the change of scale as easy as possible.

Ex. 1. To join a given point A to an inaccessible point L, the intersection of two given straight lines **b**, **c**.

This problem is solved as soon as we have an accessible point X lying on AL.

(i) One method is to take any straight line AD through A and determine a point D upon it such that A, D are harmonically separated by **b**, **c**. Take any other straight line DX through D, and determine a point X on it such that D, X are again harmonically separated by **b**, **c**. These two harmonic ranges are in perspective with L as vertex, and therefore AX passes through L.

Start by taking any two straight lines B_1AC_2, B_2AC_1 through A to meet **b**, **c** with intersections as in fig. 52. Join B_1C_1, B_2C_2 to meet in D. Then AD, which need not be drawn, is a diagonal of the quadrangle $B_1B_2C_1C_2$, and it is divided harmonically by **b**, **c**, which are a pair of opposite sides of the same quadrangle. Draw any other

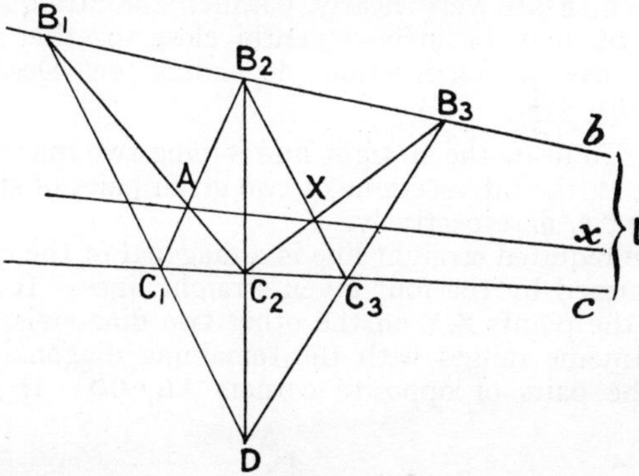

FIG. 52.

straight line DB_3C_3 through D to meet **b**, **c**, and join B_2C_3, B_3C_2 to meet in X; then DX is divided harmonically by **b**, **c**, for the same reason as before. Join AX, which is the straight line required.

This construction involves projective properties only. It may fail if A is nearly midway between **b** and **c**, for then the point D may be inaccessible for all accessible positions of B_1, B_2.

(ii) In that case we may adopt a method which depends

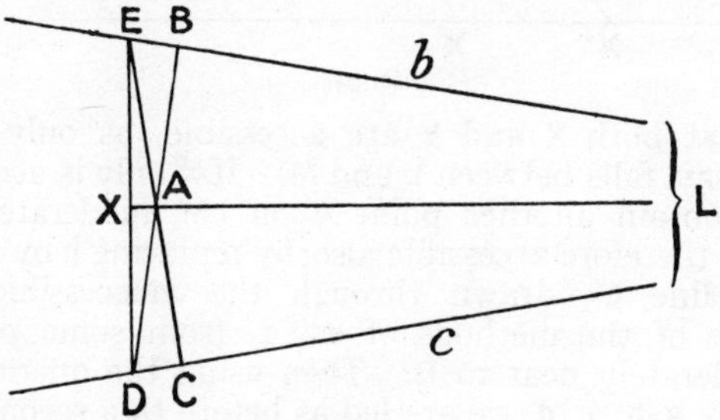

FIG. 53.

on the property of the orthocentre. Draw DAB perpendicular to **b** to meet **b** in B and **c** in D; draw EAC perpendicular to **c** to meet **c** in C and **b** in E; join DE, and draw AX perpendicular to DE. Then A is the orthocentre of the triangle LDE, and AX passes through L.

(iii) If **b**, **c** are very nearly parallel, the straight lines BD, CE, DE may be inconveniently close together; then we may use a construction depending on Desargues' theorem (p. 37).

Ex. 2. To draw the straight line joining two inaccessible points L, M, the intersections of two given pairs of straight lines **a**, **b**; **c**, **d** respectively.

(i) The required straight line is a diagonal of the quadrilateral formed by the four given straight lines. It passes through the points X, Y on the other two diagonals, which form harmonic ranges with the remaining diagonal point E and the pairs of opposite corners AB, CD. It is not

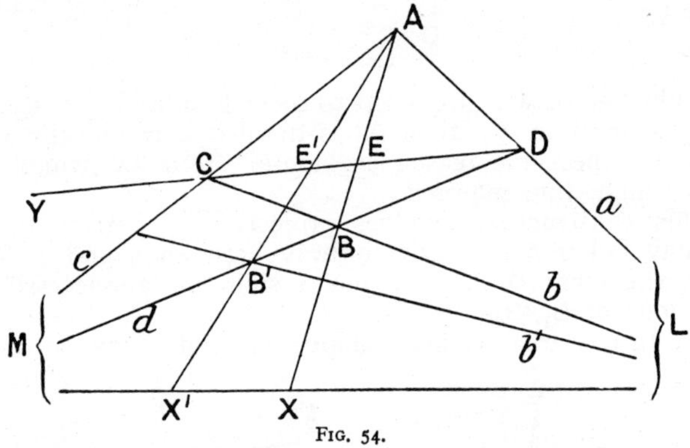

FIG. 54.

likely that both X and Y are accessible, as only one of them, X say, falls between L and M. If X only is accessible, we can obtain another point X′ on LM, moderately near to X, and therefore accessible also, by replacing **b** by another straight line **b**′, drawn through the inaccessible point L by one of the methods of ex. 1, from some point B′ of **d** moderately near to B. Then using the quadrilateral formed by **a**, **b**′, **c**, **d**, we are led as before to a second point X′ lying on the required straight line, which is therefore obtained by joining XX′.

(ii) If C or D is inaccessible, so that the point E cannot be readily obtained, we may fall back on the method of similar figures, using A as centre of similitude, for then the three straight lines **a**, **c** and AX are not altered. Take AB_1 any convenient fraction $1/\lambda$ of AB; draw B_1L_1, B_1M_1 parallel to **b**, **d** to meet **a**, **c** in the accessible points L_1, M_1

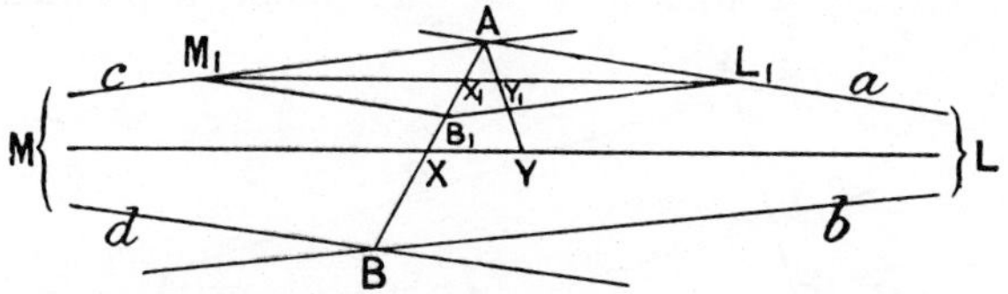

FIG. 55.

respectively. Then in the smaller figure determine the point X_1 corresponding to X, by drawing the diagonal L_1M_1 corresponding to LM to meet AB_1 in X_1. Take X on AB, making $AX = \lambda AX_1$; then X is a point on LM, and the straight line XY through X parallel to L_1M_1 is the straight line required. Or we may determine a second point Y upon it by taking any point Y_1 on L_1M_1 and producing AY_1 to Y, making $AY = \lambda AY_1$. It is convenient to take λ to be a power of 2, for then B, X and Y are found by the processes of halving and doubling repeated as often as necessary.

Ex. 3. To bisect an angle whose vertex L is inaccessible.

(i) We may use the property of the inscribed circle of a triangle. Take any points A, B on **a**, **b**; join AB,

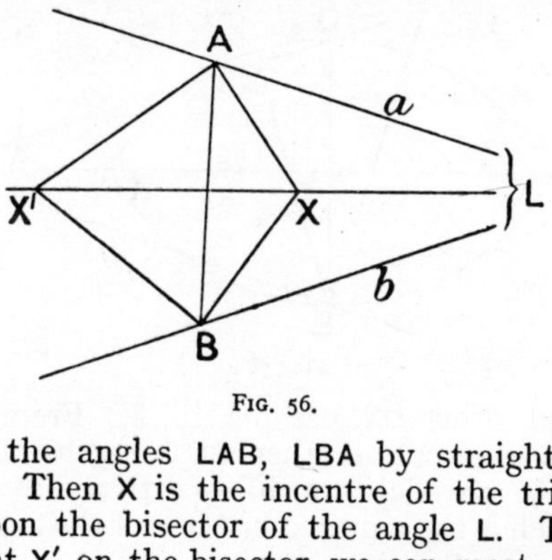

FIG. 56.

and bisect the angles LAB, LBA by straight lines that meet in X. Then X is the incentre of the triangle ALB, and lies upon the bisector of the angle L. To obtain a second point X′ on the bisector, we can most easily bisect the exterior angles at A, B by straight lines that meet in X′ the centre of the escribed circle of ALB opposite L. Then XX′ is the required bisector.

(ii) Since the bisector is the locus of a point equidistant from the given arms **a**, **b**, we can obtain a position of

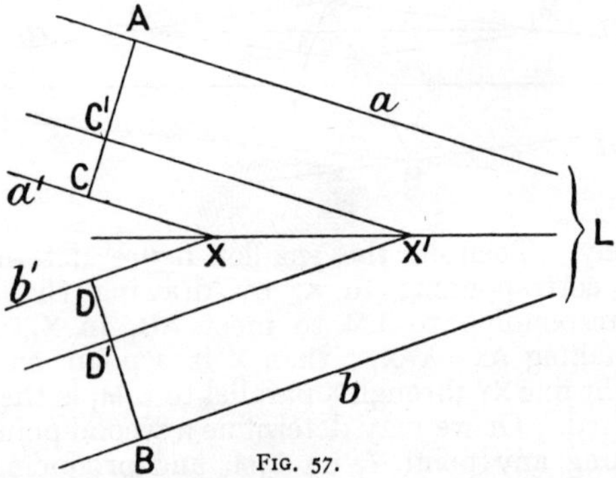

FIG. 57.

X as the intersection of parallels **a**′, **b**′ drawn at the same perpendicular distance from **a**, **b** respectively. Take any point A on **a**, draw AC perpendicular to **a** and take any point C on it. Through C draw CX parallel to **a**. Take any point B on **b**, draw BD perpendicular to **b**, making

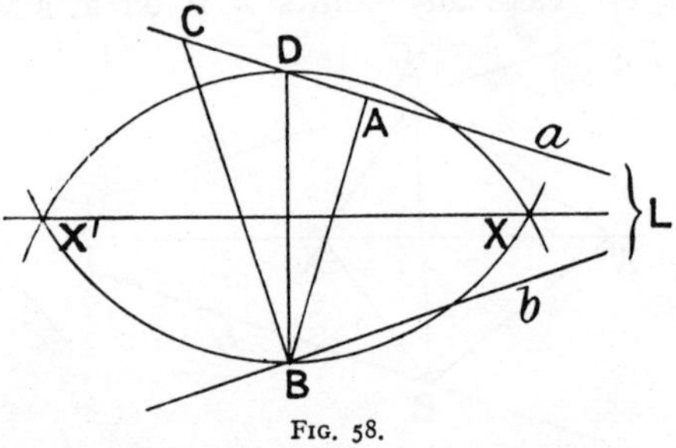

FIG. 58.

BD = AC, and draw DX parallel to **b**. From the same two perpendiculars cut off other equal lengths AC′, BD′, and draw C′X′, D′X′ parallel to **a**, **b** respectively to meet in X′. Join XX′, which is the required bisector of the angle between **a**, **b**.

Instead of constructing X′ in this way, we may use the property that XX′ is also the bisector of ∠CXD. This is the method of translation.

(iii) Since XX′ is the axis of symmetry of the pair of arms **a**, **b**, it bisects at right angles any intercept BD which is equally inclined to **a**, **b**. To construct this axis, take any point A on **a**, draw AB perpendicular to **a** to meet **b** in B; draw BC perpendicular to **b**; draw BD to bisect ∠ABC and to meet **a** in D. Then XX′, the perpendicular bisector of BD, is also the bisector of the angle between **a**, **b**; for BD, being equally inclined to AB, CB, is also equally inclined to **a**, **b**, which are perpendicular to them respectively; LBD is an isosceles triangle, and XX′, which bisects the base at right angles, also bisects the vertical angle DLB.

Ex. 4. To draw tangents to a circle from an inaccessible point.

(i) If the point L is given as the intersection of two straight lines which meet the circle in real points, let these intersections be A, B; C, D. If not, take points A, C on the circumference, and by ex. 1 draw AB, CD to pass

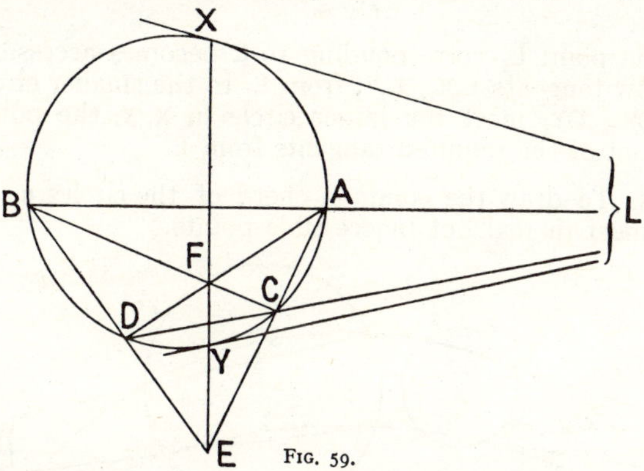

Fig. 59.

through L and meet the circumference again in B, D respectively. Let AC, BD meet in E, and AD, BC in F. If we arrange that AB is nearly a diameter and CD nearly a tangent, then both E and F are accessible. Join EF to meet the circle in X, Y. Then these are the points of contact of the tangents from L to the circle, and the tangents themselves are constructed as the perpendiculars to the radii at X, Y. For the diagonal EF of the complete quadrilateral AB, AD, BC, BD divides harmonically the two chords LAB, LCD; it is therefore the polar of L, and passes through the points of contact of the tangents from L to the circle.

(ii) Or we may use similar figures with O the centre of the circle as centre of similitude. Reduce the scale

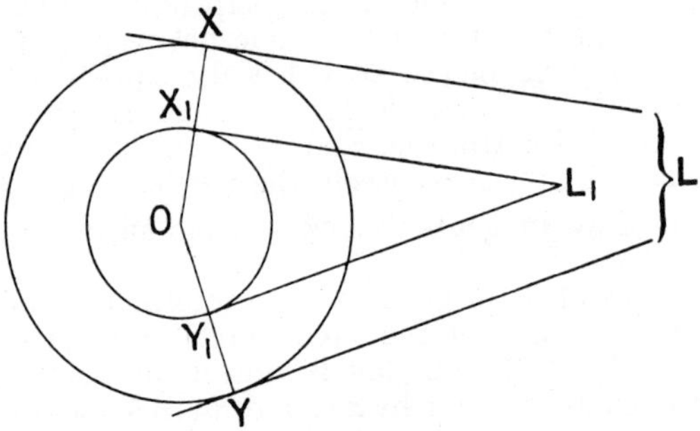

FIG. 60.

until the point L_1 corresponding to L becomes accessible, and draw tangents L_1X_1, L_1Y_1 from L_1 to the smaller circle. Then OX_1, OY_1 meet the larger circle in X, Y, the points of contact of the required tangents from L.

Ex. 5. To draw the common chord of the circles C_1, C_2, which meet in real but inaccessible points.

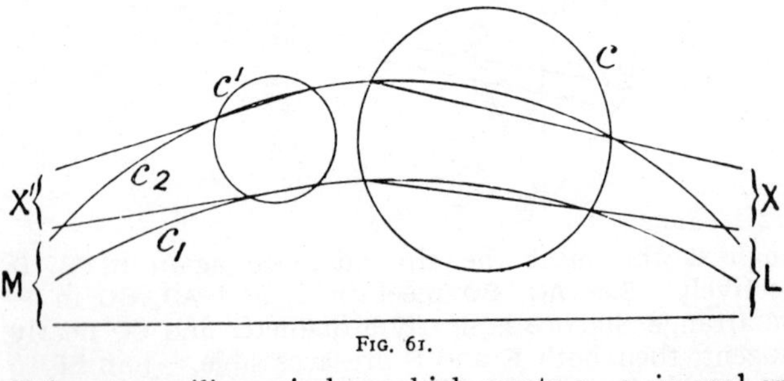

FIG. 61.

Take an auxiliary circle C which meets C_1, C_2 in real and accessible points. Then the two common chords of C, C_1;

$\mathbf{c}$, $\mathbf{c}_2$ meet in X, the radical centre of the three circles, which lies upon the required common chord of $\mathbf{c}_1$, $\mathbf{c}_2$. By using another auxiliary circle $\mathbf{c}'$, we can construct another point X′ of the common chord. Both X and X′ are likely to be inaccessible, and we may have to construct the straight line joining XX′ by the methods of ex. 2.

Ex. 6. One circle $\mathbf{c}_1$ is partly drawn, and another $\mathbf{c}_2$ is known to pass through three given points A, B, C ; both centres are inaccessible. It is required to draw the radical axis.

This can be done by the method of the last example, if $\mathbf{c}$ is taken to pass through A, B, and $\mathbf{c}'$ through B, C. The

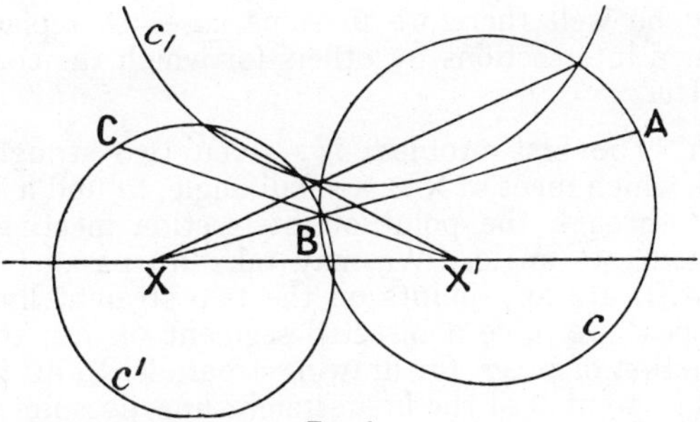

FIG. 62.

common chords of $\mathbf{c}$, $\mathbf{c}_2$ and $\mathbf{c}'$, $\mathbf{c}_2$ are then the straight lines AB and BC, so that we need not have the circle $\mathbf{c}_2$ drawn. Care must be taken that $\mathbf{c}$, $\mathbf{c}'$ meet $\mathbf{c}_1$ in points lying on the arc that is drawn.

II. Ill-defined Intersections.

It is always well to economize the chance of error: among many ways in which errors may arise, one of the most frequent is in determining the point of intersection of two lines, straight or curved, which meet at a very small angle. In an actual figure, in which lines are represented by strips of small breadth $\mathbf{b}$ say, an intersection is only determined as some point within the rhombus which is common to the two strips. If the two directions are at right angles, the rhombus is a square, and the distance of any one of its points from the centre is not greater than $\frac{1}{2}\mathfrak{b}\sqrt{2}$; but if the two directions make a very small angle ϵ, the greatest distance of a point of the rhombus from the centre is $\frac{1}{2}\mathfrak{b}\operatorname{cosec}\frac{1}{2}\epsilon$, which has no upper limit as ϵ

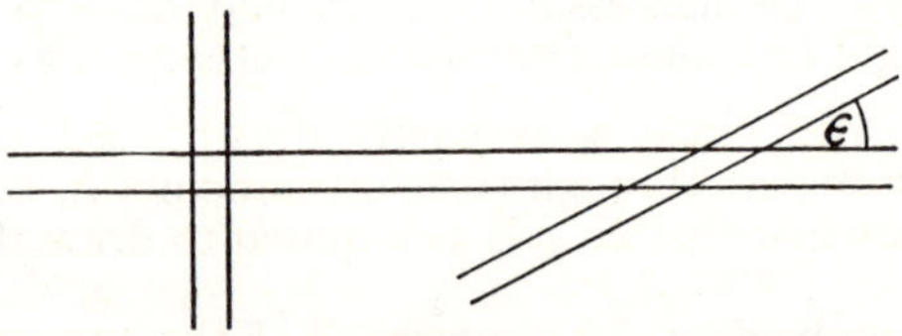

FIG. 63.

decreases, and may be a source of quite sensible error. It may be well therefore in some cases to replace such ill-defined intersections by others for which the conditions are better.

Ex. I. The first problem is: given two straight lines AB, CD, which meet at X at a small angle, to find a straight line XY through the point of intersection meeting either at a moderate angle. We may take AC parallel to BD; for if A, C are any points on the two straight lines, and we suppose we have a bisected segment on AC, the ruler construction of p. 47, for drawing a parallel to AC through any other point B of the first straight line, does not depend on any ill-defined intersections when B is very close to AC. We can take B on AX produced, and then the problem

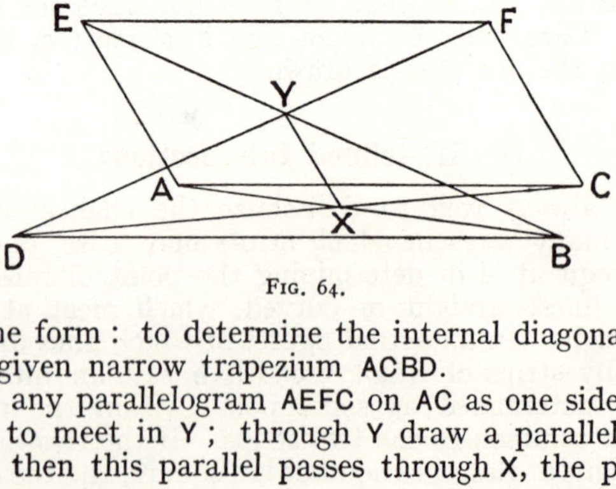

FIG. 64.

takes the form: to determine the internal diagonal point X of a given narrow trapezium ACBD.

Draw any parallelogram AEFC on AC as one side. Join BE, DF to meet in Y; through Y draw a parallel to AE or CF; then this parallel passes through X, the point of intersection of AB, CD. For if XY is joined, by similar triangles,

$$\frac{AX}{XB}=\frac{AC}{BD}=\frac{EF}{BD}=\frac{EY}{YB};$$

therefore XY is parallel to AE.

Ex. 2. To determine the two intersections X, Y of a straight line AB with a circle which it nearly touches.

Let C be the centre, and Z the pole of AB with regard to the circle. Then X, Y are the points of contact of tangents from Z to the circle, the angles CXZ, CYZ are right angles and CXZY lie on a circle whose centre is O the midpoint of CZ. Now the angle between AB and the tangent XZ to the given circle at X is equal to $\angle$ZCX between their perpendiculars CZ and CX; and similarly, the angle between AB and the circle O(X) is equal to $\angle$ZOX. But $\angle$ZOX at the centre of the circle O(X) is twice $\angle$ZCX at its circumference; hence the point X is determined by the intersection of AB with the circle O(X) with twice as much accuracy as it is determined by the intersection of AB with the given circle.

In order to construct first Z and then X, Y, take any point A on the given straight line, and on AC as diameter

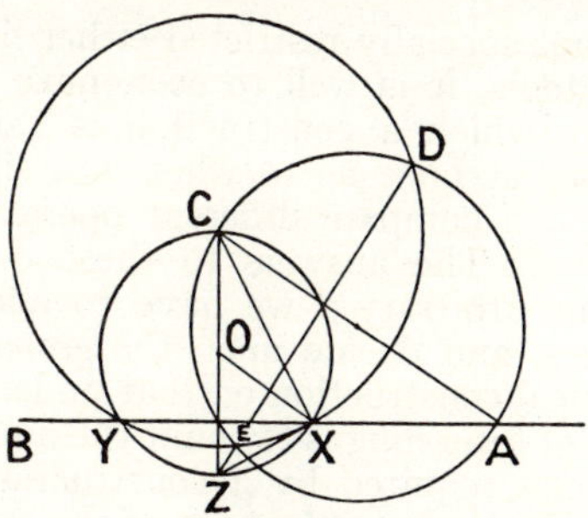

FIG. 65.

describe a circle to meet the given circle in D, E; join DE, which is the polar of A, and therefore passes through Z the pole of AB. Draw CZ perpendicular to AB to meet DE in Z; on CZ as diameter describe a circle to meet AB in the required points X, Y.

Ex. 3. To determine the second intersection X of a straight line with a circle, when one intersection A is given and the straight line is nearly a tangent.

Draw a parallel chord BC, and cut off an arc CX from the circumference equal to AB. We must take BC sufficiently far from AX for B and C to be well defined. If we

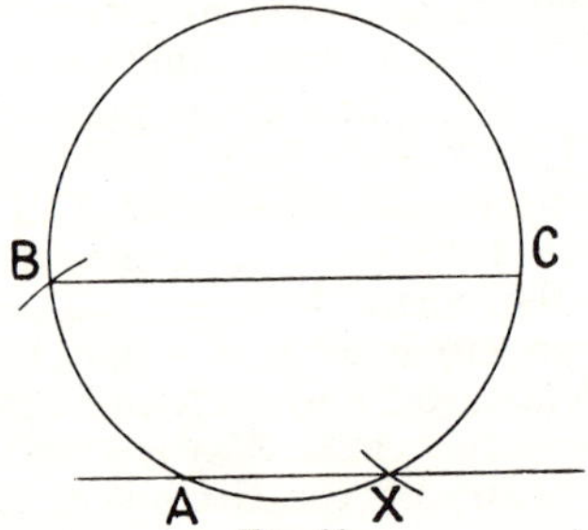

Fig. 66.

make AB and CX equal to the radius of the circle, then each of the points B, C, X is determined as the intersection of two lines that meet at an angle of nearly $\frac{1}{3}\pi$.

III. Geometrography.

If we are not specially restricted either by lack of space or by blunt tools, it is well to economize the number of operations by which a construction is carried out. The question arises at once as to what is a single operation, and how we can compare different operations with ruler and compasses. The answers to these questions are to a large extent arbitrary; we have to adopt some fixed scale of values, and decide upon the *geometrographical* or most economical construction on that understanding. The simplest way of reckoning is to count the number of straight lines and circles required by a construction, and to take this total number as a rough gauge of the amount of labour expended. Or we may consider that to describe a circle is more laborious than to rule a straight line. If we reckon that each of these two operations involves a fixed expenditure of energy, and that these are in the ratio R : C, then if a construction requires L straight lines and m circles, its measure is the quantity LR + mC, and among different constructions for the same problem, those are best for which this measure is least. If we take $\frac{C}{R}=\infty$, we try to avoid the use of compasses, so as to keep m as small as possible, however much this may entail an increase in L. We have already discussed what constructions are possible with ruler alone, which is the case $m=0$; and m need never be greater than 1 (see chapter VII). If we take $\frac{R}{C}=\infty$, and avoid as far as possible the use of a ruler, we can always have $L=0$ (see chapter VIII).

But we may distinguish between various operations with the same instrument. As long as we restrict ourselves to definite constructions, whenever we use a ruler we have to make its edge coincide with two given points and then rule along the edge; but often we only need to draw an indefinite straight line with one or no coincidence; and these are simpler operations. Again, to describe the circle $C(AB)$ requires us to make a compass leg coincide with a fixed point three times: first with the two ends A, B of the given radius and then with the centre C. But to describe the circle $C(A)$ only requires two coincidences, one with the centre C and one with the given point A of the circumference; and to describe a circle with given centre and any radius, or any circle through a given point, only requires one coincidence. Also, with modern compasses (p. 71), when we describe two circles in succession with the same radius, the second only requires one coincidence, at its centre.

Lemoine has worked out an elaborate system of geometrography. He distinguishes the following operations:

R_1 make the straight edge pass through one given point,
R_2 rule a straight line,
C_1 make one compass leg coincide with a given point,
C_2 make one compass leg coincide with any point of a given line,
C_3 describe a circle.

Then if these operations occur respectively L_1, L_2, m_1, m_2, m_3 times in a construction, its *symbol* is

$$L_1R_1 + L_2R_2 + m_1C_1 + m_2C_2 + m_3C_3\,;$$

the sum of all the coefficients $L_1 + L_2 + m_1 + m_2 + m_3$, which is the total number of operations, is the *coefficient of simplicity*, and the total number of coincidences $L_1 + m_1 + m_2$ is the *coefficient of exactitude*. As Lemoine remarks, these coefficients really measure, not the simplicity and exactitude, but the complication and the chance of error. The number of straight lines is L_2 and of circles m_3, so that the total number of lines drawn is $L_2 + m_3$, which is the difference between the coefficients of simplicity and exactitude. To join two points A, B is a complex operation, which has the symbol $2R_1 + R_2$; to describe the circle C(AB) has the symbol $3C_1 + C_3$.

With these conventions, Lemoine proceeds to give, for a large number of problems, the geometrographical constructions, those for which the coefficient of simplicity is least. It is only in easy cases that we can assert that a given construction is as simple, in the sense defined, as any other possible method ; in general we can only say that it is as simple as any other known method. Further, the construction which is simplest for a certain problem standing alone may not be the simplest when the same problem occurs as part of another ; for a construction less simple in itself may fit in better with the context, by making use of lines which are already drawn or which can be used later on in the course of the construction of the more complicated problem.

Other systems, more or less elaborate than this, could no doubt be given, and justified from different points of view. We shall keep to these definitions in the examples that follow in this chapter.

Ex. 1. Through a given point A to draw a straight line parallel to a given straight line BC.

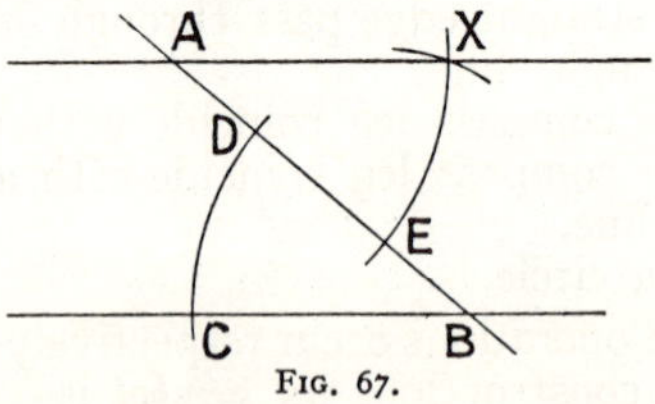

FIG. 67.

(i) A usual construction is by means of equal alternate angles. Draw any straight line through A to meet BC in B (symbol R_1+R_2). With centre B and any radius ρ describe the circle $B(\rho)$ to meet BC in C and BA in D (C_1+C_3). Describe the circle $A(\rho)$ to meet AB in E (C_1+C_3), and describe E(CD) to meet $A(\rho)$ in X ($3C_1+C_3$). Join AX ($2R_1+R_2$), which is the required parallel. The symbol of the whole construction is $3R_1+2R_2+5C_1+3C_3$; simplicity **13**, exactitude **8**.

(ii) Compare the following construction. With any centre D describe D(A) to meet the given straight line in

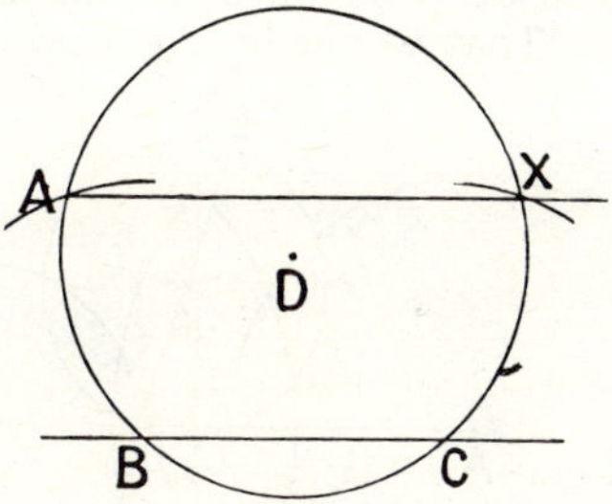

FIG. 68.

B, C (C_1+C_3). Describe C(AB) to meet D(A) in X ($3C_1+C_3$); join AX ($2R_1+R_2$). Symbol $2R_1+R_2+4C_1+2C_3$; simplicity **9**, exactitude **6**.

(iii) Or compare this. With any radius ρ describe the circles $A(\rho)$ to meet BC in B; $B(\rho)$ to meet BC in C; $C(\rho)$ to meet $A(\rho)$ in X. Join AX. Then ABCX is a rhom-

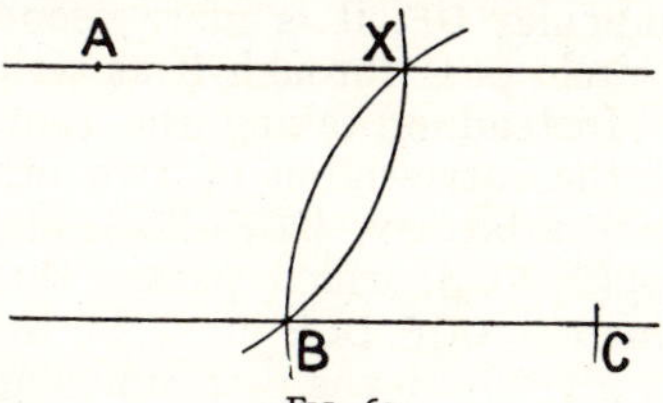

FIG. 69.

bus and AX is parallel to BC. Symbol $2R_1+R_2+3C_1+3C_3$; simplicity **9**, exactitude **5**. Though this construction has one more line than the last, it gains by using the same radius (and that arbitrary) for all the circles, instead of two different radii, one of them assigned.

Ex. 2. Let us analyse the constructions already given for the radical axis of the circles whose common points are either imaginary or inaccessible. We suppose the centres A, B given.

(i) In the construction of p. 79, ex. 2, we begin by joining AB $(2R_1+R_2)$ to meet the circles in C, D. Then we have to draw CE perpendicular to AC. The shortest method for this perpendicular is to describe any circle G(C) through C to meet AC in H, and join HG to meet G(C) in E. Then CE (which need not be drawn) is perpendicular to AC. That would be the best construction for

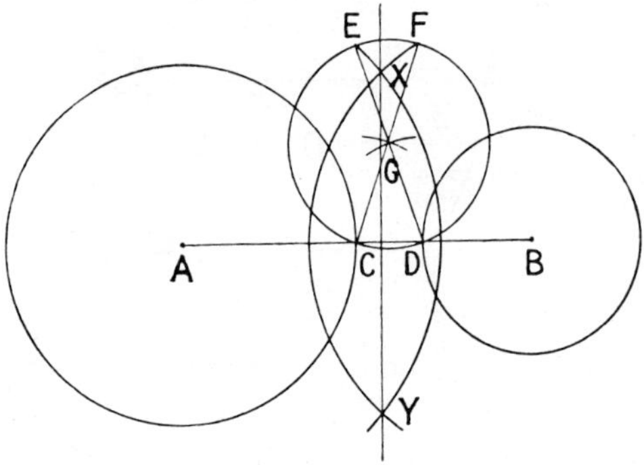

FIG. 70.

a point E on the perpendicular if there were no further construction to be made; but as we also require the second perpendicular DF, it is more economical to make the auxiliary circle pass through D as well as C, so as to do for both. Instead of taking the centre G arbitrary, we take it at the intersection of two other circles $C(\rho)$, $D(\rho)$, where ρ is arbitrary, $(2C_1+2C_3)$, and then describe the circle $G(\rho)$ (C_1+C_3), which passes through C and D. Then the diameters CGF, DGE must be drawn $(4R_1+2R_2)$, which not only gives both the perpendiculars CE, DF (which need not be drawn), but also gives them of equal length, for the figure GCDEF is symmetrical. It remains to describe the circles A(E), B(F) $(4C_1+2C_3)$ to meet in X, Y, and to join XY $(2R_1+R_2)$, which is the required radical axis. The whole symbol is $8R_1+4R_2+7C_1+5C_3$; simplicity **24**, exactitude **15**.

(ii) In the construction of p. 108, ex. 5, we obtain the point X on the radical axis by using one arbitrary circle and two common chords; then X′ is found in the same way, and XX′ is joined. Symbol $\mathbf{10}R_1+\mathbf{5}R_2+\mathbf{2}C_3$; simplicity **17**, exactitude **10**.

(iii) In this last construction, if the centres A, B are given, instead of X′ we can find the reflexion Y of X in AB, as the other intersection of the circles A(X), B(X), and then join XY. Symbol $\mathbf{6}R_1+\mathbf{3}R_2+\mathbf{4}C_1+\mathbf{3}C_3$; simplicity **16**, exactitude **10**.

Ex. 3. To bisect an angle whose vertex is inaccessible.

Compare the methods of p. 105, ex. 3, with this: With any point C of one arm as centre and any convenient radius ρ describe $C(\rho)$ to meet the arms AB, CD in A, D.

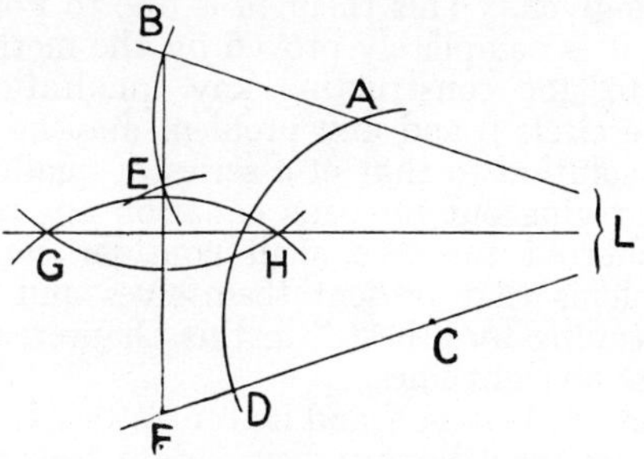

Fig. 71.

Describe $A(\rho)$ to meet AB in B, and describe $D(\rho)$ to meet $A(\rho)$ in E. Join BE to meet CD in F. Describe $B(\rho)$, $F(\rho)$ to meet in G, H; join GH, which is the required bisector.

For the figure CAED is a rhombus of side ρ, and BAE is an isosceles triangle with sides parallel to the given arms; BEF is perpendicular to the required bisector, which must also bisect BF. Symbol $\mathbf{4}R_1+\mathbf{2}R_2+\mathbf{4}C_1+C_2+\mathbf{5}C_3$; simplicity **16**, exactitude **9**.

CHAPTER VII.

ONE FIXED CIRCLE.

ANY point that can be constructed with ruler and compasses can be constructed with ruler only if one fixed circle and its centre are given. This theorem is due to Poncelet and Steiner, and it is completely proved by the methods given on pp. 60, 61 for constructing any quadratic surd by means of one circle; and any problem may be solved by reducing its solution to that of a series of quadratic equations, and carrying out for each equation one of the constructions referred to. We shall consider some of the simpler problems that present themselves and their best solutions, meaning by "best," in this chapter, those that require fewest straight lines.

We call the fixed circle Γ and its centre O. If the centre is not given, we must have sufficient data to construct it; for example, a parallelogram, for then we can draw parallel chords and bisect them, and so obtain diameters. When O is found, all the metrical data are at hand.

The methods of this chapter are appropriate in connection with the method of trial and error (p. 80), or others which require the common points of a homography, which can be found by projecting on to the fixed circle. They are also of use in problems connected with the centres of similitude of two circles, in a way which is fully illustrated below.

We shall first consider separately the best methods for drawing parallels, which are required for all the fundamental constructions.

Lemma. Through a given point R *to draw a parallel to a given straight line* PQ.

We have first to obtain a bisected segment PZQ on the

given straight line, and then construct a complete quadrilateral with one diagonal point at infinity, as on p. 47.

Case (i). PQ passes through O, the centre of Γ. The problem is to draw through R a parallel to a given diameter of Γ.

Then if P, Q are the ends of the diameter, POQ is a bisected segment already given, and we complete the construction thus. Join PR, QR; on PR take any point U. Join UO

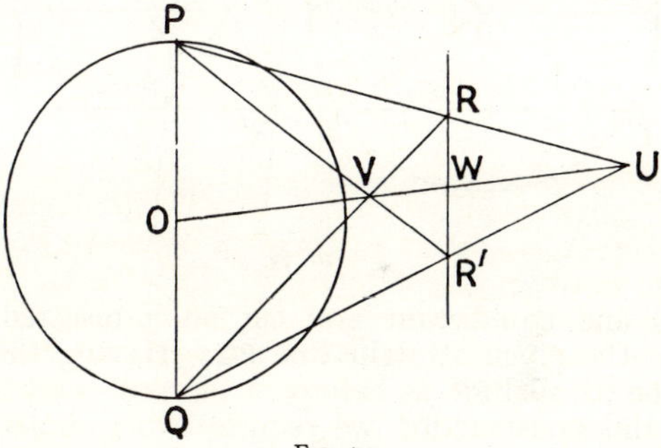

FIG. 72.

to meet QR in V; join QU, PV to meet in R′. Then RR′ is the required parallel; for UV, RR′ divide harmonically the third diagonal PQ of the quadrilateral PR, PR′, QR, QR′; and since UV bisects PQ, therefore RR′ meets it at infinity, so that PQ, RR′ are parallel. The figure consists of a quadrilateral and its three diagonals, that is, of seven straight lines, one of which is given.

An important aspect of this figure is when RR′ is regarded as the diameter of a circle γ, which is not drawn. Since the diagonals RR′, UV of the quadrilateral meet in W the midpoint of RR′, this point W is the centre of the circle γ, and U, V are the external and internal centres of similitude of Γ, γ. In the first of these similitudes, P corresponds to R and Q to R′; in the second, P corresponds to R′ and Q to R.

Case (ii). PQ and R have general positions.

In order to obtain a bisected segment PZQ, we first draw any straight line OZ through O to meet PQ in Z; then, by case (i), a parallel chord ST, taking care that its intersections S, T with Γ are real. Join SO, TO and produce them to meet the circumference again in S′, T′; then S′T′ is the reflexion of ST in OZ, so that these three straight lines are

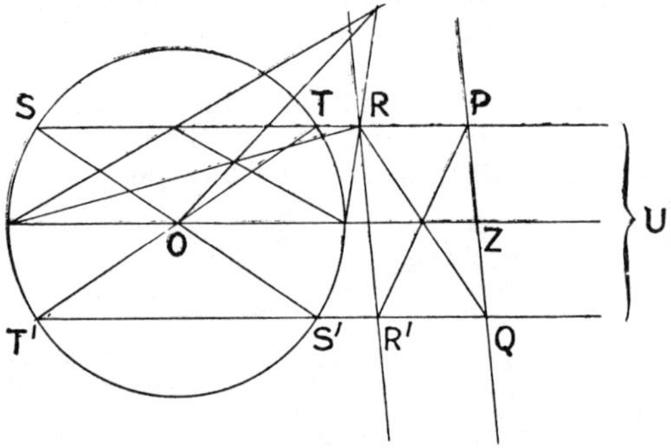

FIG. 73.

parallel and equidistant and cut off a bisected segment PZQ on the given straight line PQ. Having this we can draw the parallel RR′ as before.

For this construction we require two complete quadrilaterals, each with its three diagonals, in order to draw the two pairs of parallels OZ, ST and PQ, RR′; in addition we have the two diameters SS′, TT′ and the third parallel S′T′: if the figure is drawn without precaution, it may contain seventeen straight lines. But this may be reduced by making some lines serve two purposes. If we take OZ to pass near enough to R, the chord ST may be drawn through R and still cut Γ in real points. Then the second quadrilateral may be reduced to a parallelogram, by taking U at infinity; for then UP, UZ, UQ coincide with the three parallels already drawn; the second quadrilateral has two sides and one diagonal already in the figure, and only requires four more lines. The whole figure contains fourteen straight lines.

Case (iii). R coincides with O. The problem is to draw the diameter of Γ parallel to a given straight line.

The simplification just given fails if R coincides with O, for we cannot draw ST through the centre of Γ. The best we can do is to make one side of the second quadrilateral, QR′U say, coincide with one of the parallel chords, S′T′ say. The whole figure contains sixteen straight lines.

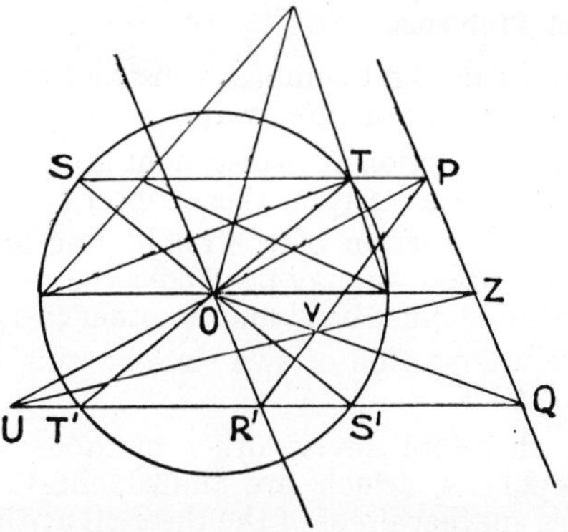

FIG. 74.

Thus when a circle γ is given by means of its centre **o** and a point **a** on its circumference, we can first draw the diameter AOB of Γ parallel to **ao**. Then to obtain other points on the circumference of γ, we make use of E, I, the external and internal centres of similitude of Γ, γ. These are the intersections of the line of centres Oo with Aa and

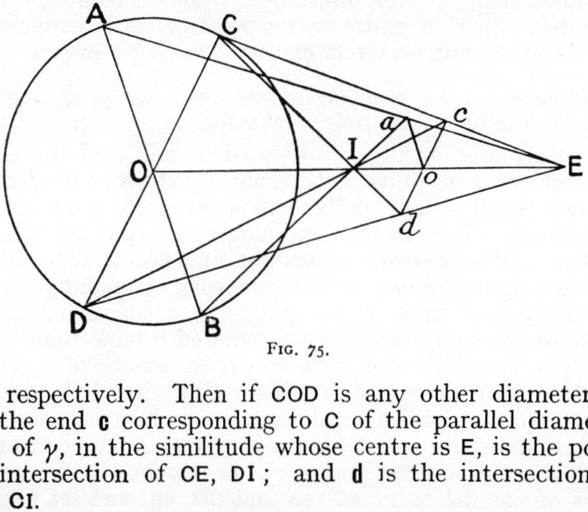

FIG. 75.

Ba respectively. Then if COD is any other diameter of Γ, the end **c** corresponding to C of the parallel diameter **cod** of γ, in the similitude whose centre is E, is the point of intersection of CE, DI; and **d** is the intersection of DE, CI.

Fundamental Problems.

In a general ruler and compass construction, new points are determined in one of three ways :

(i) as the intersection of two straight lines : this method we may use freely in this chapter ;

(ii) as the intersection of a straight line and a circle : this method we may use under the restriction that the circle must be Γ and no other ;

(iii) as the intersection of two circles : this we may not use at all.

We must therefore devise other methods for the two following problems, which are fundamental, and whose solution gives another proof of the theorem at the beginning of this chapter :

I. To determine the points of intersection of a given straight line and a given circle other than Γ.

II. To determine the points of intersection of two given circles, both in case (i) in which one circle is Γ, and is therefore already drawn, and in case (ii) in which neither circle is Γ, so that neither is to be actually drawn.

Throughout, a circle must be considered to be given or obtained when its centre and a point on its circumference are known ; but no circle except Γ must be drawn.

Problem I. To find the intersections **x**, **y** *of a given straight line* **fg** *with a circle* γ *other than* Γ.

The straight line **fg** is supposed drawn, and the circle γ given by means of its centre **o** and a point **a** on its circumference, that is, one radius **oa** is given in position and magnitude. The method employed is that of similar figures. We construct a straight line FG, which with Γ makes a figure similar to that composed of **fg** and γ ; then if FG meets Γ in X, Y, the points **x**, **y** corresponding to these in the similitude are the required intersections of **fg** and γ. The first step is to construct as above E, I the external and internal centres of similitude of Γ, γ. We use the direct similitude, which has E as centre ; then any straight line through E meets any two parallel radii of Γ, γ in corresponding points. Now suppose we have two pairs of parallel radii, AO, **ao** and CO, **co**, and let **f**, **g** be the intersections of the given straight line with **ao**, **co**. Then the corresponding points are F the intersection of **Ef**, AO, and G the intersection of **Eg**, CO. Then if FG meets Γ in X, Y, the required points **x**, **y** are the intersections of **fg** with EX, EY.

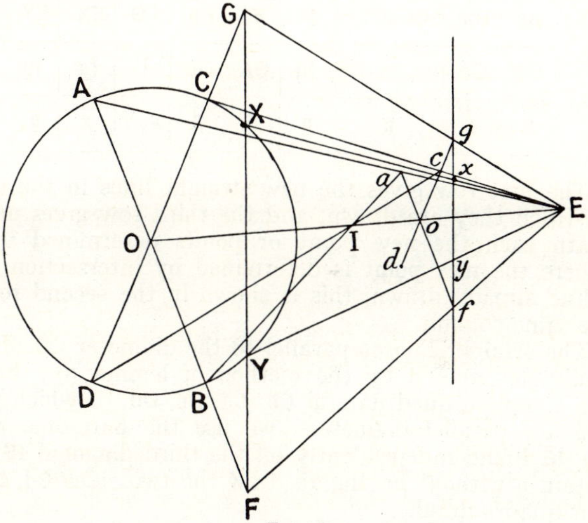

FIG. 76.

The construction succeeds or fails according as FG meets Γ in real or imaginary points; that is, according as **fg** meets γ in real or imaginary points. Hence if **x**, **y** are real, they can always be constructed by this method.

Construction in Detail.

The whole figure contains twenty-seven straight lines. We start with the one straight line **fg** already drawn. Then we carry out the construction of lemma (iii) in order to obtain the diameter AOB of Γ parallel to **ao**; this introduces sixteen straight lines. Now we can arrange that two of these sixteen, besides **ao**, AO, have positions that are used again; for the diameter of Γ first to be drawn (OZ of fig. 74) can coincide with O**o**, and one of the other diameters (SS′ say of the same figure) can be used for the arbitrary second diameter CD of Γ which is required. The point **f** is now determined as the intersections of **oa**, **fg**. The rest of the construction requires ten more lines, given by this table :

Aa	Ba	CE, DI	**oc**	Ef	E**g**	FG	EX	EY
O**o**	O**o**		**fg**	OA	OC	Γ	**fg**	**fg**
E	I	**c**	**g**	F	G	X, Y	**x**	**y**

The first row gives the new straight lines in the order in which they are drawn, and the third row gives underneath each the new point or points determined by it. Where the new point is determined by intersection with a line already drawn, this is shown in the second row in the same column.

The straight line **oc** parallel to the diameter CD of Γ is really determined by the method of lemma (i) ; but in the complete quadrilateral C**c**, C**d**, D**c**, D**d**, of which COD, **cod** are parallel diameters, we use the harmonic range O**o**, IE found independently on the third diagonal IE, and obtain **c** without having to draw the two sides C**d**, D**d** of this quadrilateral.

Problem II case (i). *To find the intersections of* Γ *and a given circle* γ.

This problem is the same as to find the radical axis of Γ, γ ; for this straight line is their common chord if they meet in real points, and it meets Γ in X, Y, the required intersections of the two circles. Now we can obtain two points on the radical axis most quickly as follows.

Let **o** be the centre and **co** the given radius of γ. Draw the parallel diameter COD of Γ and construct E, I, the external and internal centres of similitude of the two circles. Let EC, ED meet Γ again in H, K. Determine the points **d**, **h**, **k** of γ corresponding to D, H, K ; let CK, **hd** meet in L, and DH, **kc** in M. Then LM is the radical axis and meets Γ in the required points X, Y.

This follows from the fact that the four points C, K, **d**, **h** lie on an auxiliary circle, and L is the intersection of CK its common chord with Γ and **hd** its common chord with γ. For in fig. 77

$$\angle \mathrm{hCK} = \angle \mathbf{hck}, \text{ by parallels,}$$
$$= \angle \mathbf{hdk}, \text{ in the circle } \gamma,$$
$$= \pi - \angle \mathrm{hdK};$$

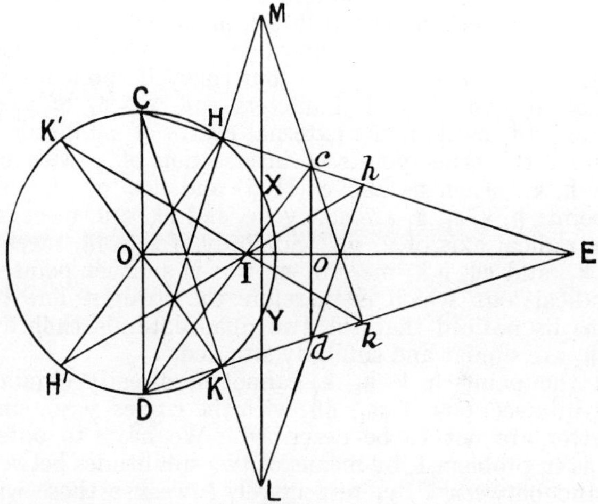

FIG. 77.

hence the points C, K, **d**, **h** are concyclic. Similarly, M lies on the radical axis of Γ, γ.

Construction in Detail.

We are given Γ and the two points **o**, **c**. Draw the diameter COD of Γ parallel to **co** by lemma (iii). The figure now contains sixteen straight lines, including **co**, COD, and one (OZ of fig. 74) which, as before, can be taken to coincide with Oo. The rest of the construction is given by the table:

Cc	DE	Dc	HO	KO	H′I	K′I	CK, **hd**	DH, **kc**	LM
Oo, Γ	**oc**, Γ	Oo	Γ	Γ	Cc	DE			Γ
E, H	**d**, K	I	H′	K′	**h**	**k**	L	M	X,Y

The whole figure contains twenty-eight straight lines.

Case (ii). *To find the intersections of two given circles* γ, γ_1.

We reduce this to depend on problem I by first finding **lm**, the radical axis of the two circles. Then the points of intersection **x**, **y** of **lm** with one of the circles, γ say, are the required points of intersection of γ, γ_1. We have therefore to determine two points **l**, **m** on this radical axis. Each is obtained as the intersection of two chords which meet the circles respectively in four concyclic points. We first obtain two parallel diameters **cod**, $\mathbf{c_1o_1d_1}$ of γ, γ_1. Then $\mathbf{cc_1}$, $\mathbf{dd_1}$ meet in the external centre of similitude of γ, γ_1; let the other points of intersection of γ with $\mathbf{cc_1}$, $\mathbf{dd_1}$ be **h**, **k**. Then as above (II (i) and figs. 77, 81), the four points **h**, **k**, $\mathbf{c_1}$, $\mathbf{d_1}$ are concyclic, and **hk**, $\mathbf{c_1d_1}$ meet in **l** on the radical axis of γ, γ_1. Similarly, if $\mathbf{cc_1}$, $\mathbf{dd_1}$ meet γ_1 in $\mathbf{h_1}$, $\mathbf{k_1}$, and **cd**, $\mathbf{h_1k_1}$ meet in **m**, this is another point on the radical axis, which is therefore the straight line **lm**. It is to be noticed that the two quadrilaterals **ckdh** and $\mathbf{c_1k_1d_1h_1}$ are similar and similarly situated.

But the points **h**, **k**, $\mathbf{h_1}$, $\mathbf{k_1}$ cannot be directly obtained as the intersections of $\mathbf{cc_1}$, $\mathbf{dd_1}$ with the circles γ, γ_1, since the latter are not to be described. We have to obtain them as in problem I, by means of two similitudes between Γ, γ and between Γ, γ_1 respectively; we use those with the external centres E, $\mathrm{E_1}$. The quadrilaterals **ckdh**, $\mathbf{c_1k_1d_1h_1}$, inscribed in γ, γ_1 respectively, that we wish to construct, are similar (in these two similitudes) to the same quadrilateral CKDH inscribed in Γ (fig. 80). We require two pairs of parallel diameters of Γ, γ, and we take them parallel to the given radii **oa**, $\mathbf{o_1c_1}$ of γ, γ_1; these same pairs of parallels can be used again later to determine the intersections of **lm** with γ, which are the required points **x**, **y**.

The construction can be divided into stages as follows:

Stage (i), fig. 78. Construct, by lemma (iii), the diameters AOB, COD of Γ parallel to the given radii **ao**, $\mathbf{c_1o_1}$ of the given circles γ, γ_1.

Stage (ii), fig. 79. Determine, by means of the centres of similitude E, I, $\mathrm{E_1}$, the other ends **b**, $\mathbf{d_1}$ of the diameters **aob**, $\mathbf{c_1o_1d_1}$ of γ, γ_1, and also the diameter **cod** of γ parallel to COD or $\mathbf{c_1o_1d_1}$. We have now the two pairs of parallel diameters AOB, **aob** ; COD, **cod** of Γ, γ, and by means of intersections with these we can find the chord of either circle corresponding to any chord of the other.

Stage (iii), fig. 80. Draw $\mathbf{cc}_1$, $\mathbf{dd}_1$, and determine their other intersections **h**, **k** with γ and $\mathbf{h}_1$, $\mathbf{k}_1$ with γ_1, by means of the corresponding chords CH, DK of Γ.

Stage (iv), fig. 81. Determine the points **l**, **m**, by means of the cyclic quadrilaterals $\mathbf{hkc}_1\mathbf{d}_1$, $\mathbf{cdh}_1\mathbf{k}_1$, and so construct the radical axis **lm** of γ, γ_1.

Stage (v), fig. 82. Determine, by problem I, the intersections **x**, **y** of **lm** with γ.

Construction in Detail.

We are given: the circle Γ fully drawn and its centre O; the centres **o**, $\mathbf{o}_1$ of the given circles γ, γ_1 and the points **a**, $\mathbf{c}_1$ on their respective circumferences.

Stage (i). Draw the diameter AOB of Γ parallel to the straight line joining **ao**, by the method of lemma (iii).

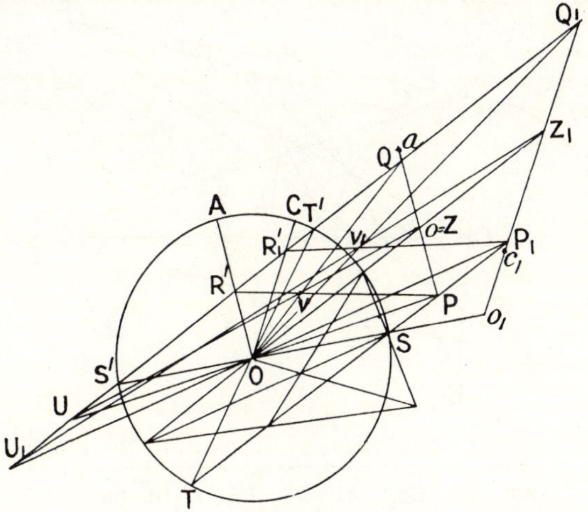

FIG. 78.

This requires the sixteen straight lines of fig. 74. Next, draw COD parallel to $\mathbf{c}_1\mathbf{o}_1$; this also requires sixteen straight lines, but we can use the same set of three parallels as before, and therefore the same set of ten lines: namely, those called OZ, ST, S′T′, SS′, TT′ in figs. 74 and 78, and the five lines, not lettered, used in the construction of ST. The number of straight lines drawn at this stage is therefore $2 \times 16 - 10 = 22$.

These include **ao**, $\mathbf{c_1o_1}$, AO, CO, which are needed later; we can also arrange for it to include Oo and Oo_1. For to start with, the point Z of fig. 74 can be taken at **o**, and then S can be taken at a point of intersection of Oo_1 with Γ, so that Oo, Oo_1 of fig. 78 coincide with OZ and SOS′ of fig. 74 respectively.

The constructions of the remaining stages are given by the following diagrams and tables. In order to keep the figure as clear and compact as possible, we use in stage (v) the inverse similitude between Γ, γ, whose centre is **I**, instead of the direct similitude, whose centre is E.

Stage (ii).

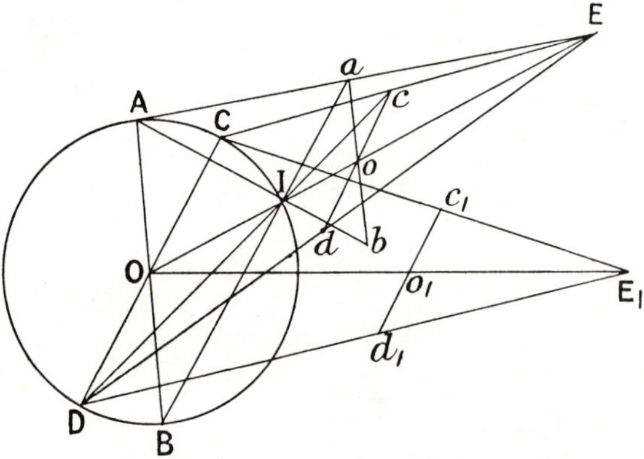

FIG. 79.

Aa	Ba	Cc_1	AI	DE_1	CE, DI	DE, **co**
Oo	Oo	Oo_1	**ao**	$\mathbf{c_1o_1}$		
E	**I**	E_1	**b**	$\mathbf{d_1}$	**c**	**d**

Stage (iii).

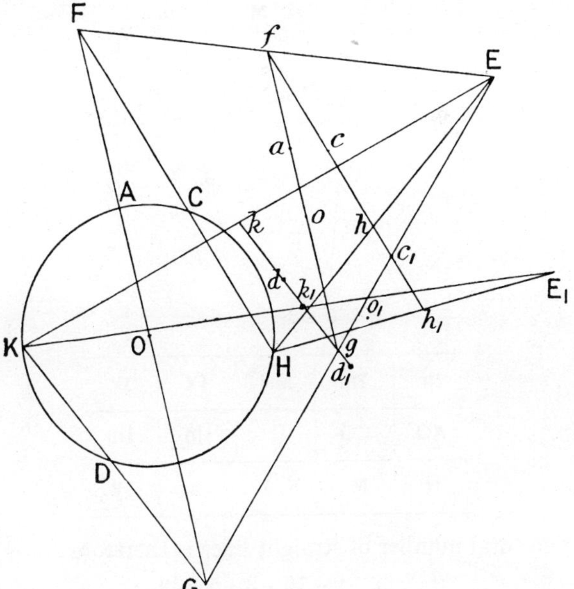

FIG. 80.

cc_1	dd_1	Ef	Eg	CF	DG	EH	EK	E_1H	E_1K
ao	ao	AO	AO	Γ	Γ	cc_1	dd_1	cc_1	dd_1
f	g	F	G	H	K	h	k	h_1	k_1

Stage (iv).

hk	h_1k_1	lm
c_1o_1	co	ao
l	m	n

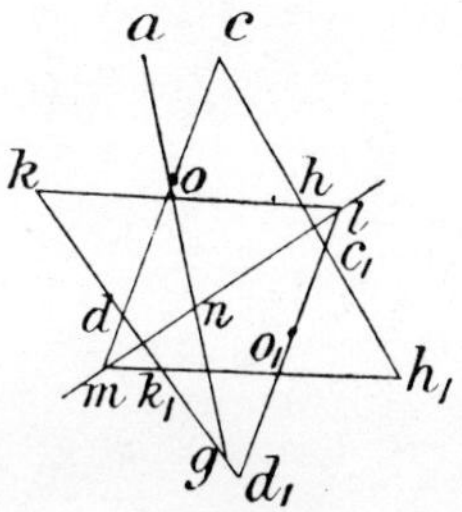

FIG. 81.

Stage (v).

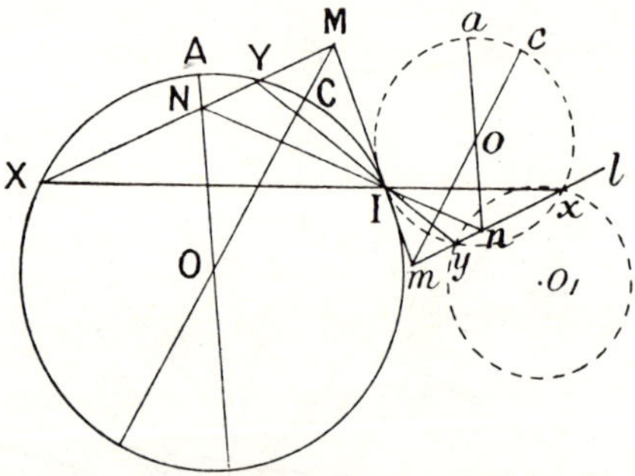

FIG. 82.

In	Im	MN	IX	IY
AO	CO	Γ	Im	Im
N	M	X, Y	x	y

The total number of straight lines is therefore

$$22 + 9 + 10 + 3 + 5 = 49,$$

but it is very possible that this might be reduced by greater ingenuity.

CHAPTER VIII.

COMPASSES ONLY.

If our only ruler were badly chipped or warped, ruling a straight line would be an altogether less desirable operation than describing any number of circles. We should then wish to avoid as far as possible having to determine a point as the intersection of two straight lines; Mascheroni and Adler have shown that we can avoid it altogether.

Of course, no straight line can be obtained in its entirety, if we make use of compasses only; it must be considered as given or obtained when two points lying on it are known. It is the main object of this chapter to show that we can then obtain, still with compasses only, any other points of the straight line that we may want, that is, its points of intersection with any given circle, or with any straight line given in the same way by means of two points lying on it.

Mascheroni's methods are the older; generally speaking, they are based on the idea of reflexion in a straight line. Some of his constructions are shorter than Adler's, mainly because Mascheroni uses his compasses freely as dividers. Adler's methods are based on the idea of inversion with regard to a circle. They are much more systematic, and more powerful as regards the theory, and they use Euclidean compasses throughout. The single operation, according to Mascheroni, of describing the circle C(AB) would have to be replaced, if we use Euclidean compasses, by a construction requiring five circles. The following is as short as any:

Describe the circles A(C), C(A) to meet in D, E. Then DE is the axis which bisects AC at right angles. Describe

D(B), E(B) to meet again in X. Then X is the reflexion of B in the axis DE, and CX is equal to its reflexion AB. Describe C(X); then this is the circle required, for its centre is the required point C and its radius CX is equal to the required radius AB.

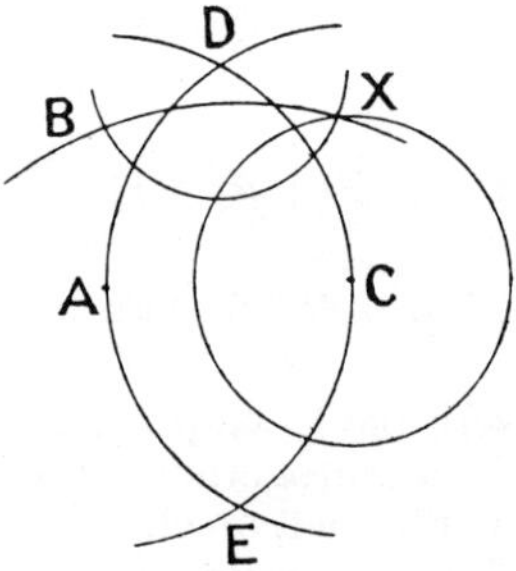

FIG. 83.

This double process, of first determining an axis, and then reflecting in it part of the figure already obtained, will occur later several times. When it forms part of another construction, we can usually arrange that some of the five circles are used either earlier or later in other parts of the construction.

Fundamental Problems.

Mascheroni's proof of the possibility of his method consists in actually giving compass constructions to replace those two, out of the three standard methods of determining points (p. 122), which involve the use of a ruler. All points, other than the data, whether required by the problem or used in the course of construction, must be determined by the third standard method only, as the intersections of pairs of circles. Just as in the last chapter, we have the two fundamental problems:

I. To determine the points of intersection of a given circle and the straight line joining two given points.

II. To determine the point of intersection of the two straight lines joining two given pairs of points.

When these are solved, we have, theoretically at least, a compass construction for any problem for which a ruler and compass construction is known. We have only to replace each step of the latter, which requires the use of a ruler, by the solution of the corresponding problem I or II.

But before giving the solutions of these fundamental problems, let us consider Adler's discussion of the matter as a whole. His proof is based on the theory of inversion. If any figure consisting of straight lines and circles is inverted with regard to a circle whose centre does not lie on any line of the figure, then the inverse figure consists entirely of circles, p. 93 (3), (4). If the original figure is that belonging to a known ruler and compass construction of any given problem, then the inverse figure consists of (i) the inverses of the given points, (ii) the inverses of the points, straight lines and circles used in the construction, and (iii) the inverses of the required points.

Now using compasses only, we can pass from the given points to their inverses (i) with regard to the auxiliary circle, whose centre is O say, by means of the first lemma given below: to find A′ the inverse of A with regard to a given circle centre O. The same proof shows that we can pass to the required points themselves as soon as we have obtained the points (iii), which are their inverses. And we shall show that we can carry out each of the steps by which we pass from (i) to (ii) and from (ii) to (iii), which correspond to the steps of the original ruler and compass construction. The latter steps are of two types: to draw a straight line AB, and to describe a circle C(D). To the first of these there corresponds in the inverse figure the step: to describe the circle A′B′O, which can be done if we first find its centre by means of the second lemma given below: to find the centre of the circle through three given points.

To a step of the second type, namely to describe the circle C(D) in the original figure, there corresponds in the inverse figure the step: to describe the circle through D′ with regard to which O and C′ are inverse points, p. 93 (3). This may be reduced to depend upon the same two lemmas. Starting with C′, D′, which we suppose already obtained, by lemma (i) we construct their inverses C, D with regard to the auxiliary circle, if they are not already in the figure, and then describe the circle C(D). If C(D) meets the auxiliary circle in real and distinct points E, F, these are unaltered by the inversion, p. 92 (2), and therefore lie on the inverse of C(D), and this inverse can be described by lemma (ii) as the circle through D'EF. Or in any case, we can find as in lemma (i) the inverse O_1 of O with regard to the circle C(D), and then the inverse O_1' of O_1 with regard to the auxiliary circle. Then O_1' is the required centre of the inverse of C(D), p. 93 (3).

If the known construction which we are inverting is a Steiner construction, the original figure contains only one circle Γ, which is arbitrary, and straight lines. Then in the inverse figure, we start with the auxiliary circle centre O and an arbitrary circle Γ', and take the inverse of Γ' to be Γ; then every other circle of the inverse figure is determined by three points on its circumference, one of them being O. It would not in general be convenient to take Γ itself as the auxiliary circle, for some of the straight lines of the original figure probably pass through its centre, and so would not invert into circles but into straight lines.

Lemma (i). *To find* A' *the inverse of a given point* A *with regard to a given circle* O(B).

Case I. $OA > \frac{1}{2}OB$.

Describe the circle A(O). Since its diameter 2OA is greater than the radius OB of the given circle, A(O) cuts

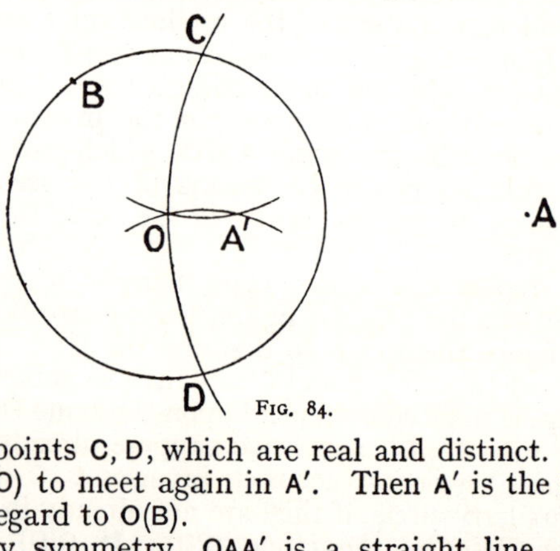

FIG. 84.

O(B) in points C, D, which are real and distinct. Describe C(O), D(O) to meet again in A'. Then A' is the inverse of A with regard to O(B). (3 circles)

For by symmetry, OAA' is a straight line. Also the triangles OAC, OCA' are both isosceles by construction, and they have a common base angle at O; they are therefore similar, and

$$\frac{OA}{OC}=\frac{OC}{OA'} \quad \text{or} \quad OA \,.\, OA'=OC^2,$$

which shows that A, A′ are inverse points.

Case 2. $OA \leqq \frac{1}{2}OB$.

The construction just given fails if C, D are coincident or imaginary. We then begin by finding the point A_1 in OA produced, so that $OA_1=2OA$. This can be done very simply by describing first the circle A(O) and then in succession O(A), P(A), Q(A) to meet A(O) in P, Q, A_1 respectively. Then O, P, Q, A_1 are successive angular points of a regular hexagon inscribed in A(O); OA_1 is a diameter, and is twice the radius OA.

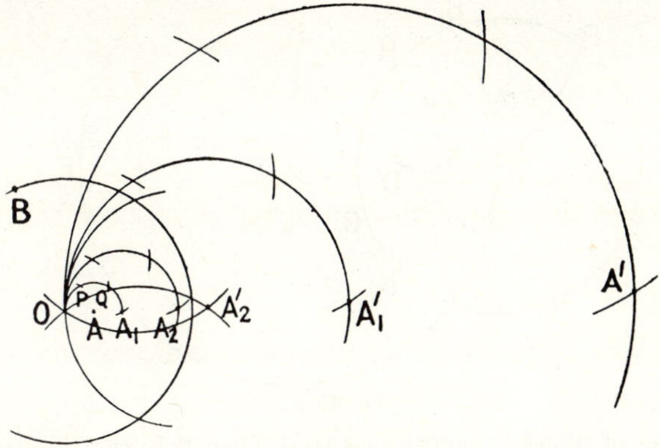

FIG. 85.

If $OA_1>\frac{1}{2}OB$, find the inverse A_1' of A_1, by case I; then, as above, find A′ in OA_1' produced, such that $OA'=2OA_1'$. Then A′ is the inverse of A. (11 circles)

For $\quad OA \,.\, OA'=\frac{1}{2}OA \,.\, 2OA_1'=OA_1 \,.\, OA_1'=OB^2.$

If $OA_1 \leqq \frac{1}{2}OB$, we must go on doubling, finding in succession A_2, A_3 ... A_p in OA produced, where

$$OA_2=2OA_1, \quad OA_3=2OA_2 \;\ldots\; OA_p=2OA_{p-1}=2^pOA,$$

until $OA_p>\frac{1}{2}OB$. We then find the inverse A_p' of A_p, and again double p times, constructing in succession the points $A'_{p-1} \ldots A_2', A_1', A'$, where

$$OA' = 2OA_1' = 2^p OA_p'.$$

Then A' is the inverse of A. ($8\mathsf{p}+3$ circles)

For $OA \cdot OA' = 2^{-p}OA_p \cdot 2^p OA_p' = OA_p \cdot OA_p' = OB^2$.

The figure is drawn for $\mathsf{p}=2$, with nineteen circles besides $O(B)$, which is given.

Lemma (ii). *To find the centre* D *of the circle through three given points* A, B, C.

Describe A(B). Find C′ the inverse of C with regard to A(B); find D′ the reflexion of A in BC′ as axis; find D the

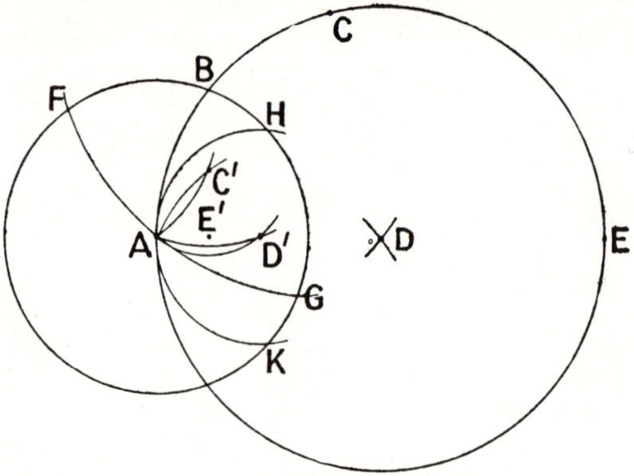

FIG. 86.

inverse of D′ with regard to A(B). Then D is the required centre of the circle ABC. (9 circles)

To prove this, let AE be the diameter through A of the circle ABC. When we invert with regard to A(B), this circle becomes a straight line, through B (which is unaltered) and C′, perpendicular to AE, so that the inverse E′ of E is the foot of the perpendicular from A on BC′, and therefore the midpoint of AD′. Now the required centre D is the midpoint of AE; hence

$$AD \cdot AD' = \tfrac{1}{2}AE \cdot 2AE' = AB^2,$$

so that D, D′ are inverse points with regard to A(B).

Or we may regard the straight line BC′ as a circle of infinite radius, with regard to which A, D′ are a pair of inverse points. When we invert with regard to A(B), the centre A becomes a point A′ at infinity, and D′ becomes the inverse of A′ with regard to the circle ABC, that is, its centre, p. 92 (1).

This construction is given by the table :

A(B), C(A)	F(A), G(A)	B(A), C′(A)	D′(A)	H(A), K(A)
			A(B)	
F, G	C′	D′	H, K	D

The first row gives all the circles of the figure in the order in which they are described, and the third row gives underneath each the new point or points determined by it. Where the new point is determined by intersection with a circle already described, this is shown in the second row in the same column.

Problem I. To determine the points of intersection X, Y *of a given circle* C(D) *and the straight line joining two given points* A, B.

(i) The construction suggested by the theory of inversion rests upon the fact that the required points X, Y, lying on C(D), are not altered by inversion with regard to that circle, p. 92 (2).

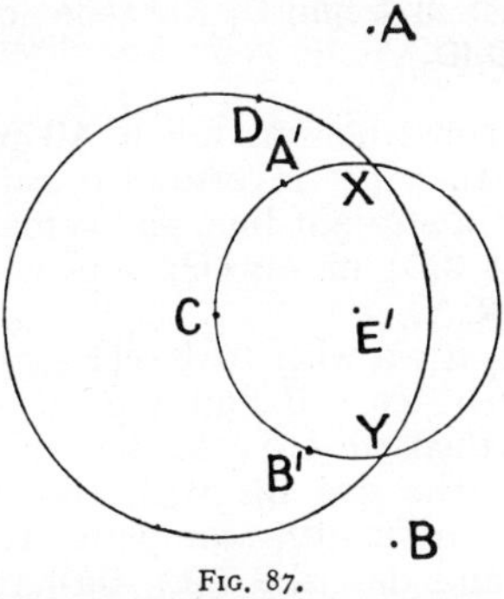

Fig. 87.

Find A′, B′ the inverses of A, B with regard to C(D). Find E′ the centre of the circle A′B′C, and describe the latter circle, which is the inverse of the straight line AB, and meets C(D) in the required points X, Y. (16 circles)

(ii) Mascheroni's construction is much shorter. The intersections of C(D) with the straight line AB are the same as its intersections with its own reflexion in AB. We therefore find C_1 the reflexion of C in AB by means of the circles A(C), B(C), and describe C_1(CD); this meets C(D) in the required points X, Y. (3 circles)

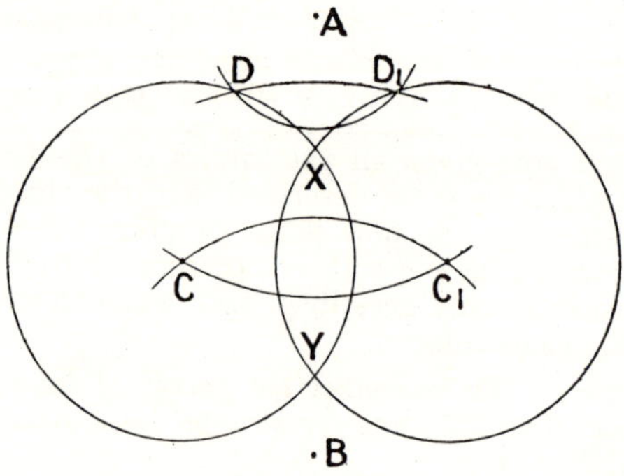

FIG. 88.

If the compasses are Euclidean, instead of describing C_1(CD), we must first find D_1 the reflexion of D in AB, and then describe $C_1(D_1)$. (5 circles)

Both these constructions fail if AB passes through the centre C; for then the inverse of the straight line ABC is not a circle but a straight line, and cannot be drawn; and the reflexion of C(D) in AB coincides with itself, and does not determine X, Y.

For example, if we wish to bisect a given arc DE whose centre C is given, we can find a point A on the bisecting diameter, and then we have to determine X the point of intersection of the arc DE with the diameter CX. To overcome this difficulty, complete the parallelograms DECF, EDCG, and describe F(E), G(D) to meet in A. By symmetry, A is on the bisecting diameter, and it can be proved that FX = CA. We have the construction given by the following table; the arc DE and the point C are given.

D(C), C(DE)	E(C)	F(E), G(D)	F(CA)
	C(DE)		C(D)
F	G	A	X

Problem II. To determine the point of intersection X *of the two straight lines joining two given pairs of points* A, B ; C, D.

(i) Describe any convenient auxiliary circle centre O; find the inverses A′, B′, C′, D′ of the given points; find the centres E′, F′ of the circles OA′B′, OC′D′ respectively, and describe these circles to meet again in X′; find X the

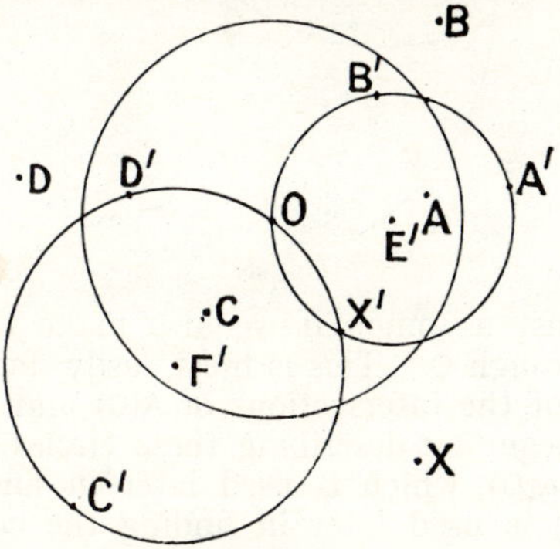

FIG. 89.

inverse of X′ with regard to the auxiliary circle. Then X is the required point of intersection of AB, CD. If carried out without precautions, this construction requires

$$\mathbf{1}+\mathbf{4}\times\mathbf{3}+\mathbf{2}\times\mathbf{10}+\mathbf{3}=\mathbf{36} \text{ circles.}$$

But this number can be very much reduced with a little care. First of all, the auxiliary circle may be taken to pass through A; then A′ coincides with A and needs no construction. Also now we need not construct B′; for in finding E′ the centre of OAB′, we require not B′, but the inverse of B′ with regard to O(A), which is B itself. In finding E′, we need only determine E the reflexion of O in AB, by means of A(O), B(O), and then E′ is the inverse of E with regard to the auxiliary circle O(A).

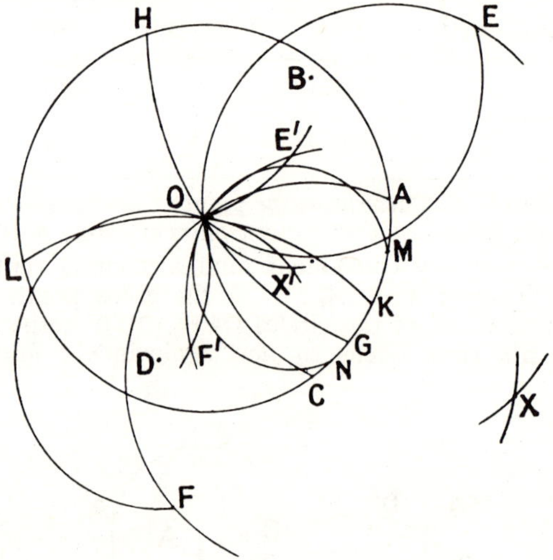

FIG. 90.

We save just as much if we also make the auxiliary circle pass through C. This is most easily done by taking O to be one of the intersections of A(C) and C(A). Then we have to begin by describing these circles, but A(C) is the same as A(O), which is used later in finding E, and similarly C(A) is used later in finding the corresponding point F.

The construction requires sixteen circles, eight of which have a radius equal to AC; it is given by the table:

A(C) C(A)	B(O)	E(O) O(A)	G(O) H(O)	D(O)	F(O)	K(O) L(O)	E′(O) F′(O)	X′(O)	M(O) N(O)
	A(C)			C(A)	O(A)			O(A)	
O	E	G, H	E′	F	K, L	F′	X′	M, N	X

We have assumed the most favourable case for each inversion, namely that each of OE, OF, OX′ is greater than

$\frac{1}{2}$OA. The first of these conditions can always be fulfilled, by choosing O to be that one of the two intersections of A(C) and C(A) which lies furthest from the straight line AB. Then the least value of the length of the perpendicular from O to AB is $\frac{1}{2}$AC, when $\angle BAC = \frac{1}{2}\pi$, and the least value

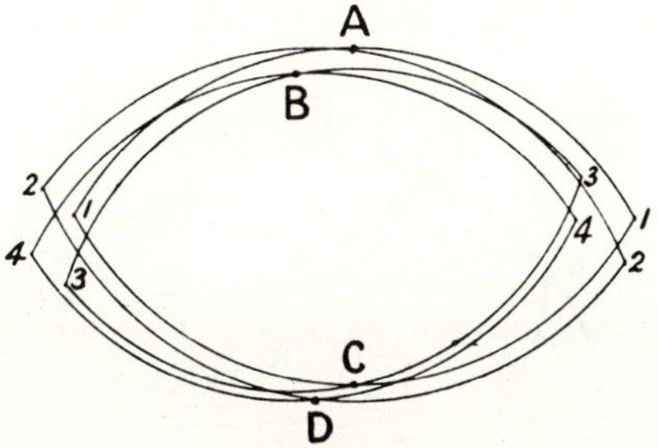

FIG. 91.

of OE is AC, so that we have OE$>\frac{1}{2}$AC. But it may happen that the position of O that is far enough from AB is too near to CD. Then instead of starting with the pair of points A, C, we may start with any other of the four pairs A, C ; A, D ; B, C ; B, D, each of which gives two possible positions of O. But there is an unfavourable case in which all the eight positions of O fail to satisfy both the conditions OE$>\frac{1}{2}$OA and OF$>\frac{1}{2}$OA ; this occurs when AB, CD are short and nearly parallel, each being inclined at an angle of nearly $\frac{1}{3}\pi$ to AC. In this case we must use case 2 of lemma (i), and the construction requires more than sixteen circles.

(ii) Mascheroni's construction is shorter if we allow the modern use of compasses, but longer if we are restricted to the Euclidean use. Find the reflexions C_1, D_1 of C, D in AB. Then the required point of intersection X of AB, CD is also the intersection of CD and C_1D_1 ; the triangles CXC_1, DXD_1 are similar, and

$$\frac{CX}{CD} = \frac{C_1C}{C_1C + DD_1}.$$

The length $C_1C + DD_1$ is constructed by completing the parallelogram $CDD_1\mathbf{d}$, where $\mathbf{d}$ is the intersection of the circles $C(DD_1)$ and $D_1(CD)$. Then $C_1\mathbf{d} = C_1C + C\mathbf{d} = C_1C + DD_1$, and we have to construct a fourth proportional to C_1d, C_1C, CD. This is done by means of similar triangles each bounded by two radii of concentric circles. Describe

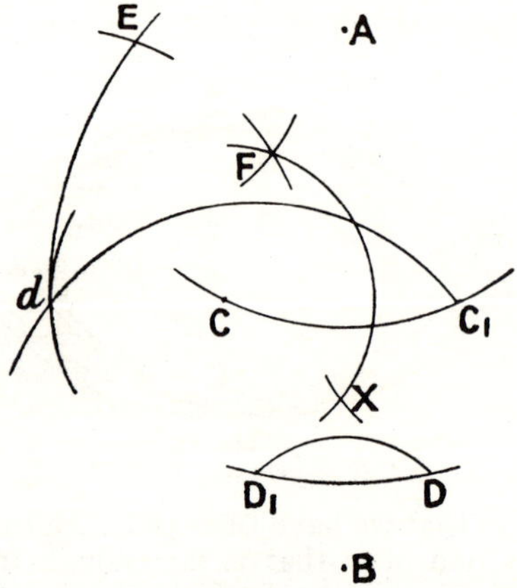

FIG. 92.

$C_1(d)$, $C_1(C)$ and place in the former a chord dE equal to CD by means of the circle $d(CD)$. The point F where the radius C_1E meets $C_1(C)$ is found as the point of contact of the latter circle and the circle $E(dC)$, or better as the point of intersection of either with $d(CE)$. Then C_1dE, C_1CF are similar isosceles triangles, and

$$\frac{CF}{dE}=\frac{C_1C}{C_1d}.$$

Hence $CF=CX=C_1X$, and X is found as the point of intersection of $C(F)$ and $C_1(CF)$.

The construction requires eleven circles as follows; three radii, namely CD, DD_1 and CF, are each used twice over.

$A(D)$ $B(D)$	$A(C)$ $D_1(CD)$	$C(DD_1)$	$C_1(d)$ $d(CD)$	$d(CE)$ $E(DD_1)$	$C(F)$ $C_1(CF)$
		$D_1(CD)$			
D_1	C_1	d	E	F	X

(iii) In adapting this construction to Euclidean compasses, I have not been able to reduce the number of circles below seventeen.

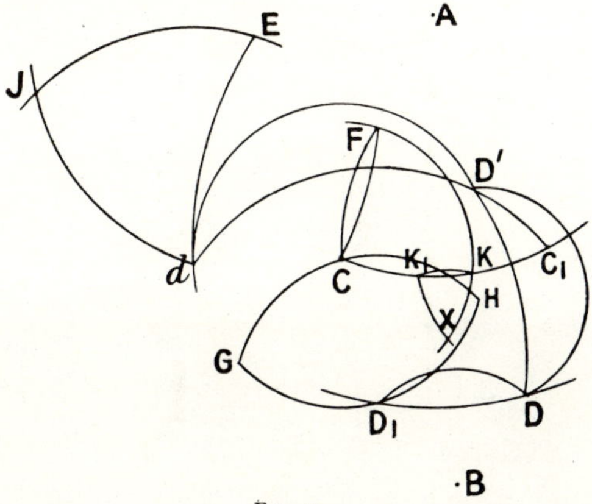

FIG. 93.

In order to construct **d**, we find D′, where CD′ is the reflexion of D_1D in a certain axis GH bisecting CD_1 at right angles, which has first to be determined; we can then describe C(D′), which is the same as $C(DD_1)$. The point D_1 is constructed as before, but C_1 is the intersection of the circle $D_1(D')$, just required in finding **d**, and A(C), which can be used again later. In order to construct F, we reflect C in an axis C_1J bisecting dE at right angles, which also bisects $\angle dC_1E$. The point J lies on **d**(E), which is the same as $\mathbf{d}(D_1)$, and is already in the figure. Finally, in order to describe $C_1(CF)$, we might first reflect CF in AB, but it is shorter to reflect another radius CK of the circle C(F), so as again to make use of A(C).

The whole construction is given by the table:

A(D) B(D)	$C(D_1)$ $D_1(C)$	G(D) H(D)	C(D′) $D_1(D')$	A(C)	$d(D_1)$ $C_1(d)$	E(d)	J(C) $C_1(C)$	C(F)	B(K)	$C_1(K_1)$
				$D_1(D')$		$d(D_1)$		A(C)	A(C)	C(F)
D_1	G, H	D′	d	C_1	E	J	F	K	K_1	X

Lucknow University Studies

Faculty of Science

Edited by **B. SAHNI.**

No. I.

A. N. SINGH

THE THEORY AND CONSTRUCTION OF NON-DIFFERENTIABLE FUNCTIONS.

PREFACE

A systematic account of the theory and construction of non-differentiable functions is not found in any published text book on the theory of functions. Hobson, in his *Theory of Functions of a Real Variable*, has given an account of Knopp's method of construction of non-differentiable functions by means of infinite series, but does not mention geometrical and arithmetical methods of obtaining such functions. Other text books contain isolated examples and do not attempt to give any general theory.

In this summary of a course of four lectures, delivered at the Lucknow University, my aim has been to include as comprehensive an account of developments relating to non-differentiable functions as is possible within a limited scope of about a hundred pages. Necessary proofs of the theorems have been given in some cases. In other cases the reader will have to consult the original sources cited.

The first lecture contains the history of attempts made by nineteenth century mathematicians to construct non-differentiable functions, as well as a brief account of the general method evolved by Dini of obtaining such functions in the form of infinite series. A proof of the non-differentiability of Weierstrass's function based on a method of M. B. Porter has been given. This proof is simpler and yet more powerful than the one ordinarily found in text books.

In the second lecture is given an account of several non-differentiable functions defined geometrically. Although some of these functions are multiple-valued yet they have been included for the sake of their historical importance, especially as they were originally given as examples of continuous nowhere differentiable functions.

The third lecture deals with the history of arithmetically-defined non-differentiable functions. A detailed discussion of the derivates of an example constructed by me in 1924 is included. Two other simple examples have also been discussed. In one of these the decimal representation of fractions is used for obtaining the definition of the function.

The fourth lecture is devoted to the discussion of the properties of non-differentiable functions, especially with regard to their oscillating nature and the existence of cusps and maxima and minima. Some theorems relating to the derivates of continuous functions, which have a direct bearing on our subject, have also been included.

A bibliography of original sources cited in the text is appended at the end. In the body of the book they are referred to by their serial numbers.

It is a pleasure to record here my indebtedness to Dr. Birbal Sahni for asking me to deliver these lectures and for making their publication possible. My thanks are also due to my friends and colleagues, Mr. M. L. Bhatia for preparing the diagrams and Mr. R. D. Misra for going through the prcofs.

LUCKNOW UNIVERSITY
June 1935.

A. N. SINGH.

CONTENTS

FIRST LECTURE

FUNCTIONS DEFINED BY SERIES

SECOND LECTURE

FUNCTIONS DEFINED GEOMETRICALLY

THIRD LECTURE

FUNCTIONS DEFINED ARITHMETICALLY

FOURTH LECTURE

PROPERTIES OF NON-DIFFERENTIABLE FUNCTIONS.

FIRST LECTURE

FUNCTIONS DEFINED BY SERIES

1. **Early Notions.** In to-day's lecture I propose to deal with the historical aspect of our subject. Non-differentiable functions have played a great part in the refinement of our geometrical intuition, and were in part, if not wholly, responsible for the critical study of the notion of "limit" made by the nineteenth century mathematicians—a study which resulted in placing mathematical analysis on a sure and sound foundation. Up to the middle of the nineteenth century the notion of "function" was connected with the geometrical notion of "curve" defined as the path traced out by a moving point. This notion of curve implies that—

(*i*) The curve is continuous, because the moving point must pass through every point between any two points P and Q on the path;

(*ii*) the curve has a determinate tangent at each point, because a moving point has at every point of its path a determinate direction of motion;

(*iii*) the arc of the curve between any two points has a finite length, because the arc is described in finite time; and (*iv*) the curve does not make an indefinitely large number of oscillations in the neighbourhood of any point.

Examples of functions which could be expressed by a simple analytical formula, but which did not satisfy one or more of the above conditions, were known to mathematicians in the early nineteenth century, so that a definition of function, free from appeal to geometrical intuition, was a desideratum. Such a definition was provided by Dirichlet (1805—1859). Analytical definitions of continuity were given by Weierstrass (1815—1896), Cauchy (1784—1857) and Hankel. The notion of "magnitude" was given an analytical garb by Cantor (1845—1918) and Dedekind (1831—1916), who, independently of each other, developed their theories of the irrational number. By the help of their theories it is now possible to interpret all mathematical processes in terms of arithmetic.

2. Early History of the Calculus. Newton and Leibnitz are said to have discovered the *Differential Calculus*, but their ideas about it were incorrect and hazy. It is said that Newton did not believe in the results that he obtained by the help of the calculus until he had proved the same results by other methods. These mathematicians seem to have stumbled upon a very powerful tool, the exact nature of which they did not know. While Newton was suspicious, Leibnitz and his followers made free and sometimes indiscriminate use of the calculus in all branches of mathematics. So far as the use of the differential calculus, either as "the rate of flow" or as "the infinitesimal increment" is concerned, Newton and Leibnitz may be said to have made more extensive use of it than their predecessors, but, certainly, they were not the first to do so. It is well known that Fermat (1608—1665) had actually obtained the equation of the tangent in the form

$$Y-y = \lim_{\substack{y' \to y \\ x' \to x}} \frac{y-y'}{x-x'} (X-x).$$

3. **The Calculus in India.** The Hindu mathematicians appear to have been familiar with the idea of the "infinitesimal increment" from very early times. Mañjula (932 A.D.) has given the formula,

$$d\,(\,Sin\,\theta\,) = Cos\,\theta\, d\,\theta$$

for the calculation of $Sin\,(\theta+\delta\theta)$, when $Sin\,\theta$ is known and $\delta\theta$ is small. Bhâskarâcârya (1150 A. D.) uses the term *tâtkâlikagati* to denote the infinitesimal increment, and has applied the conception to the evaluation of the instantaneous velocity of the moon and planets, and to a problem of maxima and minima, stating that the infinitesimal increment at a maximum is zero. Nîlakantha (c. 1500 A. D.) has shown that the increment of the increment of *Sin* θ varies as $-$ *Sin* θ, which stated in the notation of the calculus gives the formula

$$\frac{d^2}{d\theta^2}(Sin\ \theta) = -\ Sin\ \theta.$$

More extensive use of the method of the calculus seems to have been made by Hindu mathematicians during the fifteenth and sixteenth centuries. Talakulattura Nambutiri (1432 A. D.) has given the expansion of *Tan* x in ascending powers of x, which is usually attributed to Gregory (1671 A. D.). In another work, the *Sadratnamâlâ*, we find the well-known expansions in infinite series of *Sin* x and *Cos* x in powers of x.

I am sure that, if political conditions in India had been favourable, the method of the *Infinitesimal Calculus* would have been independently developed along indigenous lines in India.

4. **Continuity and Differentiability.** That continuity is necessary for the existence of a finite differential coefficient was probably known to Newton and Leibnitz, but whether it is or is not sufficient for differentiability seems to have been one of the outstanding questions till 1860, when it was finally answered in the negative by Weierstrass.

An attempt to prove that continuity was a sufficient condition for differentiability was made by Ampère ([1]) in 1806. Although Ampère's proof was defective yet his result was believed in by most mathematicians for a long time. Mention may be made of Duhamel, Bertrand and Gilbert ([30]) among those who definitely expressed their belief in Ampère's result. Even in the writings of such eminent mathematicians as Gauss, Cauchy, and Dirichlet there is nothing to show that they held a different opinion. Although they did not endorse Ampère's statement none of them seems to have had the conviction that a function which was everywhere continuous but nowhere differentiable could exist. Darboux in his memoir on "Discontinuous functions," published in 1875, mentions only one auditor, M. Bienaimé, who said that he was unconvinced by Ampère's proof.

5. **Riemann's Non-differentiable Function.** It was asserted by some pupils of Riemann that, in his lectures in the year 1861, he gave the function represented by the infinite series

$$\sum_{n=1}^{\infty} \frac{Sin\ (n^2 \pi x)}{n^2}$$

as an example of a continuous non-differentiable function. No proof of Riemann's assertion was ever published by him or any of his pupils,* while Paul du Bois-Reymond,

* In a letter to Du Bois-Reymond, dated 23rd November, 1873, Weierstrass says that Riemann was reported to have said that the proof would come from elliptic functions; see *Acta Math.*, Vol. 39, p. 199.

in a paper in *Crelle's Journal*, 1874, states without proof that Riemann's Function, for certain values of x, in any interval, ever-so-small, has no finite differential coefficient. The only writer who has considered the non-differentiability of Riemann's Function is G. H. Hardy ([32]). He has shown that Riemann's Function "is certainly not differentiable for any irrational and some rational values of x". Definite information is not available as regards the existence or non-existence of the differential coefficient at the other points. It can, however, be easily proved that Riemann's Function has an infinite differential coefficient with positive sign at the point $x = 0$. Thus the function is not *totally* non-differentiable.

6. **Condensation of Singularities.** Methods of constructing functions which do not possess a differential coefficient at an everywhere dense set of points were given by Cantor ([15]) and Hankel ([34]). These methods, however, fail to give a non-differentiable function in the strict sense of the term.

7. **Weierstrass's Discovery.** The question whether a continuous no-where differentiable function could exist was finally solved by Weierstrass's discovery of the classical example

$$f(x) = \sum_{n=1}^{\infty} a^n \, Cos\,(b^n \pi x),$$

where b is an odd integer, $0 < a < 1$, and $ab > 1 + \frac{3\pi}{2}$.

Although discovered much earlier, and communicated by Weierstrass in his lectures, the function was first published in 1874 by Paul du Bois-Reymond ([24]). Due Bois-Reymond was awe-struck by Weierstrass's discovery and terms it as 'equally too strange for immediate perception as well as for critical understanding.'

8. Attempts of Other Writers. Darboux ([20]) in 1875 gave an example of a function which does not possess a finite differential coefficient. He makes no mention of Weierstrass's Function published in 1874, and was doubtless not aware of it.

Of earlier attempts to get a non-differentiable function may be mentioned those of Cellerier and Bolzano. There is reason to believe that the function

$$\sum_{n=1}^{\infty} \frac{Sin\ (a^n \pi x)}{a^n}$$

was discovered by Cellerier before 1850, as has been pointed out by G. C. Young ([100]). The function is, however, not non-differentiable in the strict sense of the term, as it possesses infinite differential coefficients at an everywhere dense set of points.*

9. Bolzano's Function. Bolzano's non-differentiable function was brought to light in 1921, when

*Hobson (36), pp. 406 – 7; also B. N. Prasad (64). G. C. Young (100) and A. Falanga (28) thought that Cellerier's function was completely non-differentiable.

its discovery, in a manuscript of Bolzano said to date from the year 1830, was first announced by M. Jasek in the sitting of the 16th December, 1921, of the Bohemian Society of Sciences. Proofs of the non-differentiability of the function have been supplied by K. Rychlik ([70]), G. Kowalewsky ([49, 50]), and A. N. Singh ([80]). From Jasek's paper it appears that Bolzano contented himself with establishing the want of a differential coefficient at an everywhere dense but enumerable set of points. Not only was Bolzano unaware of the complete non-differentiability of his function, but that he, at one time, held the wrong opinion that 'a continuous function must be differentiable for every value of the variable with the exception of *isolated values*' is evident from a footnote to Art. 37 of his book, "*Paradoxien des Unendlichen*," published in 1847-48.

10. **The Effect of Weierstrass's Discovery.** The publication of Weierstrass's example created a sensation in mathematical circles. The discovery was hailed by men of keen acumen like Du Bois-Reymond, but there were others who could not easily bring themselves round to believe in Weierstrass's demonstration. Evidence of this tendency is to be found in a comprehensive paper published by Wiener ([97]) in which he made a detailed study of Weierstrass's Function, and sought to prove that it possessed a differential coefficient at an everywhere dense set of points.

Besides supplying the answer to a question which had long been agitating the minds of mathematicians Weierstrass's discovery opened up a new field of research—the subject of non-differentiability—a subject which has exercised great charm on the minds of mathematicians. Indeed, there are few amongst mathematicians of note who have not contributed something to the subject.

11. **Work Relating to Weierstrass's Function.** A large number of papers on the subject of non-differentiability group around Weierstrass's Function, or the generalised series

$$\sum a_n \, Cos \, (b_n \pi x) \text{ and } \sum a_n \, Sin \, (b_n \pi x),$$

where the a's and b's are positive the series $\sum a_n$ is convergent, and the b's increase steadily with more than a certain degree of rapidity.

The conditions under which the series

$$\sum a^n \, Cos \, (b^n \pi x)$$

does not possess a differential coefficient, finite or infinite, were given by Weierstrass as:

$$0 < a < 1, \; ab > 1 + \frac{3\pi}{2},$$

where b is an odd integer.

The only direct improvement known on this is Bromwich's [14]

$$0 < a < 1, \; ab > 1 + \frac{3\pi}{2}(1 - a),$$

where b is odd.

For the non-existence of a finite differential coefficient several conditions have been given. Dini gives the condition ([23]):

$$ab \geqq 1, \; ab^2 > 1 + 3\pi^2,$$

Lerch ([53]):

$$ab > 1 \quad , \; ab^2 > 1 + \pi^2$$

and Bromwich ([14]):

$$ab \geqq 1, \; ab^2 > 1 + \frac{3\pi^2}{4}\,(1-a).$$

All these conditions presuppose that b is an odd integer. But Dini has shown that if

$$ab > 1 + \frac{3\pi}{2}\cdot\frac{1-a}{1-3a} \quad , \quad a < \tfrac{1}{3}\,;$$

or

$$ab \geqq 1, \; ab^2 > 1 + 15\pi^2\,\frac{1-a}{5-21a}, \; a < \frac{5}{21},$$

this restriction may be removed.

The best result in this connection is due to G. H. Hardy ([32, 33]) who has shown that neither of the functions

$$\Sigma\, a^n \, Cos\, (b^n \pi x) \text{ or } \Sigma\, a^n \, Sin\, (b^n \pi x)$$

where $o < a < 1$, $b > 1$,

possesses a finite differential coefficient at any point in any case in which $ab \geqq 1$. It has been further shown by him that the result is untrue if the word 'finite' is omitted. It has also been shown that these functions possess cusps at everywhere dense sets of points.

12. The Series Definition. Attempts have been made by various writers to generalize Weierstrass's result by considering the function

$$f(x) = \sum_{1}^{\infty} U_n(x) \qquad (i)$$

instead of Weierstrass's function

$$W(x) = \sum_{1}^{\infty} a^n \, Cos \, (b^n \pi x). \qquad (ii)$$

Mention may be made of Faber who replaces *Cos* $(b^n \pi x)$ in (*ii*) by the function $\psi(b_n x)$, where $\psi(x)$ is a polygonal function of period 1, such that in (0, 1),

$$\psi(x)=x, \text{ for } o \leq x \leq \tfrac{1}{2};$$

$$\text{and } \psi(x)=1-x, \text{ for } \tfrac{1}{2} \leq x \leq 1.$$

The function actually considered by Faber (25, 26) is

$$\sum_{1}^{\infty} \frac{1}{10^n} \psi(2^{n!} x).$$

The general case $\sum a^n \psi(b^n x)$ has been shown by Knopp to be non-differentiable when $ab > 4$.

Van der Waerden (96) has recently given a simple definition of the above case when $a=1/10$ and $b = 10$.

Let $f_n(x)$ denote the distance between x and the nearest number of the form $m/10^n$, where m is an integer. Then, it is easy to see that

$$\sum_1^\infty \frac{1}{10^n}\psi(10^n x)= \sum_1^\infty f_n(x).$$

The above series does not give a non-differentiable function in the strict sense of the term for it can be shown to possess infinite differential coefficients at an enumerable everywhere dense set of points.*

Non-differentiable functions defined by series as in (i) above have been studied by Dini and Knopp. An account of Knopp's method of construction is given in Hobson's *Theory of Functions of a Real Variable* (Cambridge, 1927), Vol. II. I shall give a summary of Dini's method.†

13. **Dini's Method.** A general method of construction of non-differentiable functions was given by Dini (22) in 1877. He considers the general series

$$f(x)= \sum_1^\infty U_n(x)$$

* See footnote (2) to Van der Waerden's paper. Titchmarsh (92) mentions this example as a non-differentiable function although he proves only the non-existence of a finite differential coefficient.

† Dini - Lüroth (28) and Hobson's *Theory of Functions*, first edition, Cambridge, 1907.

where $U_n(x)$ is continuous in (a, b) for all values of n, and the series $\sum U_n(x)$ converges everywhere in (a, b) and defines a continuous function. It is further assumed that, for each value of n, $U_n(x)$ possesses maxima and minima, such that the interval between each maximum and the next minimum is a number δ_n which diminishes indefinitely as n is indefinitely increased; and also that $U_n(x) = -U_n(x+\delta_n)$, so that all the maxima of $U_n(x)$ are equal to one another, the maxima and minima being equal in absolute value and opposite in sign. It is also assumed that, for finite n, $U_n(x)$ possesses finite differential coefficients of the first and second orders $U'_n(x)$ and $U''_n(x)$ everywhere in (a, b); and that the upper limits of $|U'_n(x)|$ and $|U''_n(x)|$ have finite values $\overline{U}_n$ and $\overline{\overline{U}}_n$.

Let D_m denote the excess of a maximum over a minimun of $U_m(x)$. Let a neighbourhood $(x, x+\epsilon)$ or $(x-\epsilon, x)$ on either side of x be chosen, m may be chosen so great that several oscillations of $U_m(x)$ are completed in the chosen neighbourhood. Let the point $x+h$ be taken at a maximum or minimum of $U_m(x)$ in $(x, x+\epsilon)$ or in $(x-\epsilon, x)$; and let it be the first maximum or minimum of $U_m(x)$ on the right or on the left of x, of which the distance from x is $\geq \frac{1}{2}\delta_m$. The condition

$$|U_m(x+h) - U_m(x)| \geq \tfrac{1}{2} D_m$$

is satisfied. We note that $|h| \leq \frac{3}{2}\delta_m$.

Writing $U_m(x+h)-U_m(x)=\frac{1}{2}\alpha_m\nu_m D_m$, where ν_m is positive and >1, and $\alpha_m=\pm 1$, its sign depending on x and m, and possibly on h, the incrementary ratio

$$(i)\qquad \frac{f(x+h)-f(x)}{h}=\frac{\nu_m\alpha_m D_m}{2h}\left[1+\frac{2\eta'_m h}{D_m}\sum_{n=1}^{m-1}\overline{U}_n+\frac{2\alpha_m}{\nu_m}\frac{R_m(x+h)-R_m(x)}{D_m}\right],$$

where η'_m lies between 1 and -1, and $R_m(x)$ denotes the remainder after m terms of the series $\sum_1^\infty U_n(x)$. Again let $x+h_1$ be the next following extreme point of $U_m(x)$ after $x+h$, so that h and h_1 have the same sign and $|h_1|>|h|$. The difference $U_m(x+h_1)-U_m(x)$, when it is not zero, has the sign opposite to that of

$$U_m(x+h)-U_m(x),$$

and therefore $U_m(x+h_1)-U_m(x)=-\frac{1}{2}\xi'_m\alpha_m\nu_m D_m$, where $o\leq\xi'_m<1$.

Then we have

$$(ii)\qquad \frac{f(x+h)-f(x)}{h}-\frac{f(x+h_1)-f(x)}{h_1}$$

$$=\frac{\alpha_m\nu_m D_m}{2h}\left[1+\xi'_m\frac{h}{h_1}+\frac{4h\eta''_m}{D_m}\sum_{n=1}^{m-1}\overline{U}_n+\frac{2\alpha_m}{\nu_m}\frac{R_m(x+h)-R_m(x)}{D_m}-\frac{2\alpha_m h}{\nu_m h_1}\frac{R_m(x+h_1)-R_m(x)}{D_m}\right]$$

and also

$$=\frac{\alpha_m \nu_m D_m}{2h}\left[1+\varepsilon'_m \frac{h}{h_1}+\zeta''_m \frac{h(h+h_1)}{D_m} \sum_{n=1}^{m-1} \overline{U}_n + 2\alpha_m \theta_m \right.$$

$$\left. \frac{R_m(x+h)-R_m(x)}{D_m} - 2\alpha_m \theta'_m \frac{h}{h_1} \frac{R_m(x+h_1)-R_m(x)}{D_m}\right]$$

where η''_m and ζ''_m lie between 1 and -1, and θ_m, θ'_m between 0 and 1.

Dini applies (*i*) and (*ii*) to get the following four forms of sufficient conditions for the non-differentiability

of $f(x)=\sum_1^\infty U_n(x)$:

(A) If

(1) $\frac{\delta_m}{D_m}$ has the limit zero when m is indefinitely increased;

(2) $R_m(x+h)-R_m(x)$ has, for values of m greater than an arbitrarily chosen integer m', the same sign as α_m;

(3) $\frac{3\delta_m}{D_m}\sum_{n=1}^{m-1}\overline{U}_n$ remains numerically less than unity by more than some fixed difference;

then, $f(x)$ has at no point a differential coefficient, either finite or infinite.

(B) If

(1) $\frac{\delta_m}{D_m}$ has the limit zero when m is indefinitely increased;

(2) $| R_m(x+h)-R_m(x) |$ has a finite upper limit $2R'_m$ for all values of x, and $R_m(x+h)-R_m(x)$ has, for values of m greater than an arbitrarily chosen integer m', the same sign as α_m;

(3) $\frac{3\delta_m}{D_m} \sum_{n=1}^{m-1} \overline{U}_n + \frac{4R'_m}{D_m}$ remains less than unity by more than some fixed difference;

then, $f(x)$ has at no point a differential coefficient, either finite or infinite.

(C) If

(1) $\frac{\delta_m}{D_m}$ has not the limit zero but remains less than some finite number, for all values of m;

(2) $Rm(x+h)-R_m(x)$ has the same sign as α_m and $R_m(x+h_1)-R_m(x)$ has the opposite sign;

(3) $\frac{6\delta_m}{D_m} \sum_{n=1}^{m-1} \overline{U}_n$ or $6 \frac{\delta^2_m}{D_m} \sum_{n=1}^{m-1} \overline{\overline{U}}_n$ remains less than unity by more than some fixed difference;

then, $f(x)$ has nowhere a finite differential coefficient, although it may have an infinite one at some points.

(D) If

(1) $\frac{\delta_m}{D_m}$ has not the limit zero, but remains less than some finite number, for all values of m;

(2) $| R_m(x+h) - R_m(x) |$, $| R_m(x+h_1) - R_m(x) |$ never exceed a finite number $2 R'_m$;

$$(3) \quad \frac{6\ \delta_m}{D_m} \sum_{n=1}^{m-1} \overline{U}_n + \frac{32}{5} \frac{R'_m}{D_m} \text{ or } \frac{6\ \delta^2_m}{D_m} \sum_{n=1}^{m-1} \overline{\overline{U}}_n + \frac{32\ R'_m}{5\ D_m}$$

remains less than unity by more than a fixed difference; then $f(x)$ has nowhere a finite differential coefficient, although it may have an infinite one at some points.

It is easy to see that Weierstrass's Function is a particular case of the class of functions considered by Dini.

14. Proof of Weierstrass's result. I shall give a proof of the non-differentiability of Weierstrass's function, and also show that *almost everywhere* the function does not possess a progressive or a regressive derivative.

Let $W(x) = \sum_0^\infty a^n \,Cos\, (b^n \pi x)$, where $| a | < 1$, and b is integral.

Setting $\delta x = 2k/b^{m+1}$, k integral, we get by applying the mean value theorem to the first m terms and a trigonometric identity to the $(m+1)th$ term,

$$(14.1)\quad \frac{\Delta W(x)}{\delta x} = -\pi \sum_{0}^{m-1} (ab)^n \, Sin\, b^n\pi(x+\varepsilon\delta x)$$

$$-\pi\ (ab)^m \frac{Sin\ k\pi/b}{k\pi/b}\ Sin\ (b^m x+\tfrac{k}{b})\pi,$$

other terms vanishing on account of having a *sine* factor whose argument is a multiple of π.

Evidently the absolute value of the first m terms is less than

$$\pi \sum_{0}^{m-1} |\, ab\, |^n < \frac{\pi\,|\, ab\,|^m}{|\, ab\,| -1}, \text{ if } |\, ab\,| > 1.$$

Now, if we suppose that $k \leq \frac{3}{4}b$, the last term in (14.1) is in absolute value

$$(14.2)\quad \geq \pi\,|\, ab\,|^m \frac{1}{\frac{3}{4}\pi\sqrt{2}} \left|\ Sin\ (b^m x+\tfrac{k}{b})\pi\ \right|$$

Let x be expressed in the scale of b as

$$x = \frac{c_1}{b} + \frac{c_2}{b^2} + \cdots\cdots + \frac{c_n}{b^n} + \cdots\cdots\cdots\cdots,$$

where the c's are integers such that $0 \leq c_n \leq b-1$.

$$\text{Then} \left|\ Sin\ (b^m x + \tfrac{k}{b})\pi\ \right| = \left|\ Sin\ (I + \tfrac{k}{b} + \frac{c_{m+1}}{b} + \ldots)\right|$$

$$= \left|\ Sin\ (\tfrac{k}{b} + \frac{c_{m+1}}{b} + \cdots\cdots \right|$$

because I is an integer.

It is now easy te see that two values of k, k_1 and k_2, not necessarily equal, can be found such that

$$\text{(A)} \quad 1 \geq Sin\ (b^m x + \frac{k_1}{b})\,\pi \geq \frac{1}{\sqrt{2}}$$

$$\text{(B)} \quad -\frac{1}{\sqrt{2}} \geq Sin\ (b^m x + \frac{k_2}{b})\pi \geq -1,$$

where k_1 and k_2 have *in general* opposite signs.

Thus for these values of k,

$$\left| \pi\,(ab)^m \frac{1}{\frac{3}{4}\pi\sqrt{2}}\ Sin\ (b^m x + \frac{k}{b})\,\pi \right| \geq \frac{\pi\,|\,ab\,|^m}{\frac{3}{2}\pi}.$$

The last term of (14.1) will then dominate in sign and magnitude the first m terms if

$$|\,ab\,| > 1 + \frac{3\pi}{2}.$$

Hence the right and left incrementary ratios which we are considering will become infinite with m but will always have opposite signs.

This proves that $W(x)$ has at no point a differential coefficient.

15. Non-existence of the derivative.

We have $b^m x\pi = (I + \frac{c_{m+1}}{b} + \frac{c_{m+2}}{b^2} + \ldots\ldots)\pi$.

That is,

$$Sin\ (b^m x + \frac{k}{b})\pi = \pm\ Sin\ \ (\frac{k}{b} + \frac{c_{m+1}}{b} + \cdots)\pi$$

(15.1) Now, if $\frac{1}{2} < \frac{c_{m+1}}{l} + \frac{c_{m+2}}{b^2} + \cdots\cdots < \frac{3}{4}$, it is easy to see that a positive k_1 can be chosen so that (A) holds and another positive k_2 can be chosen for which (B) holds. Thus at all those points x, whose representation is such that the condition (15.1) is realized for infinitely many values of m, the right hand incrementary ratio oscillates between $+\infty$ and $-\infty$.

(15.2) Similary, if the representation of x' be such that the condition :

$$\tfrac{1}{4} \leq \frac{c_{m+1}}{b} + \frac{c_{m+2}}{b^2} + \cdots\cdots < \tfrac{1}{2}$$

is realized for infinitely many values of m, it can be shown that at each point x' the left hand incrementary ratio oscillates between $+\infty$ and $-\infty$.

Further, it can be easily shown that the set of points $C[x]$ complementary to the set $[x]$ for each point of which (15.1) holds, is a *null set*, if $b \geq 8$.* The same remark holds for the set $C[x']$ complementary to the set $[x']$.

Thus we obtain the following result :

Except at a set of measure zero both the upper derivates of $W(x)$ *are* $+\infty$ *and both the lower derivates* $-\infty$.†

* If $b=8$, the set $C[x]$ consists of points in whose representation 4 and 5 do not occur an infinite number of times. This set is made up of an enumerable number of perfect null sets, and is everywhere dense on the line.

†This result was proved by G. C. Young (100) by means of an elaborate and lengthy analysis.

16. Some Important Non-Differentiable Functions Defined by Series.

I now give a list of some of the important non-differentiable functions which have been studied.

(1) $\Sigma a^n \, Cos \, (b^n \pi x)$ where $o < a < 1$, b an odd integer, $ab > 1 + \frac{3\pi}{2}$; or $ab > 1 + \frac{3\pi}{2}(1-a)$, is non-differentiable.

(Weierstrass, Dini, Lerch, Bromwich)

(2) $\Sigma n^{-2} \, Sin \, n^2 x$ does not possess a finite differential coefficient.

(Riemann, Hardy)

(3) $\Sigma \, a^n \, Sin \, (b^n \pi x)$, where $o < a < 1$, $b = 4m$, $ab > 1 + \frac{3\pi}{2}$, is non-differentiable whatever be the signs of the individual terms.

(Dini, Porter)

(4) $\Sigma \, a^n \, Sin \, (b^n \pi x)$, where $o < a < 1$, $ab > 9$, is non-differentiable whether b be odd or even.

(Porter)

(5) $\Sigma \, a^n \, Sin \, (b^n \pi x)$, where $o < a < 1$, $b = 4m + 1$, $ab > 1 + \frac{3\pi}{2}$, is non-differentiable.

(Dini, Knopp)

(6) $\Sigma \, (-1)^n \, a^n \, Sin \, (b^n \pi x)$, where $o < a < 1$, $b = 4m + 3$, $ab > 1 + \frac{3\pi}{2}$, is non-differentiable.

(Knopp)

(7) $\Sigma \frac{a^n}{n!} \frac{Sin}{Cos} (n!\pi x)$, $| a | > 1 + \frac{3\pi}{2}$,

is non-differentiable.

(Porter)

(8) $\Sigma \frac{a^n}{1.\,3.\,5 \ldots (2n-1)} Cos\,(1.\,3.\,5 \ldots (2n-1)\pi x)$,

where $| a | > 1 + \frac{3\pi}{2}$ is non-differentiable.

(Darboux, Dini)

(9) $\Sigma \frac{a^n}{1.\,5.\,9 \ldots (4n+1)} Sin\,(1.\,5.\,9 \ldots (4n+1)\pi x)$,

$a > 1 + \frac{3\pi}{2}$, is non-differentiable.

(Darboux, Dini)

(10) If $\Sigma \frac{a^r}{10^r}$ denote any non-terminating decimal,

$\Sigma \frac{a^r}{10^r} \frac{Sin}{Cos} (10^{3^r}\pi x)$ is non-differentiable.

(Porter)

(11) $\Sigma a^n \frac{Sin}{Cos} (a^n \pi x)$ does not possess a finite differential coefficient.

(Cellerier, Hardy)

(12) If the periodic function $\psi(x) = x$, for $o \leq x \leq \frac{1}{2}$; $\psi(x) = 1 - x$ for $\frac{1}{2} \leq x \leq 1$, then $\Sigma a^n \psi(b^n x)$, where $o < a < 1$ and $ab > 4$ is non-differentiable,

(Faber, Knopp)

(13) If the periodic function $\varphi(x)=x$, for $o \leq x \leq \frac{1}{2}$, $\varphi(x)=1-x$ for $\frac{1}{2} \leq x \leq \frac{3}{2}$, $\varphi(x)=x-2$, for $\frac{3}{2} \leq x \leq 2$, then $\Sigma\, a^n\, \psi(b^n x)$, where $o < a < 1$, $b=4m+1$, $ab > 4$, is non-differentiable.

(Knopp)

(14) $\Sigma\,(-1)^n\, a^n\, \varphi(b^n x)$, where $o < a < 1$, $b=4m+3$, $ab > 4$, is non-differentiable.

(Knopp)

(15) $\Sigma\, a^n\, |\, Sin\,(b^n \pi x)\,|$, where $o < a < 1$, and $ab > 1 + \frac{3\pi}{2}$ is non-differentiable.

(Knopp)

(16) $x\, \Sigma\, a^n\, Sin\,(b^n \pi x)$, where $|\,a\,| < 1$, $|\,ab\,| > 1 + \frac{3\pi}{2}$ has a differential coefficient for $x=0$ but for no other value of x,

(Porter)

(17) $\Sigma\, \dfrac{x^n}{n!}\, {Sin \atop Cos}\,(n!\pi x)$ has differential coefficients between -1 and 1, and no differential coefficients if $|\,x\,| > 1 + \frac{3\pi}{2}$.

(Lerch, Porter)

SECOND LECTURE

FUNCTIONS DEFINED GEOMETRICALLY

1. It is well known that a continuous curve $\Phi(x)$ can be defined by a convergent sequence of polygonal curves $[\Phi_n(x)]$. This process has been used by various writers to construct non-differentiable functions. To illustrate the method I shall give the construction of some typical curves obtained by the above process, pointing out the main properties of each curve.

2. **Bolzano's Curve.** The first non-differentiable function defined geometrically is Bolzano's curve. The construction, which depends upon the successive stretching and deformation of straight lines, may be given* as follows:

Divide a straight line PQ (which we denote by F_0) into the two halves PM and MQ, and each of these again into four equal parts PP_1, P_1P_2, P_2P_3, P_3M, and MQ_1, Q_1Q_2, Q_2Q_3, Q_3Q. Now let Q_3 be carried to Q'_3, and P_3 to P'_3, as shown in Fig. 1, so that the join of $PP'_3MQ'_3Q$ gives a zig-zag consisting of four stretches. We note that the slope of each stretch is double that of the original stretch PQ. We denote

* The construction given here is a modification due to Kowalewsky (50); see Singh (80).

$PP'_3MQ'_3Q$ by F_1. Bolzano's fundamental operation consists in the stretching of the line PQ into the zig-zag form $PP'_3MQ'_3Q$ which is made up of four doubly steep stretches. By applying the fundamental operation

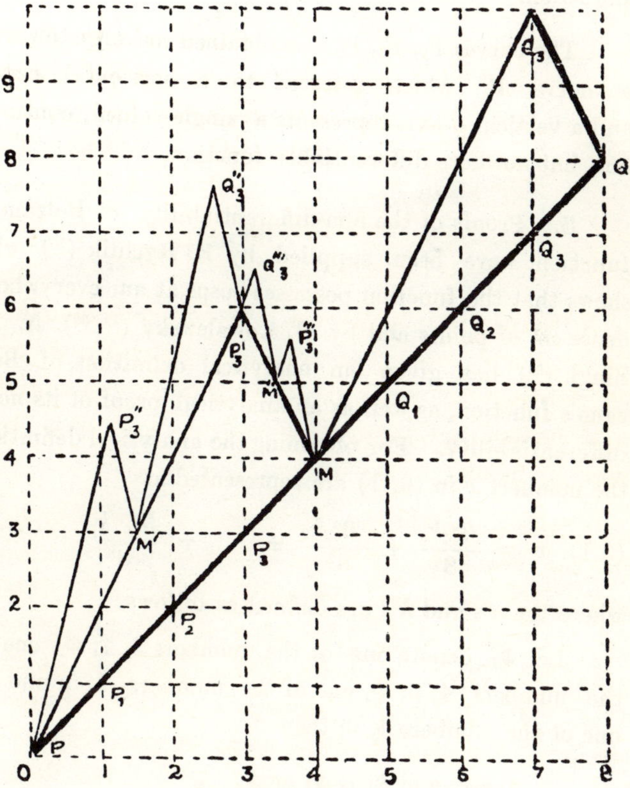

Fig. 1.

to each of the four parts of the zig-zag $PP'_3MQ'_3Q$ (curve F_1) Bolzano obtains a zig-zag line (curve F_2) consisting of sixteen parts (as shown in Fig. 1) each of which is again subjected to the fundamental operation, and so on.

The curves F_0, F_1, F_2...so obtained converge towards a curve F, which referred to a horizontal x-axis and a vertical y-axis represents a single-valued, continuous, but nowhere differentiable function.

3. Proofs of the non-differentiability of Bolzano's function have been supplied by K. Rychlik (70) who shows that the function possesses cusps at an everywhere dense set of points and by G. Kowalewsky (49, 50). A. N. Singh (80) has given an analytical definition of Bolzano's function, as well as an analytical proof of its non-differentiability. For obtaining the analytical definition the numbers x in (0, 1) are represented as

$$(3.1)\quad x = \frac{3^{a_1}k_1}{8} + \frac{3^{a_2}k_2}{8^2} + \cdots + \frac{3^{a_n}k_n}{8^n} + \cdots\cdots,$$

where the a's and k's are defined as follows:

Let Φ_1 denote one of the numbers 3, 7; Φ_2 one of the numbers 4, 0; ϕ_1 one of the numbers 1, 5; and ϕ_2 one of the numbers 4, 0; then

$$a_1 = o, \text{ and } k_1 \text{ is } \Phi_1 \text{ or } \Phi_2 \quad ,$$

and, in general,

if k_v is Φ_1, then $a_{v+1} = a_v$, and k_{v+1} is ψ_1 or ψ_2,
if k_v is Φ_2, then $a_{v+1} = a_v + 1$ and k_{v+1} is Φ_1 or Φ_2,
if k_v is ψ_1, then $a_{v+1} = a_v + 1$ and k_{v+1} is ψ_1 or ψ_2
if k_v is ψ_2, then $a_{v+1} = a_v$ and k_{v+1} is Φ_1 or Φ_2,

Considering the ending representations of x, we see that, when x consists of only one term, four points are obtained; these being $\frac{3}{8}$, $\frac{7}{8}$, or $\frac{4}{8}$, $\frac{0}{8}$ according as k_1 is Φ_1 or Φ_2. When x consists of two terms, sixteen points are obtained by giving all possible values to k_1 and k_2. Of these sixteen points four (corresponding to $k_2 = 0$) have already been obtained. Of the remaining twelve, three lie in each of the four intervals into which (0, 1) is divided by the points x consisting of one term only. Similarly it is easy to see that the aggregate of all the points consisting of n terms contains all the points whose representations run up to $(n-1)$ terms together with three points in each of the sub-intervals formed by these points. Thus it is obvious that each ending representation uniquely defines a point in (0, 1), and that the aggregate of ending representations gives an enumerable everywhere dense set of points in (0, 1), so that the ending and non-ending representations of x together give all the points in (0, 1).

A number x in $(0,1)$ being written in the form (3.1), let

$$(3.2)\qquad y = \mathrm{F}(x) = \frac{3^{a_1} q_1}{4} + \frac{3^{a_2} q_2}{4^2} + \cdots + \frac{3^{a_n} q_n}{4^n} + \cdots\cdots$$

where, if k_v is a Φ being 0, 3, 4 or 7, q_v is 0, 3, 2, or 5 respectively, and if k_v is a ψ being 0, 1, 4 or 5, q_v is 0, 1, -2 or -1 respectively.

4. Taking PQ to be the join of (0, 0) and (1, 1), F_1 is the join of $(\frac{3}{8}, \frac{3}{4})$ and $(\frac{7}{8}, \frac{5}{4})$ to the middle point and the ends of PQ. Thus F_1 consists of four lines of which the first and third are positively inclined while the second and fourth are negatively inclined. On the negatively inclined stretches the construction is carried out as shown in Fig. 1, according to Kowalewski.

It will be observed that F_1 has three edge points (for $x=\frac{3}{8}, \frac{1}{2}, \frac{7}{8}$) besides the end points, P and Q, whose ordinates are not affected by subsequent constructions. From the construction given, it follows that the edge points of F_{n-1} are also the edge points of F_n, and further that the abscissae of the edge points of F_n are all the points x whose representations in the form (3.1) run up to n places. It is also easy to verify that the corresponding values of y obtained according to the analytical definition give the ordinates of the edge points of F_n.

The function Y and Bolzano's function, therefore, agree for an everywhere dense set of values of x in (0, 1), and as both can be shown to be continuous,* they must be identical.

The analytical definition given above provides an easy method of constructing a class of non-differentiable

---

*See Singh (80).

functions of Bolzano's type. To do this we have simply to choose proper values of the a's, k's and q's in (3.1) and (3.2). It may be further pointed out that the base of representation in (3.1) and (3.2) instead of being 8 and 4, may be any other properly chosen numbers.

5. **Koch's Curve.** In two papers, ([45, 46]) published during 1903-1906, Helge von Koch developed a new method of constructing plane curves having no determinate tangent at any point. The following example will illustrate Koch's construction.

Divide the straight line (A, B) by means of points C and E into three equal parts. Construct the equilateral triangle CDE on the middle part CE (see Fig. 2).

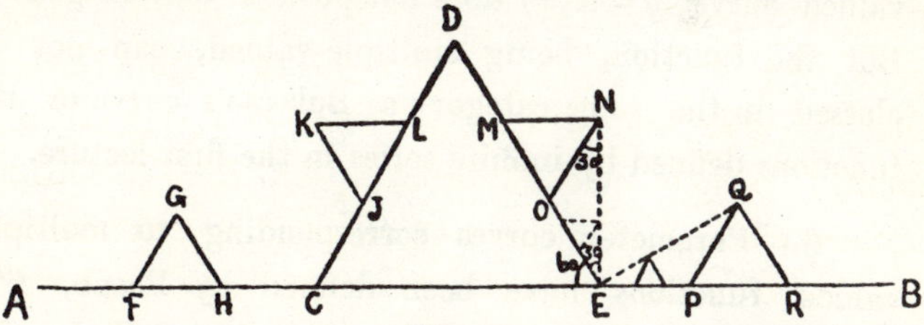

Fig. 2.

Apply the same construction to each of the four new lines AC, CD, DE and EB. Continue this construction indefinitely. The vertices of the equilateral triangles so obtained together with their limiting points form the curve of Koch.

It can be easily shown* that Koch's Curve corresponds to a multiple valued function $y=K(x)$. For

* See Singh (77).

example, the vertices of an indefinitely large number of triangles lie on the vertical through E (see Fig 2), and every vertical through a point (except the middle point) on CE cuts the curve more than once. The curve is, however, a parameter curve and can be expressed by equations of the form : †

$$x = \Phi(t)$$
$$y = \Psi(t)$$

where $\Phi(t)$ and $\psi(t)$ are single valued and continuous functions of t.

It has been shown by F. Apt (3) that the multiple-valued curve $y = K(x)$ does not possess half-tangents.‡ But the function, being multiple-valued, can not be classed in the same category as Bolzano's curve or the functions defined by infinite series in the first lecture.

6. Parameter curves corresponding to multiple-valued functions have been defined by Peano, (59), Hilbert (35), Moore (57) Schoenflies (72), Sellerio (74), Kaufmann (39) and others. All these curves can be shown to be tangentless, and even without half-tangents at any point. Although the curves are not examples of continuous non-differentiable functions (being in a sense discontinuous because they are multiple-valued), yet for

---

†See Kaufmann ( ).

‡By a half-tangent at a point P is meant the limit of the secant PQ as Q approaches P always remaining on the same side of P.

the sake of their historical importance I shall give the definitions of some of them.

7. **Peano's Curve.** Peano, in 1890, defined a parameter curve $x=\phi(t)$, $y=\psi(t)$, which passes through all the points of a unit square. The following geometrical method of obtaining the curve is based on the work of Moore ([57]) and Schoenflies ([72]).

Consider the diagonal AC of the unit square ABCD, and denote it by F_0. Divide the square into 3^2 equal parts and also the interval (0, 1) into 3^2 equal parts. Let the straight line AC be stretched and brought into the polygonal form shown in Fig. 3. Denote

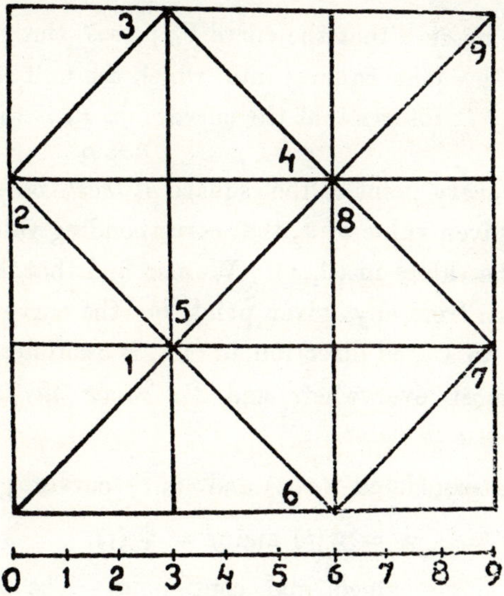

Fig. 3.

this polygonal stretch by F_1. In this manner the squares are arranged in the order $1, 2, 3, \ldots 3^2$ and are placed into correspondence with the segments of $(0, 1)$ bearing the same numbers. The stretch F_1 is made up of the diagonals of the small squares and is traversed from A to C in the order indicated by the numerals shown in the figure. The above construction is now applied to each diagonal of the small squares giving a polygonal stretch F_2, which now passes through each of the 3^4 squares into which the unit square is divided, and which goes from A to C. Continuing the construction we obtain polygonal stretches $F_3, F_4 \ldots F_n \ldots$ Peano's curve is $F = \lim_{n\to\infty} F_n$.

We observe that the curve F_n passes through each of the 3^{2n} equal squares into which the unit square is divided. It follows that the curve $\lim_{n\to\infty} F_n$ will pass through every point of the square *at least once*. Thus for any given value of x, the corresponding values of y are all the values in $(0, 1)$. We also find that the secant line drawn from any given point on the curve can not converge to a fixed direction, in fact, it oscillates through 360^0 almost everywhere and *the curve has no half-tangent at any point.*

The associated (x, t) and (y, t) curves given by

$$x = \phi(t) \text{ and } y = \psi(t),$$

are each single-valued and continuous. The functions

$\Phi(t)$ and $\psi(t)$ have been shown to be non-differentiable functions by Moore ([57]) and Banerji ([4]). These functions will be considered in the next lecture. A different method of looking at the correspondence established by such curves will be illustrated by the following:

8. **Hilbert's Curve.** Let the variable t range over the interval I (0, 1) and let the point (x, y) range over the unit square R. Single-valued continuous functions

$$x = \Phi(t), y = \psi(t)$$

can be defined, so that as t ranges over I, (x, y) ranges over the whole of the domain R. This can be done as follows:

Divide the interval I into four parts $\delta_1, \delta_2, \delta_3, \delta_4$, and the unit square R also into four parts, $\eta_1, \eta_2, \eta_3, \eta_4$.

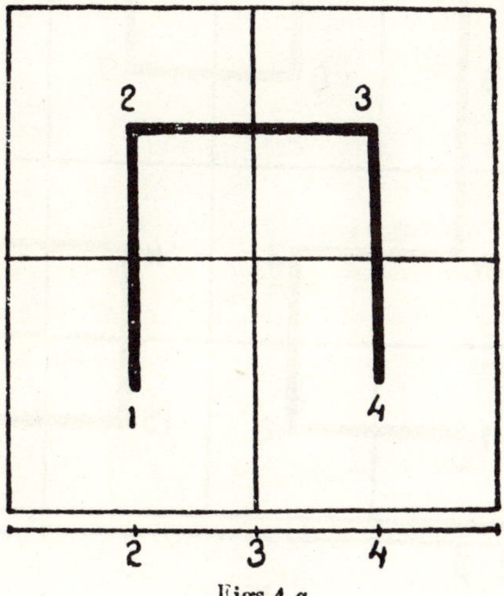

Figs 4 *a*.

We call this the first division or D_1. The correspondence between I and R is given *in the first approximation* by saying that to each point P in δ, shall correspond some point Q in η.

Let the polygonal stretch shown in Fig. 4*a* be called F_1.

We now effect a second division D_2 by dividing the intervals and the squares of D_1 each into four equal parts. Let the numbering of these parts be carried out as shown in Fig. 4*b*.

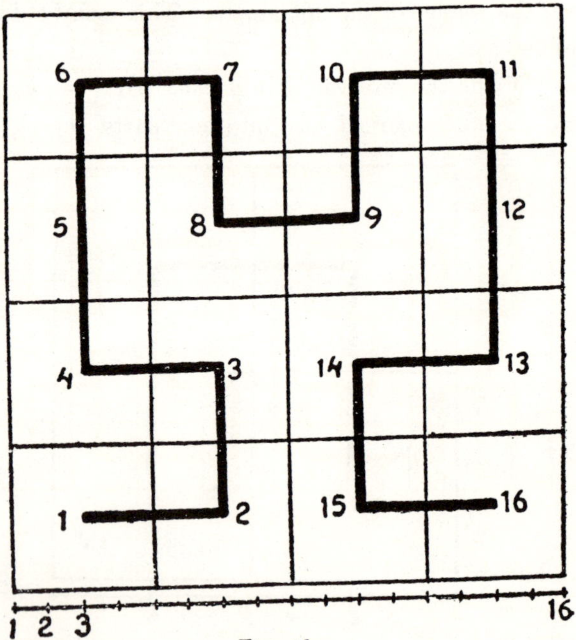

Fig. 4*b*.

Let the polygonal line shown in Fig. 4*b* be called F_2. The correspondence between I and R is given in the *second approximation* by saying that to a point P of the *r*th interval of I corresponds some point Q in the *r*th square of R.

The third division D_3 and the third polygonal stretch F_3 are illustrated in Fig. 4*c*.

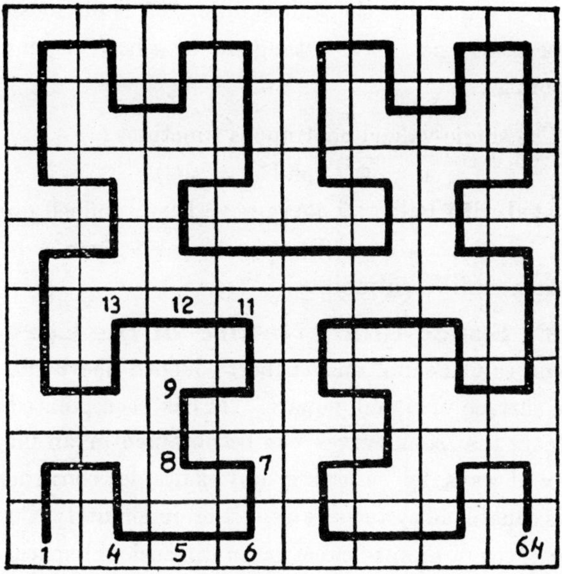

Fig. 4*c*.

The above construction is continued indefinitely. To find the point Q in R corresponding to P in I, we observe that P lies in a sequence of intervals tending to zero (in length), to which correspond uniquely a sequence of squares tending to zero (in area) and hence defining the point Q, whose co-ordinates are, therefore, single-valued functions of t.

The curves F_1, F_2, F_3......form a sequence of curves, and the limiting curve

$$F = \lim_{n \to \infty} F_n$$

is a curve which fills the unit square. It is obvious that the curve can not have a tangent or a half-tangent at any point.

The single-valued continuous functions

$$x = \Phi(t) \text{ and } y = \psi(t)$$

associated with Hilbert's curve have been defined analytically by R. D. Misra ([55]), who has also given a proof of their non-differentiability.

9. Space-filling Nature of the Curves.

The curves of Peano and Hilbert defined above fill the entire surface of a unit square. It has been pointed out by Moore that such curves can be obtained in an infinite variety of ways by assigning any suitable construction that is capable of systematic repitition indefinitely. Curves filling entirely a unit cube or n-dimensional space can also be constructed by a similar geometrical procedure. By a simple generalisation of Peano's method, Singh ([78]) has given an analytical method of obtaining curves which fill entirely a given n-dimensional space.

A curve filling the unit cube will be given analytically by the equations

$$x = \Phi_1(t), \ y = \Phi_2(t), \ z = \Phi_3(t);$$

where the point (x, y, z) ranges over the entire cube as t ranges over the linear interval (0, 1). According to Hilbert a kinematical interpretation of the functional relation between (x, y, z) and t is that a point may move so that in unit time it passes through every point of the cube. This interpretation, however, can not be realized in practice, as the length of the (x, y, z) curve is infinite.

10. **Kaufmann's Curve.**

The curve defined by Kaufmann ([39]) is a parameter curve of the same type as the curve of Koch. It corresponds to a multiple-valued function $y=f(x)$, and has at no point a half-tangent. The curve is defined as follows :

Inscribed polygonal stretches and the projection condition.—Let ABC be an isosceles triangle with the fundamental side AB and the base angle ν. Let CM be the perpendicular from C on AB. Let $A_1, A_2, \ldots \ldots A_n \ldots \ldots$ be a sequence of different points on MA converging towards A, such that A_n lies between A and A_{n-1}. Correspondingly let $A'_1, A'_2 \ldots \ldots A'_n \ldots$ lying on CA be a sequence of different points converging towards A, and likewise so arranged that A'_n lies between A and A'_{n-1}. By joining these stretches CA_1, $A_1A'_1$, A'_1A_2, $A_2A'_2$

etc., we form a broken polygonal stretch ∇_{CA}. In the same way we define in the triangle BMC a polygonal stretch ∇_{CB}. The joint ∇ of the two polygonal stretches ∇_{CA} and ∇_{CB} we call a polygonal stretch inscribed in the $\triangle$ ABC.

We now keep in mind such polygonal stretches inscribed in the triangle ABC, all of whose stretches (suitably oriented) make a constant angle with the fundamental side AB. If, now, ν is the base angle of ABC, then, there is for every value of $\theta < \frac{\pi}{2} - \nu$ one and only one polygonal stretch inscribed to the $\triangle$ ABC all of whose stretches form with the fundamental side AB of the triangle ABC a fixed angle less than $\frac{\pi}{2} - \theta$ (see Fig. 5). We call ∇ in this case a polygonal stretch inscribed in the $\triangle$ ABC with the *reflection angle* θ. In the following we shall obtain, by iteration, with the help of such inscribed polygonal stretches those parameter curves which interest us.

Construction of the Curve.

We start from an isosceles $\triangle$ ABC with the base angle $\gamma < \frac{\pi}{6}$. By ∇° we denote a polygonal stretch inscribed in ABC with the constant reflection angle $\theta < \frac{\pi}{6}$. On each stretch $\top$ of ∇° inside the reflection space θ, we construct isosceles triangles with the fundamental side T and base angle less than θ. We obtain in this manner a sequence of triangles which we denote as chain triangles of the first order. Let, again, an inscribed polygonal stretch with the reflection angle θ be given in each chain triangle of the first order. Such a polygonal stretch may be likewise called of the first order. We choose each stretch T of everyone of the polygonal stretches of the first order as a fundamental side of an isosceles triangle (inscribed in the angular space θ) *with the base angle* γ. Every one of such

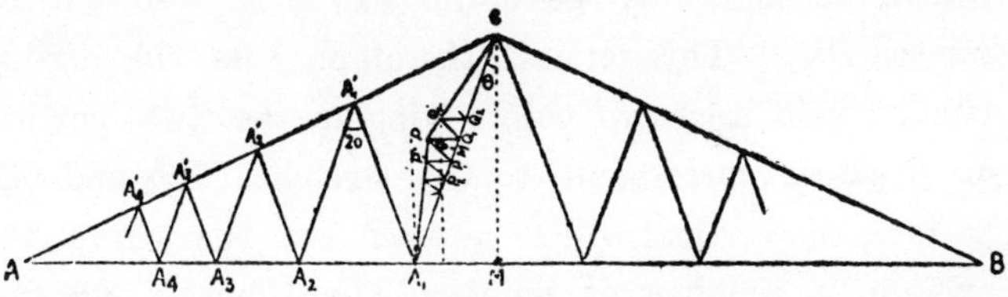

Fig. 5.

triangles we call a chain-triangle of the second order, and we determine in this an inscribed polygonal stretch of the second order with the *reflection angle* θ, and so on. In such a manner will be defined chain-triangles of the nth order ($n=1, 2, \ldots$) and corresponding inscribed polygonal stretches of the nth order. The joining together of all the inscribed polygonal stretches of any fixed order n gives a simple curvilinear arc. The sequence of these curvilinear arcs for increasing n converges as is easily seen towards a simple curve l.

11. **Kaufmann's curve is multiple-valued.** It is easy to show that Kaufmann's curve l, defined as above, corresponds to a multiple-valued function $y=f(x)$. Let $A_1P\ C$ be the triangle constructed on the first stretch CA_1 (see fig. 5). Let PP_1, $P_1P'_1$, P'_1P_2........, and PQ_1, $Q_1Q'_1$, Q'_1Q_2,......be the parts of the polygonal stretch inscribed in the triangle A_1PC. As the stretches PQ_1 and PP_1 each make an angle θ with PM_1, it is easy to see that PQ_1 is horizontal whilst PP_1 makes an angle 2θ with it. Now, as $2\theta < \frac{\pi}{3}$, therefore, the vertical drawn through any point on PP_1 must also cut the stretch PQ_1. This vertical, therefore, cuts the limiting curve l in at least two points lying on the two portions of l which correspond to the stretches PP_1 and PQ_1. In fact, the vertical will *in general* cut the curve l in an infinite number of points. The function $y=f(x)$ is, therefore, a multiple-valued function.

12. **Non-existence of half-tangents.** Kaufmann has given an indirect proof of the non-existence of half tangents at any point of the curve. The property, however, is an immediate corollary from the multiple-valued nature of the curve at an everywhere dense set of values of x. For, let (x, y) be any point on the curve and let $x_1, x_2, \ldots x_n$, ..be a sequence approaching x from the right, and let $y_1, y_2 \ldots y_n \ldots$ and $y'_1, y'_2 \ldots y'_n, \ldots$ be the corresponding values of y, lying on different parts of the

curve. It is easy to see that the angle between the two secants joining (x, y) with (x_n, y_n), and (x, y) with (x'_n, y'_n) always make with each other an angle greater than θ, so there can be no limiting secant on the right at any point. Similarly on the left there can be no limiting secant. *This shows the futility of considering parameter curves as examples of non-differentiable functions.*

13. **Modifications of Kaufmann's curve.** A single-valued function can be obtained by a suitable modification of Kaufmann's construction. It will be shown that the single-valued function, so obtained, possesses half-tangents at an everywhere dense set of points, and is differentiable there.

Let PA_1C be the first chain-triangle lying in the part AMC of the triangle ABC, and let the angle $CA_1M = a_0$.

The base angle ν_1 of the triangle PA_1C is so chosen that the inclination of each of the sides PA_1 and PC to the horizontal is less than $\frac{\pi}{2}$. Through P draw the vertical PN_1 (see fig. 6). We have now to construct a series of stretches in the triangle PA_1C, in such a way that the resulting curve, represented by these stretches shall be single-valued. For this purpose we draw through P a straight line PP_1 cutting A_1C in P_1 and lying in the angular space A_1PN_1. Let a_1 be the inclination of PP_1

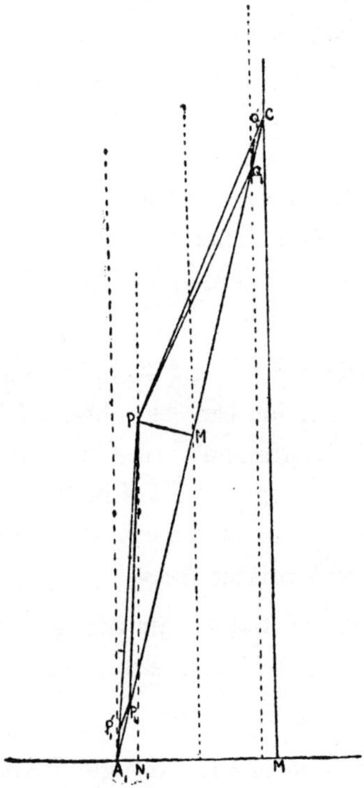

Fig 6.

to the horizontal. The other stretches $P_1P'_1, P'_1P_2 \ldots$ and PQ_1, Q_1Q_1', $Q'_1Q_2 \ldots$ are all constructed so that they are equally inclined to A_1C just as in Kaufmann's construction. Similar stretches are constructed in all the chain triangles. This gives us the stretches of the first order. On each stretch we have now to construct an isosceles triangle. For fixity of ideas we consider the stretch PP_1. Construct on it an isosceles triangle

P^1P_1P with the base angle ν_2, such that the inclination of each side P^1P_1 and P^1P is less than $\frac{\pi}{2}$. Similar triangles are constructed on all the stretches. In each chain-triangle of the second order so obtained we shall have stretches of the second order. $P^1P_1^1$ is the first stretch lying on the left of the triangle P^1P_1P, so constructed that it lies in the angular space between P^1P_1 and the vertical through P^1. The other stretches are then constructed as before. This procedure gives, at the nth stage, a curve l_n which is made up of the stretches of the nth order. The curve $l = \lim_{n\to\infty} l_n$ corresponds to a single valued function, $y=f(x)$, as is easily seen.

14. **The existence of a derivative.** Let the first chain-triangle lying in the left part of the $\triangle$ ABC be denoted by $\triangle_{0,1}$, and let its sides CA_1, A_1P and PC be denoted by $S_{0,1}$, $S_{0,2}$, $S_{0,3}$. Let the first chain-triangle constructed in $\triangle_{0,1}$ and lying in the left-part (lower part) of $\triangle_{0,1}$ be denoted by $\triangle_{1,1}$ and its corresponding sides by $S_{1,1}$, $S_{1,2}$, $S_{1,3}$. We have likewise $\triangle_{n,1}$, and its corresponding sides $S_{n,1}$, $S_{n,2}$, $S_{n,3}$. We also know that $S_{0,1}$, $S_{1,1}, \ldots \ldots S_{n,1}, \ldots \ldots$ belong to the stretches of different orders $(0, 1, 2, \ldots n, \ldots)$. Their inclinations to the horizon a_0, a_1, $a_2, \ldots a_n, \ldots$ form a monotone increasing sequence, and as each is less than $\frac{}{2}$

$$\lim_{n \to \infty} a_n = a \leq \frac{\pi}{2} \dots\dots\dots\dots\dots\dots\dots (1)$$

Let $\nu_1, \nu_2, \nu_3, \dots \nu_n, \dots$ be the base angles of $\triangle_{0,1}, \triangle_{1,1}, \dots \triangle_{n,1}, \dots$

It is easy to see that because of (1)

$$\lim_{n \to \infty} \nu_n = 0 \dots\dots\dots\dots\dots\dots\dots\dots\dots (2)$$

Moreover the inclinations of $S_{n,1}$, $S_{n,2}$ and $S_{n,3}$ are given by a_n, $a_n + \nu_n$ and $a_n - \nu_n$. It follows, therefore, that $S_{n,1}$, $S_{n,2}$ and $S_{n,3}$ each have the same limiting inclination a as n tends to infinity.

Consider now the point S which belongs to all the chain triangles $\triangle_{0,1}, \triangle_{1,1}, \dots \triangle_{n,1}, \dots$ This point is the limiting point of the vertices of the chain triangles, as n tends to infinity. Moreover, it lies in the left-part (lower part) of each chain triangle of the sequence given above. It is further easy to see that the inclination of any secant joining the point S to any point of the curve l corresponding to the side $S_{n,3}$ of $\triangle_{n,1}$ lies between a_n and a. And as

$$\lim_{n \to \infty} a_n = a,$$

therefore, the secants drawn from S to points of the curve l tend to the limiting inclination a. Thus at the point S, the curve has a progressive derivative $= \tan a$, where $a \leq \dfrac{\pi}{2}$.

It can be similarly proved that at S a regressive derivative $= \tan a$ exists.

Thus at the point S, there exists a differential coefficient whose value is equal to $\tan a$; $a \leq \frac{\pi}{2}$.

The above proof can be evidently applied to show that *there exists an everywhere dense set of points at each of which the function possesses a differential coefficient.*

15. **Besicovitch's Curve.** I shall now give the construction of the Curve of Besicovitch, which has been stated to be without half-tangents. The curve corresponds to a single-valued function and is defined as follows :

" Let us take the stretch $AB=2a$ for A $(0, 0)$ and $B(2a, 0)$, and the points $C(a, b)$ and $D(a, 0)$. On the stretch AD let us construct a stretch $l_1=\frac{a}{4}$ placing it centrally. The stretch AD is divided by the stretch l_1 into two equal parts. On each of these let us place centrally the stretches $l_2 = l_3 = \frac{a}{2^4}$. The stretches l_1, l_2, l_3 divide the stretch AD into four equal stretches. On each of these let us place centrally (calculated from left to right) the stretches $l_4=l_5=l_6=l_7=\frac{a}{2^6}$, and so on. In this manner a set L of stretches

$$l_1+l_2+l_3+\cdots = \frac{a}{2}$$

is constructed on the stretch AD.

We construct a similar system of stretches on DB. We call these stretches the first series of stretches.

Let us denote by $m(x)$ the measure* of the set of points of the interval $(0, x)$ which do not belong to the set L, and let us determine on the stretch AD a function $\varphi(x)$, whilst we assume

$$\varphi(x) = \frac{2b}{a} m(x).$$

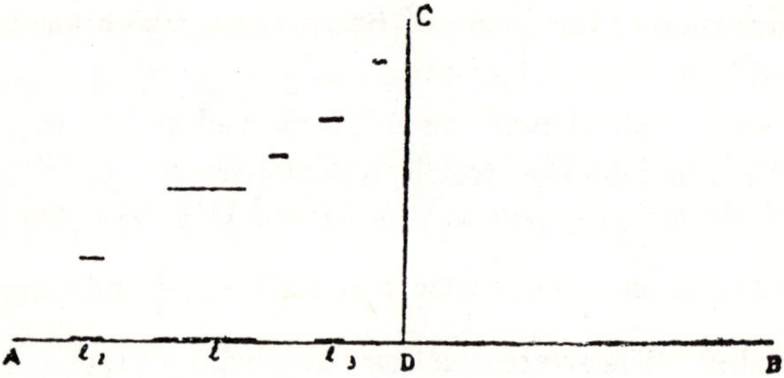

Fig 7.

The points A and D are thus connected by the curve $y=\varphi(x)$, which has a constant value on an arbitrary stretch l_r and which we call a 'ladder curve'; the points C and B are likewise connected by such a ladder curve. The figure originating in this manner is called a 'step-triangle' whose base is $2a$ and whose height is b (see Fig. 7).

---

* It has been assumed that the measure $m(x)$ exists as a unique number for every x.

On the fundamental lines corresponding to the first series of stretches of the step-triangle ABC, let us construct step-triangles directed towards below, equal on equal fundamental lines, whilst we choose the height so that the vertex of the undermost of all equal triangles lies on the side AB. The construction of all these triangles is called the operation of 'maiming' the triangle ABC towards inside. With the so obtained infinity of triangles (first series) we carry out the same operation of maiming towards inside, and thus obtain the second series of triangles; on them also perform maiming towards inside, and so on.

We now define a function $f(x)$ on the stretch AB as follows:

(1) at the points of the stretch AB, which do not belong to the first series of stretches, by the ordinates of the sides of the step-triangle ABC;

(2) at the points of the stretches of the first series, which do not belong to the stretches of the second series, by the ordinates of the sides of the triangles of the first series;

(3) at the points of the stretches of the second series, which do not belong to the stretches of the third series, by the ordinates of the sides of the triangles of the second series, and so on;

(4) at the points which belong to the stretches of all series (they form a null set) according to the principle of continuity."

It has been stated by Besicovitch ([5]) and E. D. Pepper ([62]) that the curve does not possess a half-tangent (progressive or regressive) at any point. Whilst it is certain that the curve is single-valued and non-differentiable, there is considerable doubt regarding the validity of the proof, given by Besicovitch and Pepper, for the non-existence of half-tangents.* The proof given by these writers depends upon geometrical intuition whose validity can not be defended. If an arithmetic definition of the function could be devised, it would be possible to study the derivates of Besicovitch's function in detail and to decide whether it has progressive derivatives or not.

16. **Other Curves.** In a series of papers published during 1907—1910, Faber ([25, 26, 27]) gave a geometrical method for the construction of non-differentiable functions. Faber's functions are single-valued and continuous, and have the additional advantage of being capable of expression as infinite series. It has already been pointed out that these functions are special cases of the general class considered by Knopp ([41]). Landsberg ([51]) has also constructed non-differentiable functions, which like those of Faber, are obtained by geometrical construction in a simple manner.

* See Singh (82).

THIRD LECTURE

FUNCTIONS DEFINED ARITHMETICALLY

1. **Introduction.** In the year 1890, Peano [59] defined a surface-filling curve by the help of arithmetically defined functions

$$x = \Phi(t) \text{ and } y = \Psi(t).$$

The function $\Phi(t)$ is defined as follows:

Let a point t of the interval $(0, 1)$ be represented arithmetically in radix fractions in the scale of 3 as

$$t = \frac{a_1}{3} + \frac{a_2}{3^2} + \frac{a_3}{3^3} + \cdots\cdots + \frac{a_n}{3^n} + \cdots\cdots\cdots,$$

where the a's are 0, 1 or 2.

Corresponding to t let a number $x = \Phi(t)$ be defined as follows:

$$x = \Phi(t) = \frac{b_1}{3} + \frac{b_2}{3^2} + \cdots\cdots + \frac{b_n}{3^n} + \cdots\cdots\cdots\cdots$$

where $b_1 = a_1$, and $b_n = K^{a_2 + a_4 + \cdots + a_{2n-2}}(a_{2n-1})$,

and $K^p(a) = a$ or $(2 - a)$ according as p is even or odd.

Similarly, the function $\Psi(t)$ is defined as:

$$y = \Psi(t) = \frac{c_1}{3} + \frac{c_2}{3^2} + \cdots\cdots\cdots + \frac{c_n}{3^n} + \cdots\cdots$$

where $c_n = K^{a_1 + a_3 + \cdots + a_{2n-1}}(a_{2n})$, $(n = 1, 2, \ldots\ldots)$,

and the operator K has the same meaning as before.

Peano's functions were generalized, and a proof of their non-differentiability was published by E. H.

Moore[57] in 1900. Moore's treatment is geometrical. An analytical proof of the non-differentiability of $\Phi(t)$ has been given by H. P. Banerji [4]. In to-day's lecture I shall consider a class of functions of which Peano's and Moore's functions are special cases.

A general theory of the construction of non-differentiable functions was published by Steinitz [86] in 1899. He divides the interval (a, b) into m equal parts and prescribes the 'value-difference' of a function $\Phi(x)$ for each of these m parts. He then divides each part again into m equal parts, and prescribes the 'value-differences' for each of the new m^2 parts. Proceeding in this manner he obtains a function $\Phi(x)$ valid for an everywhere dense set in (a, b). By the extension of $\Phi(x)$ he obtains a function $f(x)$ defined for the whole interval (a, b), but fails to give a sufficient condition for the non-existence of finite or infinite differential coefficients. He has, however, indicated a method of obtaining a non-differentiable function when m is greater than 5.

In a paper, published in 1918, H. Hahn [31] has considered a function, constructed according to the method of Steinitz (for $m=6$), and proves that at no point does the function possess both the progressive and regressive differential coefficients.

The advantage of the arithmetic definition over all other forms of definitions is that the numerical value of the function at each point is directly given by the definition so that the character of the derivates at any assigned point can be studied directly. That arithmetically defined functions have been attracting attention is shown by the attempts of several mathematicians to construct such functions. As early as 1904, T. Takagi ([90]) constructed an arithmetically defined function by using the representation of the points of a linear interval in the scale of 2. Takagi's function is non-differentiable at an everywhere dense set of points. E. Cesaro ([19]), in 1905, gave an arithmetically defined function which has no differential coefficient at the points of an everywhere dense set. By representing the points x of a linear interval in the scale of 5, W. Sierpinski ([76]), in 1914, obtained a function $y=f(x)$, represented in the scale of 3, which does not simultaneously possess both progressive and regressive differential coefficients at any point. A class of simple non-differentiable functions was given by K. Petr ([60]) in 1920. By using the representation of the points x of a linear interval in an *even* scale 2b, Petr obtained a function $y=f(x)$ expressed in another *even* scale $2c$ $(b>c)$.

It will be observed that radix fractions have been used for obtaining the definitions of the above functions. But the use of radix functions is not necessary as is shown by the definition of Bolzano's function given in the second lecture. In fact all that is necessary is to set up a correspondence between the points x of an everywhere dense set in an interval, say (0, 1), with those of another set y, in such a way that, (i) to every value of x there corresponds only one value of y, (ii) to every value of y, there corresponds an infinite number of values of x, (iii) to the two representations of x (ending and non-ending) there corresponds the same value of y, and (iv) the ratio $\frac{y-y'}{x-x'}$ tends to infinity as x' tends to x.

Cesaro (18) has pointed out that Koch's curve can be defined arithmetically. An arithmetic definition of Hilbert's curve has been given by R. D. Misra (55). The curves given in the second lecture can also be defined arithmetically.

I shall now give a detailed study of the derivates of a class of non-differentiable functions constructed by me. I believe that in the case of no other function have the derivates been so completely studied.

2. Definition of the functions $\Phi_{m,r,p}(x)$.

Let the numbers in the interval (0, 1) be expressed in radix fractions, in the base p, where p is an odd integer.

Then a point x in (0, 1) can be represented as

$$x = \frac{a_1}{p} + \frac{a_2}{p^2} + \frac{a_3}{p^3} + \cdots + \frac{a_n}{p^n} + \cdots.$$

where the a's are positive integers such that

$$0 \leq a \leq p - 1.$$

For a given integer r and another given integer m, let

$$Y_{m,r,p}=\Phi_{m,r,p}(x) = \frac{b_{1,r}}{p}+\frac{b_{2,r}}{p^2}+\cdots+\frac{b_{n,r}}{p^n}+\cdots.$$

where

$$b_{1,r}=k^{a_1+a_2+\cdots+a_{r-1}}(a_r),$$

$$b_{2,r}=k^{a_1+a_2+\cdots+a_{r-1}+\cdots a_{m+r-1}}(a_{m+r}),$$

$$b_{3,r}=k^{a_1+\cdots+a_{r+1}+\cdots+a_{m+r+1}+\cdots+a_{2m+r-1}}(a_{2m+r}),$$

$$\cdots \quad \cdots \quad \cdots \quad \cdots \quad \cdots \quad \cdots \quad \cdots$$

and so on; where $k^s(a)=a$ or $(p-1-a)$ according as s is even or odd.

In the above, the integer p may have any odd value 3, 5, 7,..., the integer m may have any one of the values 2, 3, 4, .. and r is an integer $\leq m$. The function $\Phi_{2,1,3}(x)$ is Peano's function $\Phi(t)$, and the function $\Phi_{2,2,3}(x)$ is Peano's function $\psi(t)$; while the functions $\Phi_{2,1,p}(x)$ and $\Phi_{2,2,p}(x)$, where p is an odd integer ≥ 3, are the functions considered by Moore.

3. **The function** $\Phi_{3,1,3}(x)$. I give below a proof of the continuity and the non-differentiability of the function $\Phi_{3,1,3}(x)$. The proof for the general case can be carried out exactly in the same manner.

We have, if

$$x=\frac{a_1}{3}+\frac{a_2}{3^2}+\cdots\cdots+\frac{a_n}{3^n}+\cdots\cdots$$

$$Y_{3,1,3}=\Phi_{3,1,3}(x)$$

$$= \frac{b_{1,1}}{3} + \frac{b_{2,1}}{3^2} + \cdots\cdots + \frac{b_{n,1}}{3^n} + \cdots\cdots$$

which, dropping out the suffixes, we write as

$$y = \Phi(x) = \frac{b_1}{3} + \frac{b_2}{3^2} + \cdots\cdots + \frac{b_n}{3^n} + \cdots\cdots,$$

where

$$b_1 = k^0(a_1) = a_1$$

$$b_2 = k^{a_2+a_3}(a_4)$$

$$b_3 = k^{a_2+a_3+a_5+a_6}(a_7)$$

$$\cdots\cdots\cdots\cdots\cdots\cdots\cdots\cdots$$

$$b_n = k^{a_2+a_3+a_5+a_6+\cdots+a_{3(n-1)-1}+a_{3(n-1)}}(a_{3(n-1)+1})$$

$$\cdots \quad \cdots \quad \cdots \quad \cdots \quad \cdots \quad \cdots \quad \cdots$$

and so on; and where $k^s(a) = a$ or $(2-a)$ according as s is even or odd.

4. $Y = \Phi(x)$ **is a continuous function of** x. The points x which have a non-ending representation such that all the a's, from and after a_n are not all 2's, are uniquely representated in a radix fraction, and thus. for all such points we have a single value of Y.

When x is representable as an ending radix fraction, it has also a non-ending representation in which all the a's, from and after some place, are equal to 2.

Three cases arise:

(1) $x_1 = \dfrac{a_1}{3} + \dfrac{a_2}{3^2} + \cdots\cdots + \dfrac{a_{3n}}{3^{3n}},$

$$= \frac{a_1}{3} + \frac{a_2}{3^2} + \cdots + \frac{(a_{3n}-1)}{3^{3n}} + \frac{2}{3^{3n+1}}$$

$$+ \frac{2}{3^{3n+2}} + \cdots\cdots$$

(2) $$x_2 = \frac{a_1}{3} + \frac{a_2}{3^2} + \cdots + \frac{a_{3n}}{3^{3n}} + \frac{a_{3n+1}}{3^{3n+1}}.$$

$$= \frac{a_1}{3} + \frac{a_2}{3^2} + \cdots + \frac{(a_{3n+1}-1)}{3^{3n+1}} + \frac{2}{3^{3n+2}}$$

$$+ \frac{2}{3^{3n+3}} + \cdots\cdots$$

(3) $$x_3 = \frac{a_1}{3} + \frac{a_2}{3^2} + \cdots + \frac{a_{3n+2}}{3^{3n+2}}.$$

$$= \frac{a_1}{3} + \frac{a_2}{3^2} + \cdots + \frac{(a_{3n+2}-1)}{3^{3n+2}} + \frac{2}{3^{3n+3}}$$

$$+ \frac{2}{3^{3n+4}} + \cdots\cdots$$

Case (1). Let the values of Y corresponding to the two representations of x be

$$Y_1 = \frac{b_1}{3} + \frac{b_2}{3^2} + \cdots + \frac{b_n}{3^n} + \frac{b_{n+1}}{3^{n+1}} + \cdots\cdots$$

and

$$Y'_1 = \frac{b'_1}{3} + \frac{b'_2}{3^2} + \cdots + \frac{b'_n}{3^n} + \frac{b'_{n+1}}{3^{n+1}} + \cdots\cdots$$

respectively.

Then as the two representations of x agree up to $3n$ places,

$$b_1 = b'_1,\ b_2 = b'_2,\ \ldots,\ b_n = b'_n,$$

and we further have

$$b_{n+1}=k^{a_2+a_3+\cdots+a_{3n-1}+a_{3n}}(0),$$

$$b_{n'+1}'=k^{a_2+a_3+\cdots+a_{3n-1}+(a_{3n}-1)}(2)$$

and in general

$$b_{n+l}=k^{a_2+a_3+\cdots+a_{3n-1}+a_{3n}}(0)$$

$$b'_{n+l}=k^{a_2+a_3+\cdots+a_{3n-1}+(a_{3n}-1)+4(l-1)}(2)$$

for $l=1, 2, 3, \ldots\ldots$

We see that if the index of k in the first case (*i.e.* b_{n+l}) is odd, it is even in the second case (*i.e.* b'_{n+l}); and if it is even in the first case, it is odd in the second.

Thus

$$b_{n+1}=b'_{n+1},\ b_{n+2}=b'_{n+2},\ \ldots,\ b_{n+l}=b'_{n+l}\ \ldots\ldots$$

$\therefore Y_1=Y'_1$, and Y is uniquely determined for the points of *Case* (1).

Case (2). Let the two representations of Y corresponding to the two representations of x be

$$Y_2=\frac{b_1}{3}+\frac{b_2}{3^2}+\cdots+\frac{b_n}{3^n}+\frac{b_{n+1}}{3^{n+1}}+\ldots\ldots$$

and

$$Y'_2=\frac{b'_1}{3}+\frac{b'_2}{3^2}+\cdots+\frac{b'_n}{3^n}+\frac{b'_{n+1}}{3^{n+1}}+\ldots\ldots$$

We have as before

$$b_1=b'_1,\ b_2=b'_2,\ \ldots\ldots\ b_n=b'_n.$$

Now let

$$a_2+a_3+a_5+a_6+\cdots\cdots+a_{3n-1}+a_{3n}=\lambda$$

Then

$$b_{n+1}=k^{\lambda}(a_{3n+1}),\ b'_{n+1}=k^{\lambda}(a_{3n+1}-1)$$

and in general

$$b_{n+l}=k^{\lambda}(0),\ b'_{n+l}=k^{\lambda+4(l-1)}(2)$$

(for $l=2, 3, 4, \ldots\ldots$).

Let λ be even, then

$$b_{n+1}=a_{3n+1} \text{ and } b'_{n+1}=(a_{3n+1}-1)$$

while $b_{n+l}=0$, and $b'_{n+l}=2$ (for $l=2, 3, \ldots$).

$$\therefore Y_2=\frac{b_1}{3}+\frac{b_2}{3^2}+\cdots+\frac{b_n}{3^n}+\frac{a_{3n+1}}{3^{n+1}}$$

$$\text{and}\quad Y'_2=\frac{b_1}{3}+\frac{b_2}{3^2}+\cdots+\frac{b_n}{3^n}+\frac{(a_{3n+1}-1)}{3^{n+1}}+\frac{2}{3^{n+2}}+\frac{2}{3^{n+3}}+\cdots$$

Hence $Y_2=Y'_2$.

Similarly, if λ be odd, it can be proved that

$$Y_2=Y'_2.$$

Therefore, Y is uniquely determined for the points of *Case* (2).

Case (3). Proceeding as in *Case* (1) we can show that Y is uniquely determined for the points of *Case* (3).

Thus Y is a single-valued function.

We observe that if two points x and x' agree up to the first $3n$ places in their representations, the corresponding values of Y agree up to the first n places.

Therefore, Y is a continuous function of x in $(0, 1)$.

5. $Y=\Phi(x)$ is nowhere differentiable. Non-existence of cusps.

For the purpose of proving the non-differentiability of $\Phi(x)$ we divide the numbers of the interval (0, 1) into the following four classes.

Class I (a). A number belongs to this class, if in its representation $(a_{3n-1}+a_{3n}) \leq 2$ and a_{3n+1} is not equal to 1, for an infinite number of values of n.

Class I (b). A number belongs to this class, if in its representation $(a_{3n-1}+a_{3n}) \leq 2$ and $a_{3n+1}=1$ for an infinite number of values of n.

Class II (a). A number belongs to this class, if in its representation $(a_{3n-1}+a_{3n}) > 2$ and a_{3n+1} is not equal to 1, for an infinite number of values of n.

Class II (b). A number belongs to this class, if in its representation $(a_{3n-1}+a_{3n}) > 2$ and $a_{3n+1}=1$ for an infinite number of values of n.

It is easy to see that a given number in (0, 1) must belong to at least* one of these four classes.

Class I(a). Let $h_n=\dfrac{2}{3^{3n}}$ [n having a value for which a_{3n-1}, a_{3n} and a_{3n+1} satisfy the conditions of I (a)].

Then

$$\Phi(x)=\Phi(x+h_n),$$

* A number may evidently belong to more than one class.

$$\therefore \lim_{h_n=0} \frac{\Phi(x+h_n)-\Phi(x)}{h_n}=0.$$

Again, let $h'_n=\frac{1}{3^{3n}}$, n having the same value as above, then

$$\Phi(x+h'_n)-\Phi(x) \,| > \frac{1}{3^{n+1}}$$

$$\therefore \lim_{h'_n=0} \left| \frac{\Phi(x+h'_n)-\Phi(x)}{h'_n} \right| > \lim_{n=\infty} \frac{3^{3n}}{3^{n+1}} = \infty.$$

$\therefore$ The progressive derivative does not exist at the points belonging to *Class* I (a).

Class I (b). Taking $h_n=\frac{2}{3^{3n}}$, we see as before that

$$\lim_{h_n=0} \frac{\Phi(x+h_n)-\Phi(x)}{h_n}=0.$$

Taking $h'_n=\frac{1}{3^{3n+1}}$, we see that

$$\left| \Phi(x+h'_n)-\Phi(x) \right| = \frac{1}{3^{n+1}}$$

$$\therefore \lim_{h'_n=0} \left| \frac{\Phi(x+h'_n)-\Phi(x)}{h'_n} \right| = \infty.$$

$\therefore$ The progressive derivative does not exist at the points belonging to *Class* I (b).

Class II (a). Taking h_n and h'_n as in I (a), we see that

$$\underset{h_n=0}{Lim} \frac{\Phi(x-h_n)-\Phi(x)}{-h_n} = 0$$

and

$$\underset{h'_n=0}{Lim} \left| \frac{\Phi(x-h'_n)-\Phi(x)}{-h'_n} \right| = \infty .$$

∴ The regressive derivative does not exist at the points belonging to *Class* II (a).

Class II (b). Taking h_n and h'_n as in I(b), we see that

$$\underset{h_n=0}{Lim} \frac{\Phi(x-h_n)-\Phi(x)}{-h_n} = 0$$

and

$$\underset{h'_n=0}{Lim} \left| \frac{\Phi(\mathrm{x}-h'_n)=\Phi(x)}{-h'_n} \right| = \infty .$$

∴ The regressive derivative does not exist at the points belonging to class II (b).

We have thus shown that at the points belonging to *Class* I, the progressive derivative does not exist, while at the points belonging to *Class* II, the regressive derivative does not exist. As any point in (0, 1) must belong to at least one of these classes, therefore, at no point x in (0, 1) do both the derivatives exist, *i.e.* $\Phi(x)$ is nowhere differentiable in (0, 1). Also *the function* $\Phi(x)$ *does not possess a cusp at any point* *.

---

* *Cf.* Hobson; *Theory of functions*, etc., Vol. II. (1926) p., 405, where it is stated, " It does not appear to be definitely known whether a non-differentiable function can exist which has no cusps.' *The analysis of the derivates given above definitely answers this question.*

6. Existence of the derivative of $\Phi(x)$.

(6.1) It has already been shown that at the points belonging to *Class* I, $\Phi(x)$ does not possess a right hand derivative; hence, in order to get the set of those points where $\Phi(x)$ possesses a right hand derivative, we search among those points x in whose representation $(a_{3n-1}+a_{3n})>2$ from and after some value v of n. We can easily show that if, for a point x, $(a_{3n-1}+a_{3n})=3$ for an infinite number of values of n, then the right hand derivative is non-existent at x. Thus the points where a right hand derivative may exist are such that in their representation $(a_{3n-1}+a_{3n})=4$ for all values of n greater than or equal to some number v.

(6.2) Let x be such that in its representation $(a_{3n-1}+a_{3n})=4$ for all values of $n\geq v$ and a_{3n+1} is either 0 or 1 for all values of $n\geq v$. Further let

$$\sum_{1}^{v-1}(a_{3r-1}+a_{3r}) \text{ be even.}$$

Then a point x of this type has the representation

$$(6.21)\quad x=\frac{a_1}{3}+\frac{a_2}{3^2}+\cdots+\frac{2}{3^{3v-1}}+\frac{2}{3^{3v}}+\frac{a_{3v+1}}{3^{3v+1}}$$
$$+\frac{2}{3^{3v+2}}+\frac{2}{3^{3v+3}}+\frac{a_{3v+4}}{3^{3v+4}}+\cdots+\frac{2}{3^{3m-1}}+\frac{2}{3^{3m}}$$
$$+\frac{a_{3m+1}}{3^{3m+1}}+\cdots$$

A point x' lying on the right of x sufficiently near it, must have the representation

$$(6.22)\quad x' = \frac{a_1}{3} + \cdots + \frac{a_{3v+1}}{3^{3v+1}} + \cdots + \frac{2}{3^{3m-1}} + \frac{2}{3^{3m}}$$

$$+ \frac{a'_{3m+1}}{3^{3m+1}} + \frac{a'_{3m+2}}{3^{3m+2}} + \frac{a'_{3m+3}}{3^{3m+3}} + \cdots\cdots,$$

where

$$(6\cdot23)\quad a'_{3m+1} > a_{3m+1}$$

The corresponding representations of $\Phi(x)$ and $\Phi(x')$ agree up to m places. After the mth place the representations are: for $\Phi(x)$

$$(6.24)\quad \frac{a_{3m+1}}{3^{m+1}} + \frac{a_{3(m+1)+1}}{3^{m+2}} + \cdots + \frac{a_{3l+1}}{3^{l+1}} + \cdots\cdots,$$

and for $\Phi(x')$

$$(6.25)\quad \frac{a'_{3m+1}}{3^{m+1}} + \frac{b_{m+2}}{3^{m+2}} + \cdots + \frac{b_{l+1}}{3^{l+1}} + \cdots\cdots$$

Therefore,

$$(6.26)\quad \Phi(x') - \Phi(x) = \frac{(a'_{3m+1} - a_{3m+1})}{3^{m+1}}$$

$$+ \frac{(b_{m+2} - a_{3(m+1)+1})}{3^{m+2}} + \cdots\cdots + \frac{(b_{l+1} - a_{3l+1})}{3^{l+1}} + \cdots\cdots$$

Now, the least value that the b's can have is *zero*, while $(a'_{3m+1} - a_{3m+1}) \geq 1$, by (6.23).

$$(6.27)\quad \therefore \Phi(x') - \Phi(x) \geq \frac{1}{3^{m+1}} - \left\{ \frac{1}{3^{m+2}} \right.$$

$$\left. + \frac{1}{3^{m+3}} + \cdots\cdots \right\} \geq \tfrac{1}{2}\,\frac{1}{3^{m+1}}\,.$$

for the a_{3r+1}'s are either 0 or 1.

(6.28) Also $$(x'-x) \leq \frac{1}{3^{3m}}.$$

Therefore,

$$\lim_{x'=x} \frac{\phi(x')-\phi(x)}{x'-x} > \lim_{m=\infty} \frac{\frac{1}{2}\cdot\frac{1}{3^{m+1}}}{\frac{1}{3^{3m}}}$$

$$> \lim_{m=\infty} \tfrac{1}{2}\, 3^{2m-1} = \infty.$$

Therefore, at the points of this type, there exists a right hand derivative equal to ∞.

(6.3) Taking the a's up to a_{3v-2} to be fixed, the points of the type (6.2) (for this fixed v) are obtained by giving to the $a_{3(v+r)+1}$'s the values 0 or 1 (where $r=0,1,2,3,\ldots$). Thus these numbers can be placed into one to one correspondence with the numbers of the continuum (0, 1) expressed as radix fractions in the scale of 2. The cardinal number of these points is, therefore, c. As v can have all finite integral values, the set of all the points of the type (6.2) is everywhere dense in (0,1).* Moreover, the set of points of the type (6.2), for a fixed v, form a non-dense set. Giving to v the values 1, 2, 3,...we get a sequence of such sets. It follows, therefore, that the set of all the points of the type (6.2) form an unenumerable, everywhere dense set of the first category in (0, 1).

* For, given any interval (α,β) in (0, 1), we can easily find a point of the type (6. 2) in (α, β). To do this we have simply to choose the a's properly.

(6.4) Let $(a_{3n-1}+a_{3n})=4$ for all values of $n \geq v$. If μ_m denote the number of a_{3n+1}'s which are equal to 2, $n=m+1, m+2\ldots\ldots\ldots$, immediately following a_{3m+1}(not equal to 2), let

$$\underset{m=\infty}{Lim} \quad (2m-\mu_m)=\infty,$$

and further, let $\sum_{1}^{v-1} (a_{3n-1}+a_{3n})$ be even.

Then reasoning as before, corresponding to the inequality (6.27), we get

$$\begin{aligned}
\phi(x')-\phi(x) &\geq \frac{1}{3^{m+1}} - \left\{ \frac{2}{3^{m+2}} + \frac{2}{3^{m+3}} + \cdots \right. \\
&\qquad\qquad \left. + \frac{2}{3^{m+\mu_m+2}} \right\} \\
&\geq \frac{1}{3^{m+1}} - \frac{2}{3^{m+2}} \left\{ 1 + \frac{1}{3} + \cdots + \frac{1}{3^{\mu_m}} \right\} \\
&\geq \frac{1}{3^{m+1}} - \frac{2}{3^{m+2}} \cdot \frac{1-\frac{1}{3^{\mu_m}}}{1-\frac{1}{3}} \\
&\geq \frac{1}{3^{m+1}} - \frac{3^{\mu_m}-1}{3^{m+\mu_m+1}} \\
&\geq \frac{1}{3^{m+\mu_m+1}}.
\end{aligned}$$

Therefore,

$$\frac{\phi(x')-\phi(x)}{x'-x} \geq \frac{\dfrac{1}{3^{m+\mu_m+1}}}{\dfrac{1}{3^{3m}}} \geq 3^{2m-\mu_m-1}.$$

Hence

$$\lim_{x'=x} \frac{\Phi(x')-\Phi(x)}{x'-x} > \lim_{m=\infty} 3^{2m-\mu_m-1} = \infty,$$

for $\lim_{m=\infty} (2m-\mu_m)=\infty$ by supposition.

Combining the results of (6.2), (6.3) and (6.4), we can assert that there exists an everywhere dense set of points at which $\Phi(x)$ has a right hand derivative equal to ∞.

(6.5) If $\sum_{1}^{v-1} (a_{3r-1}+a_{3r})$ is odd, while the other conditions of (6.2) and (6.4) are satisfied by the representation of a point x, then it can be easily shown that there exists a right hand derivative at x, which has the value $-\infty$.

It follows, therefore, that there exists an everywhere dense set of points at each of which $\Phi(x)$ has a right hand derivative equal to $-\infty$.

(6.6) The preceding results can now be summarized as below:

(1) *There exists an everywhere dense set of points* S_1 *at each of which* $\Phi(x)$ *has a progressive derivative equal to* ∞. *A point* x *of this set is such that* (*a*) *in its representation,* $(a_{3n-1}+a_n)=4$ *for all values of* $n \geq v$, *and if* μ_m *denote the greatest number of* a_{3n+1}'*s,* $(n=m+1, m+2,\ldots)$ *immediately following* a_{3m+1} (*not equal to* 2), *which are equal to* 2, *then*

$$\lim_{m=\infty} (2m-\mu_m)=\infty;$$

$$\textit{and } \sum_{1}^{v-1} (a_{3r-1}+a_{3r}) \textit{ is even.}$$

(2) *There exists an everywhere dense set of points* S_2 *at each of which* $\Phi(x)$ *has a progressive derivative equal to* $-\infty$. *A point* x *of this set is such that the condition* (*a*) *of* (1) *relating to the representation of* x *is satisfied, while*

$$\sum_{1}^{v-1} (a_{3r-1}+a_{3r}) \textit{ is odd.}$$

Similarly it can be proved that:

(3) *There exists an everywhere dense set of points* S_3 *at each of which* $\Phi(x)$ *has a regressive derivative equal to* ∞. *A point* x *of this set is such that* (*a*) *in its representation* $(a_{3n-1}+a_{3n})=0$ *for all values of* $n \geq v'$, *and if* μ'_m *denote the greatest number of* a_{3n+1}'s *immediately succeeding* a_{3m+1} (*not equal to 0*), *which are all equal to 0, then*

$$\lim_{m=\infty} (2m-\mu'_m)=\infty;$$

$$\textit{and } \sum_{1}^{v'-1} (a_{3r-1}+a_{3r}) \textit{ is even.}$$

(4) *There exists an everywhere dense set of points* S_4 *at each of which* $\Phi(x)$ *has a regressive derivative equal to* $-\infty$. *A point* x *of this set is such that the condition* (*a*) *of* (3) *relating to its representation is satisfied, while* $\sum_{1}^{v'-1} (a_{3r-1}+a_{3r})$ *is odd.*

7. Second Example.

Let a point t in the interval $(0, 1)$ be represented as

$$t=\frac{a_1}{3}+\frac{a_2}{3.5}+\frac{a_3}{3^2.5}+\frac{a_4}{3^2.5^2}+\cdots,$$

where the a's are zero or positive integers such that $a_{2r} \leqq 4$ and $a_{2r+1} \leqq 2$ $(r=0, 1, 2,...)$.

Corresponding to t let a number x be defined as

$$x=\varphi(t)=\frac{c_1}{3}+\frac{c_2}{3^2}+\frac{c_3}{3^3}+\cdots,$$

where

$$c_1=a_1,\ c_2=\mathrm{P}^{a_2}(a_3), \ldots, c_n=\mathrm{P}^{a_2+a_4+\cdots+a_{2n}}(a_{2n+1}), \ldots$$

and $\mathrm{P}^k(a)$ denotes a or $(2-a)$ according as k is even or odd.

I shall first prove that $\varphi(t)$ is a continuous function of t and then prove that it is non-differentiable.

(7.1) $x=\varphi(t)$ is a continuous function.

The numbers t may be divided into two classes:

(1) Those which are capable of double representation.

(2) Those which have a single representation only.

If t be a number of the second class, x is uniquely defined.

If t be a number of the first class, whose ending representation runs up to an odd number of terms, then, since,

$$\frac{a_1}{3}+\frac{a_2}{3.5}+\cdots+\frac{a_{2n+1}}{3^{n+1}.5^n}+\left(\frac{4}{3^{n+1}.5^{n+1}}+\frac{2}{3^{n+2}.5^{n+1}}\right)$$

$$+\left(\frac{4}{3^{n+2}.\,5^{n+2}}+\frac{2}{3^{n+3}.5^{n+2}}\right)+\ \ldots \text{ to infinity}$$

$$=\frac{a_1}{3}+\frac{a_2}{3.5}+\cdots+\frac{a_{2n+1}+1}{3^{n+1}.5^n},$$

t can be represented by the finite series or by the infinite series; and if $a_2+a_4+\cdots+a_{2n}$ is even, we see that the values obtained by applying the definition of x to the two modes of representation of t are

$$\frac{c_1}{3}+\frac{c_2}{3^2}+\cdots+\frac{c_n}{3^n}+\frac{c_{n+1}}{3^{n+1}}+\frac{2}{3^{n+2}}+\frac{2}{3^{n+3}}+\cdots$$

and

$$\frac{c_1}{3}+\frac{c_2}{3^2}+\cdots+\frac{c_n}{3^n}+\frac{c'_{n+1}}{3^{n+1}},$$

where $c_{n+1}=a_{2n+1}$, and $c'_{n+1}=a_{2n+1}+1$, so that the same value of x is obtained for both representations of t. Similarly if $a_2+a_4+\cdots+a_{2n}$ is odd, it can be seen that x has the same value for both representations of t. The case when the ending representation of t runs up to an even number of terms may be similarly treated.

$x=\varphi(t)$ is thus a single-valued function of t. It is continuous, for if t and t' are identical as regards their first $2n$ terms, the corresponding x and x' are identical as regards their first n terms, and, therefore, when t' tends to t with increasing n, x' tends to x.

(7.2) $x = \varphi(t)$ is a non-differentiable function.

For proving the non-differentiability of $\varphi(t)$, the numbers t may be divided into two classes (A) and (B).

First consider class (A), i.e., the class of numbers in which $a_{2r} \leqq 2$ for infinitely many values of r.

If t_1 be a point of this class, we see that the addition of $\dfrac{2}{3^r.5^r}$ to t_1 does not make any change in the value of $\varphi(t_1)$, since a_{2r} becomes $a_{2r}+2$ so that k remains even or odd as before; while the addition of $\dfrac{1}{3^{r+1}.5^r}$ to t_1 does so. Therefore,

$$\lim \left| \frac{\varphi\left(t_1 + \dfrac{2}{3^r.5^r}\right) - \varphi(t_1)}{\dfrac{2}{3^r.5^r}} \right| = 0$$

and

$$\overline{\lim} \left| \frac{\varphi\left(t_1 + \dfrac{1}{3^{r+1}.5^r}\right) - \varphi(t_1)}{\dfrac{1}{3^{r+1}.5^r}} \right| = \infty$$

where the limits are taken as r tends to infinity assuming those values for which the inequality $a_{2r} \leqq 2$ is satisfied.

Thus at the points of class (A) the differential coefficient is non-existent.

Now consider class (B), i.e., those numbers t in which $a_{2r} > 2$ for infinitely many values of r.

If t_2 be a point of this class, we see that the subtraction of $\frac{2}{3^r.5^r}$ from t_2 does not change the value of $\varphi(t_2)$, while the subtraction of $\frac{1}{3^{r+1}.5^r}$ from t_2 does so, so that

$$\lim \left| \frac{\varphi\left(t_2 - \frac{2}{3^r.5^r}\right) - \varphi(t_2)}{\frac{2}{3^r.5^r}} \right| = 0$$

and

$$\overline{\lim} \left| \frac{\varphi\left(t_2 - \frac{1}{3^{r+1}.5^r}\right) - \varphi(t_2)}{\frac{1}{3^{r+1}.5^r}} \right| = \infty,$$

where the limits are taken as r tends to infinity assuming those values for which the inequality $a_{2r} > 2$ is satisfied.

Thus at the points of class (B) the differential coefficient is non-existent. $x = \varphi(t)$ is, therefore, a non-differentiable function in the interval (0, 1).

Every function of this type* can, by a similar treatment, be shown to be devoid of a differential coefficient at each point in (0, 1).

---

*For the general class of functions of which the above is a particular case see Singh (79).

8. **Third Example.** Let the numbers of the interval (0, 1) be expressed in the decimal scale as

$$x = \frac{a_1}{10} + \frac{a_2}{10^2} + \cdots + \frac{a_{2n-1}}{10^{2n-1}} + \frac{a_{2n}}{10^{2n}} + \cdots$$

where every a is one of the numbers 0, 1, 2,...or 9. Corresponding to x we define a number

$$y = f(x) = \frac{b_1}{20} + \frac{b_2}{20^2} + \frac{b_3}{20^3} + \cdots + \frac{b_n}{20^n} + \cdots$$

where $b_n = \pm(2a_{2n-1} + c_n)$, and $c_n = 0$, 1 or 2 according as a_{2n} is 0, 2, 4 or 1, 3, 5, 7, 9 or 6, 8 respectively, and b_n‡ has the same sign as b_{n-1}, if a_{2n-2} is 0, 2, 4, 5, 7 or 9, otherwise it has the opposite sign; and b_1 is always positive.

(8.1) $y=f(x)$ **is a continuous function.** y is evidently continuous at all the points x in (0, 1) which have a unique non-terminating representation in the scale of ten. Those points x which have a terminating representation have also a non-terminating representation. To prove the continuity of y in (0, 1) it will be enough to show that y is uniquely determined at all the points where x has a double representation.

Let x be a point whose representation runs up to $(2n-1)$ places. Then

$$x = \frac{a_1}{10} + \frac{a_2}{10^2} + \cdots + \frac{a_{2n-1}}{10^{2n-1}}$$

$$= \frac{a_1}{10} + \frac{a_2}{10^2} + \cdots + \frac{(a_{2n-1} - 1)}{10^{2n-1}} + \frac{9}{10^{2n}} + \frac{9}{10^{2n+1}} + \cdots$$

‡ It will be observed that b_n can have the value 20.

And according to our definition, corresponding to the two representations of x

$$y = \frac{b_1}{20} + \frac{b_2}{20^2} + \cdots + \frac{b_{n-1}}{20^{n-1}} + \frac{b_n}{20^n}$$

and

$$= \frac{b_1}{20} + \frac{b_2}{20^2} + \cdots + \frac{b_{n-1}}{20^{n-1}} + \frac{b'_n}{20^n}$$

$$\pm\left(\frac{19}{20^{n+1}} + \frac{19}{20^{n+2}} + \cdots\right).$$

We have $b_n = \pm\,(2a_{2n-1}+0)$ and
$b'_n = \pm\,[2(a_{2n-1}-1)+1]$, i.e., $|\,b'_n\,| = |\,b_n\,| - 1$.
But the terms that follow b'_n have the same sign as b'_n and their sum is $\frac{1}{20^n}$. Hence the same value of y corresponds to the two representations of x.

Similarly, if x has a terminating representation running up to $2n$ places, we can show that the same value of y is obtained for both representations of x.

It follows that y is a continuous function of x in $(0, 1)$.

(8.2) y **is a non-differentiable function.**
(a) Let x be a point in whose representation an infinite number of a_{2n}'s are 0, 1 or 2. Then

$$f\left(x + \frac{2}{10^{2n}}\right) - f(x) = 0$$

while

$$\left| f\left(x+\frac{5}{10^{2n}}\right)-f(x)\right| = \frac{1}{20^n},$$

for an infinite number of values of n tending to infinity, so that one of the derivates at x is zero, whilst another is indefinitely great* (numerically). Thus the differential coefficient is non-existent at all such points.

(b) Let x be a point in whose representation an infinite number of a_{2n}'s are 3 or 4. Then

$$f\left(x-\frac{2}{10^{2n}}\right)-f(x) = 0$$

while

$$\left| f\left(x+\frac{5}{10^{2n}}\right)-f(x)\right| = \frac{1}{20^n},$$

and, therefore as before, the differential coefficient is non-existent at all such points.

(c) Let x be a point in whose representation an infinite number of a_{2n}'s are 5, 6 or 7. Then

$$f\left(x+\frac{2}{10^{2n}}\right)-f(x) = 0$$

while

$$\left| f\left(x-\frac{5}{10^{2n}}\right)-f(x)\right| = \frac{1}{20^n},$$

*For $\displaystyle \lim_{h=0}\left|\frac{f(x+h)-f(x)}{h}\right| \geq \lim_{n=\infty}\frac{\frac{1}{20^n}}{\frac{5}{10^{2n}}}$

and, therefore as before, the differential coefficient is non-existent at all such points.

(d) Let x be a point in whose representation an infinite number of a_{2n}'s are 8 or 9. Then

$$f\left(x-\frac{2}{10^{2n}}\right)-f(x)=0$$

while

$$\left|f\left(x-\frac{5}{10^{2n}}\right)-f(x)\right|=\frac{1}{20^n},$$

and, therefore as before, the differential coefficient is non-existent at all such points.

Now, any point in (0, 1) comes under at least one of the four heads enumerated above, and hence at no point x in (0, 1) does there exist a differential coefficient.*

* For the general class of functions of which the above is a particular case see Singh (83).

FOURTH LECTURE

PROPERTIES OF NON-DIFFERENTIABLE FUNCTIONS

1. In to-day's lecture I shall give an account of the recent work relating to the study of the properties of non-differentiable functions, especially with regard to the existence of cusps and derivatives. In this connection I shall also enumerate some of the important results of the theory of derivates which have a direct bearing on our subject. An account will also be given of the character of the oscillations of non-differentiable functions. It will be shown by means of an example that the oscillations of a non-differentiable function may be *unenumerable* in every interval how-so-ever small taken in the domain of the function. Whether or not this is a general characteristic of all non-differentiable functions is unknown.

2. **Upper and lower derivates.** When a continuous function $f(x)$ does not possess a right (left) derivative, the incrementary ratio

$$R(x, h) = \frac{f(x+h) - f(x)}{h},$$

$h > 0$ (< 0), does not tend to any limit as h tends to

zero. In such a case $R(x, h)$ is associated with its upper and lower limits which are called the upper and lower derivates of $f(x)$ on the right (left), and are denoted respectively by the symbols $D^+f(x)$, $D_+f(x)$ $[D^-f(x), D_-f(x)]$. In recent years, these derivates have been closely studied and a number of very interesting results obtained. Whenever the derivates are finite *almost everywhere*, they can be used in the place of the differential coefficient. The derivates of non-differentiable functions, however, can not be used as they are infinite *almost everywhere*. In fact, in the case of every non-differentiable function, it can be shown that

$$(i)\ D^+ = D^- = \infty$$

and
$$(ii)\ D_+ = D_- = \infty$$

almost everywhere.

3. **Sufficient condition for differentiability.** The discovery of continuous non-differentiable functions brought to the fore-front the question: "In what cases and for what aggregates of values of x can we assert that a function $f(x)$ possesses a differential coefficient?" Perhaps the most important answer to this question has been given by the following theorem of Lebesgue [52]:

A continuous function of bounded variation has a finite differential coefficient at every point which does not belong to a set of measure zero.

This theorem was extended to the case of monotone functions (not necessarily continuous) by W. H. Young and G. C. Young ([104]), who gave a proof independent of the notion of integration and of transfinite numbers. Elementary demonstrations of the theorem have also been given by Faber ([26]) and Tonelli ([93]). Proofs which hold for any function of bounded variation have been given by Carathéodory ([16]) Steinhaus ([88]) and Rajchman and Saks ([68]). It has been shown by Singh ([87]) that the set of points of differentiability of a function of bounded variation is of the second category.

It follows from Lebesgue's theorem that a non-differentiable function is *not a function of bounded variation* and is consequently not *rectifiable.* This leads us to the following three properties of a non-differentiable function:

(i) it is everywhere oscillating;

(ii) the length of the arc between any two points on the curve is infinite; and

(iii) the geometrical graph of the function can not be drawn.

4. **Existence of cusps.** The question arises whether a non-differentiable function can possess proper maxima and minima. As the function is non-differentiable, the proper maxima and minima, if they exist, must be either cusps or edge-points (*i.e.*, points at which the derivates on one side are positive whilst those on the

other side are negative; the derivates on one side at least being finite). Hardy ([32]) and G. C. Young ([100]) have shown that Weierstrass's function

$$W(x) = \sum_{0}^{\infty} a^n \, Cos \, (b^n \pi x)$$

has at the everywhere dense set of points $x = \frac{2r}{b^m}$ right-hand derivatives $= -\infty$, and left-hand derivatives $= \infty$. Thus these points are cusps whose edges point upwards and are proper maxima of Weierstrass's function. It has also been shown that at the set of points $x = \frac{2r+1}{b^m}$, the function has proper minima, with cusps pointing downwards. Bhar ([8]) has found similar everywhere dense set of cusps on the curve

$$D(x) = \sum_{1}^{\infty} a^n \, Sin \, (b^n \pi x).$$

All the functions defined by series and given in the first lecture, can be shown to possess cusps at everywhere dense sets of points. Hobson* has remarked, "It does not appear to be definitely known whether a non-differentiable function can exist which has no cusps." This question was answered by me in the third lecture, where it was shown that the function $\varphi_{3,1,3}(t)$ has no cusps, because at no point does the function possess

* See Hobson (36) p. 405.

both a progressive and a regressive derivative. Functions which do not possess cusps were considered by Moore ([57]) Sierpinski ([76]) and Hahn ([31]), but their investigations seem to have escaped notice.

It does not appear to be known whether a non-differentiable function can exist which has edge-points at an everywhere dense set.

(4.1) **Theorems about cusps and derivates.** B. Levi ([54]) has proved the theorem:

The aggregate of points x, of an interval (a, b), at which a function $F(x)$, continuous in (a, b), possesses a progressive and a regressive derivative, which are different from each other, is enumerable.

This theorem has been generalised by Rosenthal ([69]) and G. C. Young ([98]).

From Levi's theorem it follows that the set of points at which a function has cusps is enumerable. In connection with the existence of an everywhere dense set of cusps may also be mentioned the following theorems due to Koenig ([48]) and Rosenthal ([69]) respectively.

(1) *If the continuous function $f(x)$ possess cusps at an everywhere dense set of points, there exists an everywhere dense set of points at each of which the differential coefficient has the prescribed value c, or is*

indeterminate and such that c lies between its upper and lower limits.

(2) *If the continuous function $f(x)$ has cusps at an everywhere dense set of points, three exists a set of the second category at each point of which the function lacks a progressive and a regressive derivative.*

(4.2) **Denjoy's Theorem.** The relations which subsist between the four derivates of a continuous function at a point, if one disregards sets of measure zero, have been systematically investigated by Denjoy(21). He has obtained the following theorem :

If $f(x)$ be a continuous function, finite at each point, and if a set of measure zero be left out of account, then, at the various points x only the following four cases are possible.*

(1) $D^+ = D^- = D_+ = D_- =$ finite,

(2) $D^+ = D^- = \infty$; $D_+ = D_- = -\infty$,

(3) $D^+ = \infty$, $D_- = -\infty$; $D_+ = D^- =$ finite,

(4) $D^- = \infty$, $D_+ = -\infty$; $D_- = D^+ =$ finite.

Each of the above four cases can be individually realized, *i.e.*, a function can be constructed for which a definite one of the four cases occurs. Denjoy has

* The above result has been shown to hold for a finite and measurable function by G. C. Young (101). Saks (71) has given a proof applicable to non-measurable functions, and an extension has been made by G. C. Young to the case in which the function is infinite at the points of a set of positive measure.

constructed a function for which every one of these four cases occurs.

(4.3) **Derivates of non-differentiable functions.** The properties of the derivates of non-differentiable functions have been studied by W. H. Young ([103]), and the following result has been established by him:

If a function $f(x)$ is non-differentiable in an interval, then

(1) there is necessarily a distinction of right and left in the values of the derivates at a set of points which is everywhere dense and is of the first category ;

(2) the upper and lower bounds of the values of the derivates at the points of this set are respectively ∞ and $-\infty$;

(3) at the remaining points of the interval, both the upper derivates are ∞ and both the lower derivates are $-\infty$, exception being made of at most another set of the first category.

Young's theorem is supplemented by the following result due to Singh ([81]) :

Every finite and continuous non-differentiable function has associated with it

(1) an everywhere dense set of points S_1, at which $D_+ > M$,

(2) *an everywhere dense set of points* S_2, *at which* $D^+ < -M$,

(3) *an everywhere dense set of points* S_3, *at which* $D_- > M$,

(4) *an everywhere dense set of points* S_4, *at which* $D^- < -M$,

where M is any positive number, however large.

Although the above result does not establish the existence of points at which infinite derivatives exist, it nevertheless shows that, in the case of every continuous non-differentiable function, there exist everywhere dense sets of points at each point of which there is an arbitrarily near approximation to the existence of a determinate (right and left) derivative which is numerically infinite.

5. **Study of Derivates.** G. C. Young ([100]) has made a detailed study of the derivates of Weierstrass's function

$$W(x) = \sum a^n \operatorname{Cos}(b^n \pi x),$$

where b is an odd integer and $ab > 1 + \frac{3\pi}{2}$.

The result obtained by her may be summarised as follows:

Let a point x in (0, 1) be expressed in the scale of b, as

$$x = \frac{c_1}{b} + \frac{c_2}{b^2} + \cdots + \frac{c_n}{b^n} + \cdots\cdots$$

where the c's are 0, 1, 2,.., $b-1$, then,

(1) at the set of points S_1 represented by an ending series of the above form $W(x)$ has cusps;

(2) at the set of points S_2 given by the points x in whose representation $c_{m+1}, c_{m+2}, \ldots\ldots$ are all even, the derivates are non-symmetrical;

(3) at the set of points S_3 which is complementary to (S_1+S_2), the derivates are symmetrical, both the upper derivates being ∞ and both the lower derivates $-\infty$.

The set S_1 is enumerable and the set S_2 can be easily proved to be of zero measure. W. H. Young's theorem is thus verified. A closer study of the derivates of Weierstrass's function seems to be necessary in order to find the actual values of the derivates at the various points of the set S_2.

The investigations of Porter (63) and Bhar (8), although incomplete, show that the derivates of Dini's function

$$\sum_{1}^{\infty} a^n \, Sin \, (b^n \pi x),$$

where b is an even integer, and $ab > 1 + \frac{3\pi}{2}$, have the same general character as those of Weierstrass's function. Let the numbers x in $(0, 1)$ be expressed as radix fractions in the scale of b as

$$x = \frac{c_1}{b} + \frac{c_2}{b^2} + \cdots + \frac{c_n}{b^n} + \cdots\cdots .$$

According to Porter, except at the points of the null set $[x]$ for which $c_n = o$ and $c_n = b-1$ fail to occur infinitely often, the right and left incremental ratios of Dini's function have each for their upper and lower limits $+\infty$ and $-\infty$.

Bhar has studied the function

$$D(x) = \Sigma \frac{Sin\ (16^n \pi x)}{2^n}.$$

His method can be modified to prove that the function

$$\Sigma\ a^n\ Sin\ (b^n \pi x)$$

possesses cusps at an everywhere dense set of points in $(0, 1)$.

6. **The Derivates of $\Phi_{m,r,p}(x)$.** In the third lecture I showed how the set of points where the function $\Phi_{3,1,3}(x)$ possesses one-sided differential co-efficients may be obtained. I shall now state the similar result which can be obtained for the general class of functions $\Phi_{m,r,p}(x)$.

$\Phi_{m,r,p}(x)$ is nowhere differentiable in $(0, 1)$, has no cusps, and

(1) there exists an everywhere dense set S_1, at which $\Phi_{m,r,p}(x)$ has a right hand derivative equal to ∞.

A point x of this set is such that (a) in its representation

$$a_{m(n-1)+r+1}+a_{m(n-1)+r+2}+\cdots+a_{mn+r-1} = (m-1)(p-1)$$

for all values of $n \geq \nu$ *(a fixed number), and if* μ_q *denote the greatest number of* a_{mn+r}*'s* $(n=q+1,\ q+2,\ \ldots\ldots)$ *immediately succeeding* a_{mq+r} *(not equal to* $p-1$*) which are all equal to* $p-1$*, then*

$$\lim_{q\to\infty} (m-1)q-\mu_q=\infty,$$

and (*b*) $\sum\limits_{n=o}^{\nu-1} (a_{m(n-1)+r+1}+\cdots+a_{mn+r-1})$ *is even;*

(2) *There exists an everywhere dense set* S_2 *at which* $\Phi_{m,r,p}(x)$ *has a right hand derivative equal to* $-\infty$. *A point x of this set is such that the condition (a) of (1) is satisfied while,*

$$\sum_{n=o}^{\nu-1} (a_{m(n-1)+r+1}+\cdots+a_{mn+r-1}) \text{ is odd;}$$

(3) *There exists an everywhere dense set* S_3 *at which* $\Phi_{m,r,p}(x)$ *has a left hand derivative equal to* ∞. *A point x of this set is such that (a) in its representation* $(a_{m(n-1)+r+1}+a_{m(n-1)+r+2}+\cdots\cdots+a_{mn+r-1})=o$ *for all values of* $n \geq \nu$*, and if* μ'_q *denote the greatest number of* a_{mn+r}*'s* $(n=q+1,\ q+2,\ldots)$ *immediately succeeding* a_{mq+r} *(not equal to 0) which are all equal to 0, then*

$$\lim_{q\to\infty} (m-1)q-\mu_q'=\infty$$

and (*b*) $\sum_{n=o}^{\nu-1} (a_{m(n-1)+r+1} + \cdots + a_{mn+r-1})$ *is even ;*

(4) *there exists an everywhere dense set* S_4 *at which* $\Phi_{m,r,\nu}(x)$ *has a left hand derivative equal to* $-\infty$. *A point x of this set is such that the condition* (*a*) *of* (*3*) *is satisfied, while*

$$\sum_{n=o}^{\nu-1} (a_{m(n-1)+r+1} + \ldots + a_{mn+r-1}) \text{ is odd.}$$

7. **The Oscillations of non-differentiable functions.** Attempts were made by Wiener (97), Klein (40) and G. C. Young (100) to evolve a "graphical representation" of Weierstrass's function. As the function does not possess a graph, exact information as to the nature of the singularity at a point on the curve is obtained mainly by the study of the values of the derivates at that point. For instance, if we know that at a point P (x_1, y_1) on a continuous curve $y = f(x)$, $D^+ = \infty$ and $D_+ = -\infty$, we can at once say that the curve cuts the line $y = y_1$ at an indefinitely large number of points in any neighbourhood on the right of P, and so makes an infinite number of oscillations. Similar will be the case on the left of P, if $D^- = \infty$ and $D_- = -\infty$. It has been shown by G. C. Young (100) that for a set of points $[x]$ whose measure is 1, in (0, 1), both the upper derivates of Weierstrass's function are ∞ and both the

lower derivates are $-\infty$. It follows, therefore, that a point P on the curve which corresponds to a point x belonging to the set $[x]$ has the property that any line through P cuts the curve an infinite number of times on both sides of P.

(7.1) It is evident that Weierstrass's function

$$W(x)=\sum a^n \, Cos\,(b^n \pi x)$$

(where b is an odd interger) is zero at $x=\frac{1}{2}$. At this point, the derivates on the right as well as on the left oscillate between ∞ and $-\infty$. It follows that the point $x=\frac{1}{2}$ is a limiting point of the zeros of the function $W(x)$. A set of zeros of $W(x)$ with $x=\frac{1}{2}$ as a limiting point has been actually located by G. Prasad(67).

His result in the case of the function

$$W(x)=\sum \frac{Cos\,(13^n \pi x)}{2^n}$$

may be stated as follows:

There is a zero of $W(x)$ between

(*i*) $\left(\frac{1}{2}\pm\frac{1}{13^k}\right)$ and $\left(\frac{1}{2}\pm\frac{3/2}{13^k}\right)$;

(*ii*) another between

$$\left(\frac{1}{2}\pm\frac{3/2}{13^k}\right) \text{ and } \left(\frac{1}{2}\pm\frac{2}{13^k}\right);$$

(*iii*) a third between

$$\left(\frac{1}{2}\pm\frac{3}{13^k}\right) \text{ and } \left(\frac{1}{2}\pm\frac{7/2}{13^k}\right);$$

---

* See also Sharma (75). For the zeros of Dini's function see Mookerji (56).

(*iv*) a fourth between

$$\left(\frac{1}{2}\pm\frac{7/2}{13^k}\right) \text{ and } \left(\frac{1}{2}\pm\frac{4}{13^k}\right);$$

(*v*) a fifth between

$$\left(\frac{1}{2}\pm\frac{5}{13^k}\right) \text{ and } \left(\frac{1}{2}\pm\frac{11/2}{13^k}\right);$$

(*vi*) and a sixth between

$$\left(\frac{1}{2}\pm\frac{11/2}{13^k}\right) \text{ and } \left(\frac{1}{2}\pm\frac{6}{13^k}\right).$$

The above is not a complete list of all the zeros of the function W(x). The determination of all the zeros is probably not practically possible. Bhar ([9]) has given a list of some of the zeros of the function

$$\sum \frac{16^n \begin{matrix} Cos \\ Sin \end{matrix} \left\{ 1.\ 3.\ 5.\ \ldots\ (2n-1) \right\} \pi x}{1.\ 3.\ 5.\ \ldots\ (2n-2)}.$$

Prasad's result quoted above is a verification of the conclusion that we can arrive at by a consideration of the values of the derivates at the point $x=\frac{1}{2}$. It would be interesting to find some special character of the set of zeros, *e.g.*, whether they form an enumerable set or not, or whether the set is closed or open.

(7.2.) Very interesting results regarding the nature of the oscillations of a non-differentiable function have been recently obtained* by studying the intersections of the line $y=c$ with the curve $y=\Phi_{m,r,p}(x)$ given in the third lecture. It has been found that the roots of

---

*See Singh ([85]). For a similar study of another function see Singh ([84]).

the equation $\Phi_{m,r,p}(x) = c$ $(o \leq c \leq 1)$, form a set S_c which is perfect and is of zero measure. Thus the oscillations of the function $\Phi_{m,r,p}(x)$ in every interval, ever-so-small, are unenumerable. The function is thus much more complicated than ordinary transcendental functions which may have an enumerably infinite number of oscillations in an interval. It may be pointed out that by the help of such a function we can easily express the continuum in (0, 1) as an unenumerable aggregate of unenumerable aggregates.

8. The Set S_c of the roots of $\Phi_{3,1,3}(x) = c$.

I shall now find the roots of the equation $\Phi_{3,1,3}(x) = c$ and prove that they form a perfect set S_c of measure zero in the interval (0, 1). The proof for the general case can be similary carried out. Taking the definition of $\Phi_{m,r,p}(x)$ given in the third lecture, we have, for values of x, in (0, 1), expressed in the scale of 3 as

$$x = \frac{a_1}{3} + \frac{a_2}{3^2} + \frac{a_3}{3^3} + \cdots + \frac{a_{3n-1}}{3^{3n-1}} + \frac{a_{3n}}{3^{3n}} + \frac{a_{3n+1}}{3^{3n+1}} + \cdots,$$

$$\Phi_{3,1,3}(x) = \frac{b_1}{3} + \frac{b_2}{3^2} + \cdots + \frac{b_n}{3^n} + \cdots,$$

where

$$b_1 = k^o(a_1) = a_1,$$

$$b_2 = k^{a_2 + a_3}(a_4),$$

$$b_3 = k^{a_2+a_3+a_5+a_6}(a_7),$$

$$\cdots\cdots\cdots\cdots\cdots\cdots\cdots\cdots\cdots$$

$$b_n = k^{a_2+a_3+a_5+a_6+\cdots+a_{3(n-1)-1}+a_{3(n-1)}}(a_{3(n-1)+1}),$$

and so on; and where $k^s(a) = a$ or $(2-a)$ according as s is even or odd.

Let the constant c have the following representation when expressed in the scale of 3:

$$c = \frac{c_1}{3} + \frac{c_2}{3} + \frac{c_3}{3^3} + \cdots + \frac{c_n}{3^n} + \cdots\cdots$$

Then the a's in the representation of

$$x_c = \frac{a_1}{3} + \frac{a_2}{3^2} + \frac{a_3}{3^3} + \cdots + \frac{a_{3n-1}}{3^{3n-1}} + \frac{a_{3n}}{3^{3n}} + \frac{a_{3n+1}}{3^{3n+1}} + \cdots$$

which corresponds to the c_n's must satisfy the following conditions:

$$a_1 = c_1;$$

and if $c_{n+1} = 0$, then $a_{3n+1} = 0$ or 2 according as

$$\sum_{r=1}^{n} (a_{3r-1} + a_{3r}) \text{ is even or odd;}$$

if $c_{n+1} = 2$, then $a_{3n+1} = 2$ or 0 according as

$$\sum_{r=1}^{n} (a_{3r-1} + a_{3r}) \text{ is even or odd;}$$

and if $c_{n+1} = 1$, then $a_{3n+1} = 1$ whatever

$$\sum_{r=1}^{n} (a_{3r-1} + a_{3r}) \text{ may be.}$$

It is obvious that there are an infinite number of x_c's, the a's in whose representation satisfy the conditions given above. The points x_c form a set S_c, which is the set of the roots of $\Phi_{3,1,3}(x)=c$ and which we now propose to study.

(8.1) *S_c is unenumerable.* This follows from the fact that S_c is perfect.

(8.2) *S_c is perfect.* Let

$$x_c = \frac{a_1}{3}+\frac{a_2}{3^2}+\frac{a_3}{3^3}+\cdots+\frac{a_{3n-1}}{3^{3n-1}}+\frac{a_{3n}}{3^{3n}}+\frac{a_{3n+1}}{3^{3n+1}}+\cdots$$

be a root of $\Phi_{3,1,3}(x)=c$.

Then the point

$$x'_c=\frac{a_1}{3}+\frac{a_2}{3^2}+\cdots\cdots+\frac{a'_{3n-1}}{3^{3n-1}}+\frac{a'_{3n}}{3^{3n}}+\frac{a_{3n+1}}{3^{3n+1}}+\cdots\cdots,$$

which differs from x_c at the $(3n-1)$th and $3n$th places only is also a root of $\Phi_{3,1,3}(x)=c$, if

$$|(a_{3n-1}+a_{3n})-(a'_{3n-1}+a'_{3n})| = 2 \text{ or } 4.$$

By letting n tend to infinity, we see that points x'_c belonging to S_c can be found as near to x_c as we please. Thus x_c is a limiting point of the points of S_c. *It follows that the set S_c is dense-in-itself.*

That the set S_c is closed follows from the continuity of $\Phi_{3,1,3}(x)$.

Therefore, the set S_c is perfect.

(8.3) *S_c has zero measure.* In the representation of c, if $c_1=0$, then there are 2 intervals

$$[\tfrac{1}{3}, \tfrac{2}{3}] \text{ and } [\tfrac{2}{3}, 1]$$

in each of which there is no point of S_c; if $c_1 = 1$, then there are two intervals

$$[0, \tfrac{1}{3}] \text{ and } [\tfrac{2}{3}, 1]$$

in each of which there is no point of S_c; and similarly if $c_1 = 2$, we find that there are two intervals

$$[0, \tfrac{1}{3}] \text{ and } [\tfrac{1}{3}, \tfrac{2}{3}]$$

in each of which there is no point of S_c.

We thus find that, *whatever c_1 may be,* there are two intervals that do not contain points of S_c, and that the sum of their lengths is $\frac{2}{3}$.

Again, *whatever c_1 and c_2 may be,* there are besides the two intervals of the above type, 2.3^2 more intervals, each of length $\frac{1}{3^4}$ which do not contain points of S_c. For supposing $c_2 = 2$, a_2 and a_3 must respectively have the values

$$1, 0 \text{ or } 0, 1 \text{ or } 1, 2 \text{ or } 2, 1,$$

whilst $a_4 = 0$, otherwise a_2 and a_3 are respectively

$$0, 0 \text{ or } 0, 2 \text{ or } 2, 0 \text{ or } 1, 1 \text{ or } 2, 2,$$

whilst $a_4 = 2$. Thus (for the case $c_1 = 1$, $c_2 = 2$) there can not be points of the set S_c in the two intervals

$$\left\{ \frac{1}{3} + \frac{1}{3^2} + \frac{0}{3^3} + \frac{1}{3^4}, \ \frac{1}{3} + \frac{1}{3^2} + \frac{0}{3^3} + \frac{2}{3^4} \right\}$$

and

$$\left\{ \frac{1}{3} + \frac{1}{3^2} + \frac{0}{3^3} + \frac{2}{3^4}, \ \frac{1}{3} + \frac{1}{3^2} + \frac{1}{3^3} \right\},$$

or again in the two intervals

$$\left\{\frac{1}{3}+\frac{0}{3^2}+\frac{1}{3^3}+\frac{1}{3^4},\ \frac{1}{3}+\frac{0}{3^2}+\frac{1}{3^3}+\frac{2}{3^4}\right\}$$

and

$$\left\{\frac{1}{3}+\frac{0}{3^2}+\frac{1}{3^3}+\frac{2}{3^4},\ \frac{1}{3}+\frac{0}{3^2}+\frac{2}{3^3}\right\},$$

and so on for all the nine cases, thus making a total of 2.3^2 intervals that do not contain points of S_c. It follows that *whatever* c_1 *and* c_2 *may be*, there are a set of intervals whose total length is

$$\frac{2}{3}+\frac{2}{3^2},$$

which do not contain points of S_c.

Similarly, it can be shown that *whatever* c_1, c_2 *and* c_3 *may be*, there are a set of intervals which do not contain points of S_c, and that the sum of the lengths of these intervals is

$$\frac{2}{3}+\frac{2}{3^2}+\frac{2}{3^3}.$$

Proceeding in this manner, we find that *whatever the c's may be*, there are a set of intervals which do not contain points of S_c, and whose measure is

$$\frac{2}{3}+\frac{2}{3^2}+\frac{2}{3^3}+\cdots+\frac{2}{3^n}+\cdots\cdots=1.$$

Therefore, S_c *has zero measure.*

9. The Mean Differential Coefficient. If $f(x)$ be continuous at the point x, the mean differential coefficient at x is the limit, if it exists, of the ratio

$$\frac{f(x+h)-f(x-h)}{2h},$$

as h tends to zero.

G. Prasad (65) has shown that Weierstrass's function possesses finite mean differential coefficients at an everywhere dense set of points. The functions defined by Darboux (20), Lerch (53), Faber (25, 26, 27), Landsberg (51), Steinitz (89), Singh (78, 79, 83, 85) and Hahn 31) possess mean differential coefficients. It has been stated by Bhar (7) that Dini's function $\sum a^n Sin\ (b^n \pi x)$ does not possess a mean differential coefficient.*

10. **Remarks**. We have traced the gradual development of our knowledge regarding the nature and properties of non-differentiable functions, and have discussed some of the problems that have arisen during the course of the study of such functions. We have also pointed out some of the advances that have been made in other branches of the theory of functions due to that study. But the fundamental question which the discovery of non-differentiable functions has raised remains still unanswered. It is this: 'What minimum restrictions should be placed on a function so that it will possess a differential coefficient at each point of its domain of definition?' Or, in other words, 'Does there

* Bhar (9), p. 80, points out a slip in his paper (7), which will require a recasting of his proof for the non-existence of the mean differential coefficient.

exist a necessary and sufficient condition for the differentiability of a function in an interval ?' The question has been engaging the attention of mathematicians since the publication of Weierstrass's example of a non-differentiable function in 1874, and yet we are no nearer its solution.

It is well-known that continuity is *necessary* for differentiability, but it is not *sufficient* as is shown by the existence of continuous non-differentiable functions. The restriction of *bounded variation* has also proved insufficient. Although a continuous function of bounded variation must possess a differential coefficient *almost everywhere*, yet there are examples of such functions which do not possess differential coefficients at unenumerable everywhere dense sets of points. The same remark applies to absolutely continuous functions. The discovery of a necessary and sufficient condition for differentiability will no doubt be a great advance and will, I believe, find immediate application in geometry and in physics. But with the present state of our knowledge it does not appear to be possible to discover any such condition. Further study of the theory of aggregates and perhaps a closer classification of functions may give us the key to the solution.

It has been suggested by W. H. Young and G. C. Young that efforts should be made to evolve a definition of differentiation, according to which all continuous functions would be differentiable, at least *almost everywhere*, and correspondingly to find a definition of integration that would always lead back from the derivate to the primitive function. We have seen that in the case of non-differentiable functions the upper and lower right (left) derivates have the values $+\infty$ and $-\infty$ respectively *almost everywhere*. At such points, therefore, by choosing a suitable method of approach the incrementary ratio can be made to converge to any desired value between $+\infty$ and $-\infty$. The problem, then, is to devise a method of choosing h's which would not only provide finite derivatives *almost everywhere*, but would also suggest a corresponding integrating process, leading from such a derivative back to the primitive function.

A suggestion has been made by G. C. Young to use (what she calls) the *mean symmetric derivate*, which she has defined and shown to exist in the case of a continuous function, except at an enumerable set of points; but there is no evidence as yet to show that it has any practical value. The *mean symmetric derivate* of Weierstrass's function is 1 *almost everywhere*, and is clearly of no particular use. Likewise the notion of the generalised Riemann derivative of the nth order does not appear to take us appreciably nearer to the solution of the problem.

The successful generalisations of the notion of integration and that of summation have shown that it is possible to select and utilize one out of an infinite number of limits. In fact, if by the application of the theory of sequences, or otherwise, we can devise some method of selection, the plurality of limits can be turned to an advantage. For the solution of our problem, however, we cannot hope much from the theory of sequences alone. The success of the method of sequences is intrinsically due to the mechanism of monotony, but monotony in such a connection as this can serve no useful purpose. It is hoped that a more profound study of the theory of aggregates may lead to the solution of the problem, or at least to the determination of the limits between which the solution should lie.

BIBLIOGRAPHY

1. **Ampère.** Recherches sur quelques points de la théorie des fonctions dérivées, qui conduisent, etc. *Ecole Polytechnique*, Bd. 6, Cah. 13. (1806).

2. **Apt, F.** Die Tangentenlosigkeit der von Kochschen Kurve, *Math. Zeit.*, Vol. 13, (1922), pp. 217-222.

3. —Nicht differentiierbare Funktionen und tangentenlose Kurven (Geschichte und Problem), *Sitzungsber. Berl. Math. Ges.*, Vol. 19, (1921), pp. 85-88.

4. **Banerji, H. P.** On Peano's Function, *Bull. Cal. Math. Soc.*, Vol. 16, (1924), pp. 23-30.

5. **Besicovitch, A. S.** Discussion der stetigen Funktionen im Zusammenhang mit der Frage über ihre Differenzierbarkeit, *Bull. de l'Acad. des Sciences de Russie*, Vol. 19, (1925), pp. 527-540.

6. — On Lipschitz numbers, *Math. Zeit.*, Vol. 30, (1929), pp. 514-519.

7. **Bhar, S. K.** On a continuous function which has no mean differential co-efficient for any value of *x*, *Bull. Cal. Math. Soc.*, Vol. 23, (1931), pp. 210-230.

8. — Some remarks on the "graph" of Dini's non-differentiable function, *Bull. Cal. Math. Soc.*, Vol. 25, (1933), pp. 59-80.

9. — On the zeros of non-differentiable functions of Darboux's type, *Bull. Cal. Math. Soc.*, Vol. 22, (1930), pp. 61-90.

10. **Bieberbach, L.** *Differential-und Integral-rechnung*, Leipzig, (1917), Vol. I.

11. **Broden, T.** Beträge zur Theorie der stetigen Funktionen einer reellen Veränderlichen, *Journ. für Math.*, Vol. 118, (1897), pp. 1-60.

12. — Ueber die elementare Konstruction sogenannter Kurven ohne Tangente, *Arkiv för Math. Astr. och Physik*, Vol. 2, (1905), p. 12.

13. **Broglio, A.** Studio analitico della Curve del Sig. Helge von Koch, *Giorn. di Mat.*, Vol. 44, (1906), pp. 168-180, 354-369.

14. **Bromwitch, T. J. I'a.** *Introduction to the theory of infinite series*, London, (1908).

15. **Cantor, G.** Ueber ein neues und allgemeines Condensationsprincip der Singularitäten von Funktionen, *Math. Ann.*, Vol. 19, (1882), pp. 588-594.

16. **Caratheodory C.** *Reelle Funktionen*, Berlin (1927).

17. **Cellerier, C.** Note sur les principes fondamentaux de l'analyse, *Bull. d. Science Math.*, Ser. (2), Vol. 14, (1890), pp. 142-160.

18. **Cesaro, E.** Remarques sur la courbe de von Koch, *Atti Accad. Napoli* (2), Vol. 10, (1905), No. 15, pp. 1-12.

19. — Fonctions continues sans dérivées, *Arch. für Math. und Physik* (3), Vol. 10, (1906), pp. 57-63.

20. **Darboux, G.** Mémoire sur les fonctions discontinues, *Ann. Ecole Normale* (2), Vol. 4, (1875), pp. 57-112, & Addition, etc., Vol. 8, (1897), pp. 195-202.

21. **Denjoy, A.** Mémoire sur les nombres dérivés des fonctions continues, *Journ. de Math.* (7), Vol. 1, (1915), pp. 105-240.

22. **Dini, U.** Sur alcuni funzioni che tutto un intervallo non hanno mai derivata, *Annali di Mat.* (2), Vol. 8, (1877), pp. 121-137.

23. **Dini, U. & Lüroth-Schepp,** *Grundlagen für eine Theorie der Funktionen einer Veränderlichen reellen Grösse*, Leipzig, (1892).

24. **Du Bois-Reymond, P.** Versuch einer Klassification der willkürlichen Funktionen reeller Argumente, *Journ. für Math.*, Vol. 79, (1875), pp. 21-37.

25. **Faber, G.** Ueber stetige Funktionen I, *Math. Ann.*, Vol. 66, (1908), pp. 81-93.

26. — Ueber stetige Funktionen II, *Math. Ann.*, Vol. 69, (1910), pp. 372-243.

27. — Einfaches Beispiel einer stetigen nirgends differenzierbaren Funktion, *Deutsch. Math. Ver.*, Vol. 16, (1907), pp. 538-540.

28. **Falanga, A.** La funzioni di Cellerier, *Giorn. di Mat.*, Vol. 59, (1921), pp. 137-146.

29. **Fukuzawa, S.** Erweiterung des Satzes, betreffend die Weierstrasssche Funktion, *Proc. Tokyo Math. Phys. Soc.* (2), Vol. 4, (1907/8), pp. 202-207.

30. **Gilbert, P.** Mémoire sur l'existence de la dérivée dans les fonctions continues. Mémoires couronnés et autres mémoires publiés par l' Académie des Sciences de Belgique, Bruxelles, (1873), Vol. 3, pp. 1-31.

31. **Hahn, H.** Ueber stetige Funktionen ohne Ableitung, *Deutsch. Math. Ver.*, Vol. 26, (1918), pp. 281-284.

32. **Hardy, G. H.** Weierstrass' non-differentiable function, *Trans. American Math. Soc.*, Vol. 17, (1906), pp. 301-325.

33. **Hardy, G. H. & Littlewood, J. E.** A further note on the converse of Abel's Theorem, *Proc. Lond. Math. Soc.* (2), Vol. 25, (1926), pp. 219-236.

34. **Hankel, H.** Untersuchungen über die unendlich oft oscillierenden und unstetigen Funktionen, *Math. Ann.*, Vol. 20, (1882), pp. 63-112.

35. **Hilbert, D.** Ueber stetige Abbildung einer Linie auf ein Flächenstück, *Math. Ann.*, Vol. 38, (1891), p. 459.

36. **Hobson, E. W.** *The Theory of functions of a Real Variable*, Vol. II, Cambridge, (1926).

37. **Jasek, M.** Aus dem handschriftlichen Nachlass Bernard Bolzano, *Sitzungsber. d. kön. Böhm. Ges. d. Wiss.*, Cl. 2, (1920-21), pp. 1-32.

38. — Bolzano's Function, *Casôpis*, Vol. 51, (1922), pp. 69-76.

39. **Kaufmann, B.** Parameterkurven ohne Halbtangenten, *Sitzungsber. der Heidelberger Akademie der Wissenschaften*, Vol. 6, (1931), pp. 3-9.

40. **Klein, F.** *Anwendungen der Differentialrechnung und Integralrechnung auf die Geometrie*, Leipzig, (1902).

41. **Knopp, K.** Einfaches Beispiel einer stetigen nirgends differenzierbaren Funktion, *Deutsch. Math. Ver.*, Vol. 26, (1917/18), pp. 278-280.

42. — Einheitliche Erzeugung und Darstellung der Kurven von Peano, Osgood und V. Koch, *Arch. Math. Phys.* (3), Vol 26, (1917), pp. 103-115.

43. Eine einheitliche Erzeugungsart stetiger nirgends differenzierbarer Funktionen, *Sitzungsber. d. Berl. Math. Ges.*, Vol. 16, (1917), pp. 97-106.

44. — Ein einfaches Verfahren zur Bildung stetiger nirgends differenzierbarer Funktionen, *Math. Zeit.*, Vol. 2, (1918), pp. 1-26.

45. **Koch, H. von.** Sur une courbe continue sans tangente, obtenue par une construction géométrique élémentaire, *Arkiv för Mat. Astr. och Fysik*, Vol. 1, (1903/4), pp. 681-702.

46. — Une méthode géométrique élémentaire pour l'étude de certaines questions de la théorie de courbes planes, *Acta Math.*, Vol. 30, (1906), pp. 145-174.

47. — Remarque sur une communication de M. Brodén, *Arkiv. för Mat. Astr. och Fysik*, Vol. 2, (1905-6), pp. 1-2.

48. **Koenig, J.** Ueber stetige Funktionen die innerhalb jedes Intervalls extreme Werte besitzen, *Monat. Math. Phys.*, Vol. 1, (1890), pp. 7-12.

49. **Kowalewski, G.** Bolzanos Verfahren zur Herstellung einer nirgends differentiierbaren stetigen Funktion, *Berichte Ges. Wiss.*, Leipzig. Vol. 74, (1922), pp. 91-95.

50 —Ueber Bolzanos Nichtdifferenzierbare stetige Funktion, *Acta Math.*, Vol. 44, (1923), pp. 315-319.

51. **Landsberg, G.** Ueber die Differenzierbarkeit stetiger Funktionen, *Deutsch. Math. Ver.*, Vol. 17, (1908), pp. 46-51.

52. **Lebesgue, H.** *Leçons sur l'intégration et la recherche des fonctions primitives*, Paris, (1928).

53. **Lerch, M.** Ueber die Nichtdifferenzierbarkeit gewisser Funktionen, *Journ. für Math.*, Vol. 103, (1888), pp. 126-138.

54. **Levy, B.** Ancora alcune osservazioni sulle funzioni derivata, *Atti accad. Linc. Rend.* (Rome) (5), Vol. 15, (1906), pp. 358-368.

55. **Misra, R. D.** On Hilbert's Curve, *Proc. Benares Math. Soc.*, Vol. 16, (1934).

56. **Mookerji, B.** On the limiting points of the zeros of a non-differentiable function first given by Dini, *Bull. Cal. Math. Soc.*, Vol. 22, (1930).

57. **Moore, E. H.** On certain crinkly curves, *Trans. Amer. Math. Soc.*, Vol. 1, (1900), pp. 72-90.

58. **Paley, R. E. A. C.** On some problems connected with Weierstrass's non-differentiable function, *Proc. Lond. Math. Soc.* (2), Vol. 31, (1930), pp. 301-328.

59. **Peano, G.** Sur une courbe, qui remplit toute une aire plane, *Math. Ann.*, Vol. 36, (1890), p. 157.

60. **Petr, K.** Priklad funkee spojite nemajici v zadnem bode derivace, *Casopis pro pestovani Math. fys.*, Vol. 49, (1920), pp. 25-31.

61. **Pierpont, J.** *Lectures on the Theory of Functions of a Real Variable*, Vol. II, New York, (1911).

62. **Pepper, E. D.** On continuous functions without a derivative, *Fundamenta Math.*, Vol. 12, (1928), pp. 244-253.

63. **Porter, M. B.** Derivateless continuous functions, *Bull. Amer. Math Soc.*, Vol. 25, (1919), pp. 176-180

64. **Prasad, B. N.** Remark on infinite derivates, *Proc. Benares Math. Soc.*, Vol. 3, (1922), pp. 25-28.

65. **Prasad, G.** On the existence of the mean differential coefficient of a continuous function, *Bull. Cal. Math. Soc.*, Vol. 3, (1911-12), pp. 53-54.

66. — Weierstrass, *Proc. Benares Math. Soc.*, Vol. 5, (1923), pp. 35-42.

67. — On the zeros of Weierstrass's non-differentiable function, *Proc. Benares Math. Soc.*, Vol. 11, (1929). pp. 1-8.

68. **Rajchman, A. & Saks, S.** Sur la dérivabilité des fonctions monotones, *Fund. Math.*, Vol. 4, (1923), pp. 204-215.

69. **Rosenthal, A.** Ueber die Singularitäten der reellen ebenen Kurven, *Math. Ann.*, Vol. 73, (1912), pp. 480-521.

70. **Rychlik, K.** Ueber eine Funktion aus Bolzanos handschriftlichen Nachlasse, *Sitzungsber. der k. Böhm. Ges. der Wiss.*, (1921-22).

71. **Saks, S.** Sur les nombres dérivés des fonctions, *Fundamenta Math.*, Vol. 5, (1924), pp. 98-104.

72. **Schoenflies, A.** *Bericht über die Mengenlehre*, Vol. 1, Leipzig, (1900).

73. **Schwarz, H. A.** Neues Beispiel einer stetigen nicht differenzierbaren Funktion, *Verhandlungen der Gesellschaft Schweizerischer Naturforscher*, Vol. 56, (1873), pp. 252-258; *Arch. des sciences naturelles*, (1873), pp. 33-38.

74. **Sellerio, A.** Le curve limiti di Poligonali che si deformano con leggi assegnata, *Rend. Circ. Mat. di Palermo*, Vol. 28, (1909), pp. 153-184.

75. **Sharma, J. L.** On the zeros of Weierstrass's non-differentiable function, *Proc. Ben. Math. Soc.*, Vol. 11, (1929).

76. **Sierpinski, W.** Arytmetyczny priklad funkcyi ciaglej nicrozniczkowalnej, *Wektor*, Vol. 3, (1914), pp. 337-343.

77. **Singh, A. N.** A note on Koch's Curve, *Tôhoku Math. Journ.*, Vol. 27, (1927), pp. 302-305.

78. — On space-filling curves, *Proc. Benares Math. Soc.*, Vol. 6, (1924), pp. 1-6.

79. — On some new types of non-differentiable functions, *Annals of Mathematics*, Vol. 28, (1927), pp. 472-476.

80. — On Bolzano's non-differentiable function, *Bull. de l'Acad. Polonaise des Sciences et des Lettres*, (1928), pp. 191-220.

81. — On the derivates of non-differentiable functions, *Proc. Benares Math. Soc.*, Vol. 11, (1929), pp.9-10.

82. — Some remarks concerning a result of Besicovitch, *Bull. Cal. Math. Soc.*, Vol. 20, (1928-29), pp. 239-244.

83. — On a class of arithmetic non-differentiable functions, *Annals of Mathematics*, Vol. 31, (1930), pp. 657-659.

84. — On the unenumerable zeros of Singh's non-differentiable function, *Bull. Cal. Math. Soc.*, Vol. 22, (1930), pp. 91-98.

85. — On the method of construction and some properties of a class of non-differentiable functions, *Proc. Ben. Math. Soc.*, Vol. 13, (1931), pp. 1-17.

86. — Remarks on a paper of Boris Kaufmann relating to Parameter Curves, *Bull. Cal. Math. Soc.*, Vol. 24, (1932), pp. 103-108.

87. — On a certain lemma with applications to the theory of derivates, *Bull. Cal. Math. Soc.*, Vol. 26, (1934), pp. 15-34.

88. **Steinhaus, H.** Sur l'existence de la dérivée, *Bull. Acad. Polonaise*, (1920/21), pp. 62-65.

89. **Steinitz, E.** Stetigkeit und Differentialquotient, *Math. Ann.*, Vol. 52, (1899), pp. 59-69.

90. **Takagi, T.** A simple example of continuous function without derived function, *Journ. of the Physical School in Tokyo*, Vol. 14, (1904), pp. 1-2.

91. **Tauber, A.** Ueber die Weierstrasssche Funktion, *Monat. Math. Phys.* Vol. 8, (1897), pp. 333-340.

92. **Titchmarsh, E. C.** *The Theory of Functions*, Oxford, (1932).

93. **Tonelli, L.** Sulla ricerca delle funzioni primitive, *Rend. Acad. Linc.* (5), Vol. 25. (1916), pp. 163-170.

94. **Townsend, E. J.** *Functions of Real Variables*, New York, (1928).

95. **Vollee, Poussin, Ch. J. d. la.** Sur la série de Weierstrass représentant une fonction sans dérivée *Ann. Soc. Scientif. Bruxelles*, Vol. 16A, (1891-92), pp. 57-62.

96. **Van der Waerden, B. L.** Ein einfaches Beispiel einer nichtdifferenzierbaren stetigen Funktion, *Math. Zeit.*, Vol. 32, (1930), pp. 474-5.

97. **Wiener, C.** Geometrische und analytische Untersuchung der Weierstrassschen Funktion, *Journ. für Math.*, Vol. 90, (1881), pp. 221-252.

98. **Young, G. C.** A note on derivates and differential coefficients, *Acta Math.*, Vol. 37, (1914), pp. 141-154.

99. — Sur les courbes sans tangentes, *Ens. Math.*, Vol. 17, (1915), p. 348.

100. — On infinite derivates, *Quart. Journ. Math.*, Vol. 47, (1916), pp. 148-163.

101. — On the derivates of a function, *Proc. Lond. Math. Soc.* (2), Vol. 15, (1916), pp. 360-384.

102. — Sur les nombres dérivés d'une fonction, *C. R.* Paris, Vol. 162, (1916), pp. 380-382.

103. **Young, W. H.** On the derivates of non-differentiable functions, *Messenger of Math.*, Vol. 38. (1908/9), pp. 65-69.

104. **Young, W. H. and G. C.** On the existence of a differential coefficient, *Proc. Lond. Math. Soc.* (2), Vol. 9, (1910/11), pp. 325-335

HOW TO DRAW A STRAIGHT LINE;

A

LECTURE ON LINKAGES.

HOW TO DRAW A STRAIGHT LINE;

A

LECTURE ON LINKAGES.

BY

A. B. KEMPE, B.A.,

OF THE INNER TEMPLE, ESQ.;
MEMBER OF THE COUNCIL OF THE LONDON MATHEMATICAL SOCIETY;
AND LATE SCHOLAR OF TRINITY COLLEGE, CAMBRIDGE.

WITH NUMEROUS ILLUSTRATIONS.

NOTICE.

THIS Lecture was one of the series delivered to science teachers last summer in connection with the Loan Collection of Scientific Apparatus. I have taken the opportunity afforded by its publication to slightly enlarge it and to add several notes. For the illustrations I am indebted to my brother, Mr. H. R. KEMPE, without whose able and indefatigable co-operation in drawing them and in constructing the models furnished by me to the Loan Collection I could hardly have undertaken the delivery of the Lecture, and still less its publication.

7, CROWN OFFICE ROW, TEMPLE,
January 16*th*, 1877.

HOW TO DRAW A STRAIGHT LINE:

A LECTURE ON LINKAGES.

THE great geometrician Euclid, before demonstrating to us the various propositions contained in his *Elements of Geometry,* requires that we should be able to effect certain processes. These *Postulates,* as the requirements are termed, may roughly be said to demand that we should be able to describe straight lines and circles. And so great is the veneration that is paid to this master-geometrician, that there are many who would refuse the designation of "geometrical" to a demonstration which requires any other construction than can be effected by straight lines and circles. Hence many problems—such as, for example, the trisection of an angle—which can readily be effected by employing other simple means, are said to have no geometrical solution, since they cannot be solved by straight lines and circles only.

It becomes then interesting to inquire how we can effect these preliminary requirements, how we can describe these circles and these straight lines, with as much accuracy as the physical circumstances of the problems will admit of.

As regards the circle we encounter no difficulty. Taking Euclid's definition, and assuming, as of course we must, that our surface on which we wish to describe the circle is a plane, (1)[1] we see that we have only to make our tracing-point preserve a distance from the given centre of the circle constant and equal to the required radius. This can readily be effected by taking a flat piece of any form, such as the piece of cardboard I have here, and passing a pivot which is fixed to the given surface at the given centre through a hole in the piece, and a tracer or pencil through another hole in it whose distance from the first is equal to the given radius; we shall then, by moving the pencil, be able, even with this rude apparatus, to describe a circle with considerable accuracy and ease; and when we come to employ very small holes and pivots, or even large ones, turned with all that marvellous truth which the lathe affords, we shall get a result unequalled perhaps among mechanical apparatus for the smoothness and accuracy of its movement. The apparatus I have just described is of course nothing but a simple form of a pair of compasses, and it is usual to say that the third Postulate postulates the compasses.

But the straight line, how are we going to describe that? Euclid defines it as "lying evenly between its extreme points." This does not help us much. Our text-books say that the first and second Postulates postulate a ruler (2). But surely that is begging the question. If we are to draw a straight line with a ruler, the ruler must itself have a straight edge; and how are we going to make the edge straight? We come back to our starting-point.

Now I wish you clearly to understand the difference

[1] These figures refer to Notes at the end of the lecture.

between the method I just now employed for describing a circle, and the ruler method of describing a straight line. If I applied the ruler method to the description of a circle, I should take a circular lamina, such as a penny, and trace my circle by passing the pencil round the edge, and I should have the same difficulty that I had with the straight-edge, for I should first have to make the lamina itself circular. But the other method I employed involves no begging the question. I do not first assume that I have a circle and then use it to trace one, but simply require that the distance between two points shall be invariable. I am of course aware that we do employ circles in our simple compass, the pivot and the hole in the moving piece which it fits are such; but they are used not because they are the curves we want to describe (they are not so, but are of a different size), as is the case with the straight-edge, but because, through the impossibility of constructing pivots or holes of no finite dimensions, we are forced to adopt the best substitute we can for making one point in the moving piece remain at the same spot. If we employ a very small pivot and hole, though they be not truly circular, the error in the description of a circle of moderate dimensions will be practically infinitesimal, not perhaps varying beyond the width of the thinnest line which the tracer can be made to describe; and even when we employ large pivots and holes we shall get results as accurate, because those pivots and holes may be made by the employment of very small ones in the machine which makes them.

It appears then, that although we have an easy and accurate method of describing a circle, we have at first sight no corresponding means of describing a straight line; and there would seem to be a substantial difficulty in producing what mathematicians call the simplest curve, so that the question how to get over that difficulty becomes one of a decided theoretical interest.

Nor is the interest theoretical only, for the question is one of direct importance to the practical mechanician. In a large number of machines and scientific apparatus it is requisite that some point or points should move accurately in a straight line with as little friction as possible. If the ruler principle is adopted, and the point is kept in its path by guides, we have, besides the initial difficulty of making the guides truly straight, the wear and tear produced by the friction of the sliding surfaces, and the deformation produced by changes of temperature and varying strains. It becomes therefore of real consequence to obtain, if possible, some method which shall not involve these objectionable features, but possess the accuracy and ease of movement which characterise our circle-producing apparatus.

Turning to that apparatus, we notice that all that is requisite to draw with accuracy a circle of any given radius is to have the distance between the pivot and the tracer properly determined, and if I pivot a second "piece" to the fixed surface at a second point having a tracer as the first piece has, by properly determining the distance between the second tracer and pivot, I can describe a second circle whose radius bears any proportion I please to that of the first circle. Now, removing the tracers, let me pivot a third piece to these two *radial* pieces, as I may call them, at the points where the tracers were, and let me fix a tracer at any point on this third or *traversing* piece. You will at once see that if the radial pieces were big enough the tracer would describe circles or portions of circles on *them*, though they are in motion, with the same ease and accuracy as in the case of the simple circle-drawing apparatus ; the tracer will not however describe a circle on the *fixed* surface, but a complicated curve.

Fig. 1.

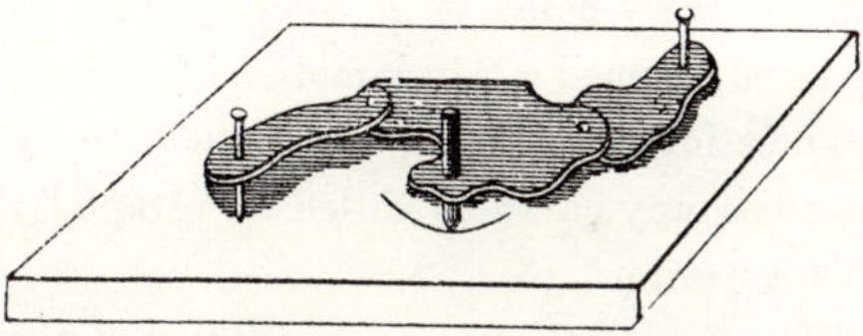

This curve will, however, be described with all the ease and accuracy of movement with which the circles were described, and if I wish to reproduce in a second apparatus the curves which I produce with this, I have only to get the distances between the pivots and tracers accurately the same in both cases, and the curves will also be accurately the same. I could of course go on adding fresh pieces *ad libitum*, and I should get points on the structure produced, describing in general very complicated curves, but with the same results as to accuracy and smoothness, *the reproduction of any particular curve depending solely on the correct determination of a certain definite number of distances.*

These systems, built up of pieces pointed or pivoted together, and turning about pivots attached to a fixed base, so that the various points on the pieces all describe definite curves, I shall term "link-motions," the pieces being termed "links." As, however, it sometimes facilitates the consideration of the properties of these structures to regard them apart from the base to which they are pivoted, the word "linkage" is employed to denote any combination of pieces pivoted together. When such a combination is pivoted in any way to a fixed base, the motion of points on it not being necessarily confined to fixed paths, the link-structure is called a "linkwork:" a "linkwork" in which the motion of every point is in some definite path being, as before stated, termed a "link-motion." I shall only add to these expressions two more: the point of a link-motion which describes any curve is called a "graph," the curve being called a "gram" (3).

The consideration of the various properties of these "linkages" has occupied much attention of late years among mathematicians, and is a subject of much complexity and difficulty. With the purely mathematical side of the question I do not, however, propose to deal to-day, as we shall have quite enough to do if we confine our attention to the practical results which mathematicians have obtained, and which I believe only mathematicians could have obtained. That these results are valuable cannot I think be doubted, though it may well be that their great beauty has led some to attribute to them an importance which they do not really possess; and it may be that fifty years ago they would have had a value which, through the great improvements that modern mechanicians have effected in the production of true planes, rulers and other exact mechanical structures, cannot now be ascribed to them. But linkages have not at present, I think, been sufficiently put before the mechanician to enable us to say what value should really be set upon them.

The practical results obtained by the use of linkages are but few in number, and are closely connected with the problem of "straight-line motion," having in fact been discovered during the investigation of that problem, and I shall be naturally led to consider them if I make "straight-line motion" the backbone of my lecture. Before, however, plunging into the midst of these linkages it will be useful to know how we can practically construct such models as we require; and here is one of the great advantages of our subject—we can get our results visibly before us so very easily. Pins for fixed pivots, cards for links, string or cotton for the other pivots, and a dining-room table, or a drawing-board if the former be thought objectionable, for a fixed base, are all we require. If something more artistic be preferred, the plan adopted in the models exhibited by me in the Loan Collection can be employed. The models were constructed by my brother, Mr. H. R. Kempe, in the following way. The bases are thin deal boards painted black; the links are neatly shaped out of thick cardboard (it is hard work making them, you have to sharpen your knife about every ten minutes, as the cardboard turns the edge very rapidly); the pivots are little rivets made of catgut, the heads being formed by pressing the face of a heated steel chisel on the ends of the gut after it is passed through the holes in the links; this gives a very firm and smoothly-working joint. More durable links may be made of tin-plate; the pivot-holes must in this case be punched, and the eyelets used by bootmakers for laced boots employed as pivots; you can get the proper tools at a trifling expense at any large tool shop.

Now, as I have said, the curves described by the various points on these link-motions are in general very complex. But they are not necessarily so. By properly choosing the distances at our disposal we can make them very simple. But can we go to the fullest extent of simplicity and get a point on one of them moving accurately in a straight line? That is what we are going to investigate.

To solve the problem with our single link is clearly impossible: all the points on it describe circles. We must therefore go to the next simple case—our three-link motion. In this case you will see that we have at our disposal the distance between the fixed pivots, the distances between the pivots on the radial links, the distance between the pivots on the traversing link, and the distances of the tracer from those pivots; in all six different distances. Can we choose those distances so that our tracing-point shall move in a straight line?

The first person who investigated this was that great man James Watt. "Watt's Parallel Motion" (4), invented in 1784, is well known to every engineer, and is employed in nearly every beam-engine. The apparatus, reduced to its simplest form, is shown in Fig. 2.

The radial bars are of equal length,—I employ the word "length" for brevity, to denote the distance between the pivots; the links of course may be of any length or shape,—and the distance between the pivots or the traversing link is such that when the radial bars are parallel the line joining those pivots is perpendicular to the radial bars. The tracing-point is situate half-way between the pivots on the traversing piece. The curve described by the tracer is, if the apparatus does not deviate much from its mean position, approximately a straight line. The reason of this is that the circles described by the extremities of the radial bars have their concavities turned in opposite directions, and the tracer being half-way between, describes a curve which is concave neither one way nor the other, and is therefore a straight line. The curve is not, however, accurately straight, for if I allow the tracer to describe the whole path it is capable of describing, it will, when it gets some distance from its mean position, deviate consider-

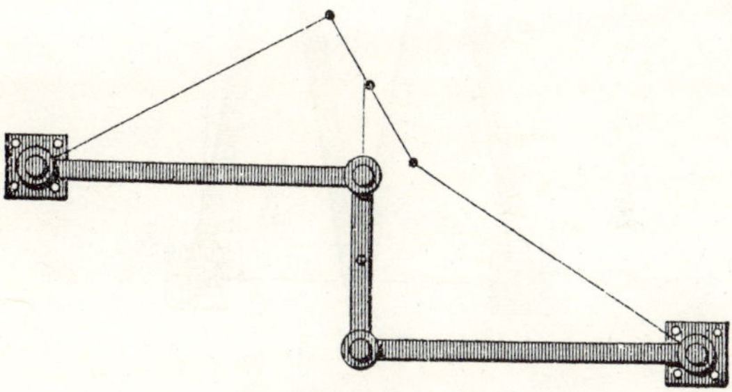

Fig. 2.

ably from the straight line, and will be found to describe a figure 8, the portions at the crossing being nearly straight. We know that they are not quite straight, because it is impossible to have such a curve partly straight and partly curved.

For many purposes the straight line described by Watt's apparatus is sufficiently accurate, but if we require an exact one it will, of course, not do, and we must try again. Now it is capable of proof that it is impossible to solve the problem with three moving links; closer approximations to the truth than that given by Watt can be obtained, but still not actual truth.

I have here some examples of these closer approximations. The first of these, shown in Fig. 3, is due to Richard Roberts of Manchester.

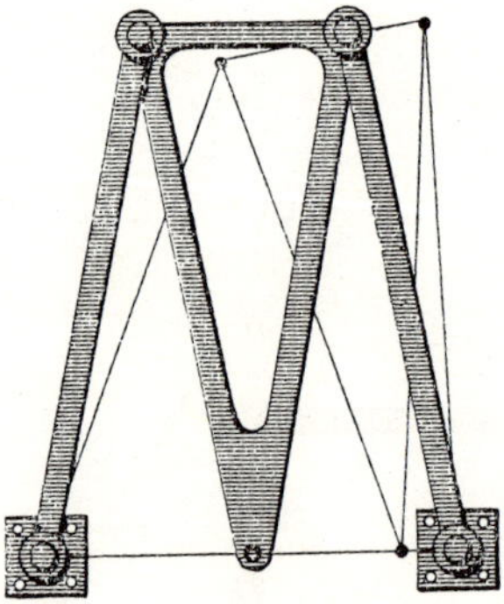

Fig. 3.

The radial bars are of equal length, the distance between the fixed pivots is twice that of the pivots on the traversing piece, and the tracer is situate on the traversing piece, at a distance from the pivots on it equal to the lengths of the radial bars. The tracer in consequence coincides with the straight line joining the fixed pivots at those pivots and half-way between them. It does not, however, coincide at any other points, but deviates very slightly between the fixed pivots. The path described by the tracer when it passes the pivots altogether deviates from the straight line.

The other apparatus was invented by Professor Tchebicheff of St. Petersburg. It is shown in Fig. 4. The radial bars are equal in length, being each in my little model five inches long. The distance between the fixed pivots must then be four inches, and the distance between the pivots or the traversing bar two inches. The tracer is taken half-way between these last. If now we draw a straight line—I had forgotten that we cannot do that

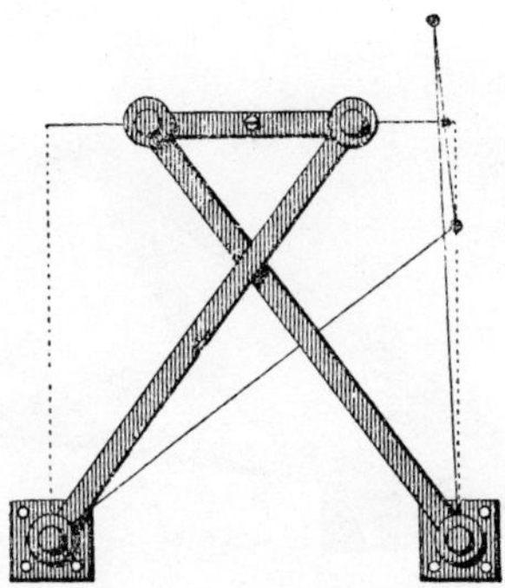

Fig. 4.

yet, well, if we draw a straight line, popularly so called —through the tracer in its mean position, as shown in the figure, parallel to that forming the fixed pivots, it will be found that the tracer will coincide with that line at the points where verticals through the fixed pivots cut it as well as at the mean position, but, as in the case of Roberts's parallel motion, it coincides nowhere else, though its deviation is very small as long as it remains between the verticals.

We have failed then with three links, and we must go on to the next case, a five-link motion—for you will observe that we must have an odd number of links if we want an apparatus describing definite curves. Can we solve the problem with five? Well, we can; but this was not the first accurate parallel motion discovered, and we must give the first inventor his due (although he did not find the simplest way) and proceed in strict chronological order.

In 1864, eighty years after Watt's discovery, the problem was first solved by M. Peaucellier, an officer of Engineers

Fig. 5.

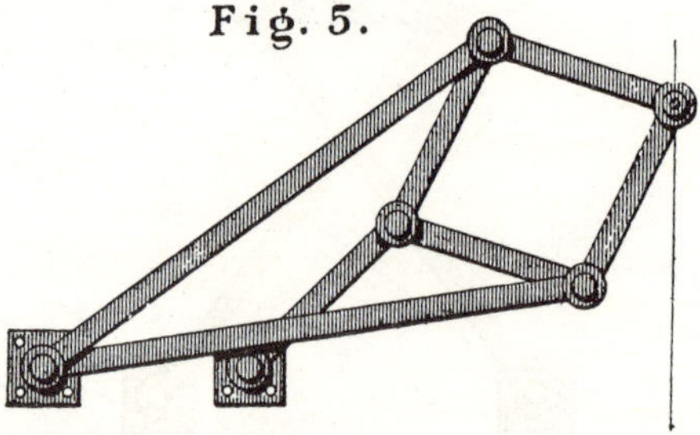

in the French army. His discovery was not at first estimated at its true value, fell almost into oblivion, and was rediscovered by a Russian student named Lipkin, who got a substantial reward from the Russian Government for his supposed originality. However, M. Peaucellier's merit has at last been recognized, and he has been awarded the great mechanical prize of the Institute of France, the "Prix Montyon."

M. Peaucellier's apparatus is shown in Fig. 5. It has, as you see, seven pieces or links. There are first of all two long links of equal length. These are both pivoted at the same fixed point; their other extremities are pivoted to opposite angles of a rhombus composed of four equal shorter links. The portion of the apparatus I have thus far described, considered apart from the fixed base, is a linkage termed a "Peaucellier cell." We then take an *extra* link, and pivot it to a fixed point whose distance from the first fixed point, that to which the cell is pivoted, is the same as the length of the extra link; the other end of the extra link is then pivoted to one of the free angles of the rhombus; the other free angle of the rhombus has a pencil at its pivot. That pencil will accurately describe a straight line.

I must now indulge in a little simple geometry. It is absolutely necessary that I should do so in order that you may understand the principle of our apparatus.

In Fig. 6, $Q\,C$ is the extra link pivoted to the fixed point Q, the other pivot on it C, describing the circle $O\,C\,R$. The straight lines $P\,M$ and $P'\,M'$ are supposed to be perpendicular to $M\,R\,Q\,O\,M'$.

Now the angle $O\,C\,R$, being the angle in a semicircle, is a right angle. Therefore the triangles $O\,C\,R$, $O\,M\,P$ are similar. Therefore,

$$O\,C : O\,R :: O\,M : O\,P.$$

Therefore,

$$OC \cdot OP = OM \cdot OR,$$

wherever C may be on the circle. That is, since $O\,M$ and $O\,R$ are both constant, if while C moves in a circle P moves

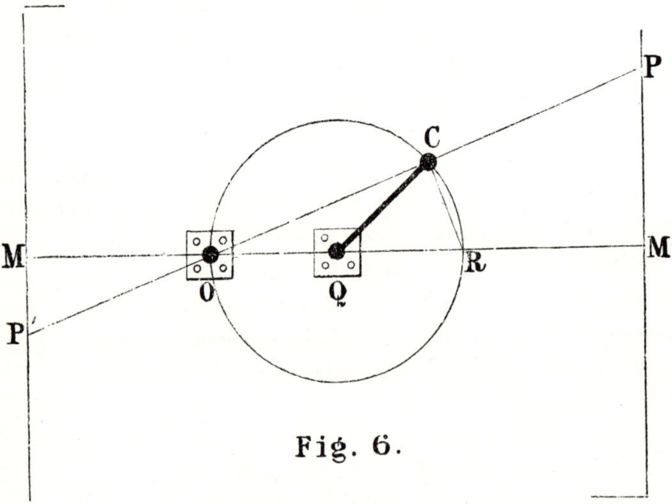

Fig. 6.

so that O, C, P are always in the same straight line, and so that $OC \cdot OP$ is always constant; then P will describe the straight line $P\ M$ perpendicular to the line $O\ Q$.

It is also clear that if we take the point P' on the other side of O, and if $OC \cdot OP'$ is constant P' will describe the straight line $P'M'$. This will be seen presently to be important.

Now, turning to Fig. 7, which is a skeleton drawing of the Peaucellier cell, we see that from the symmetry of the construction of the cell, O, C, P, all lie in the same straigh line, and if the straight line $A\ n$ be drawn perpendicular to $C\ P$—it must still be an imaginary one, as we have not proved yet that our apparatus does draw a straight line—Cn is equal to nP.

Now,

$$O\,A^2 = On^2 + An^2$$
$$A\,P^2 = Pn^2 + An^2$$

therefore,

$$\begin{aligned} O\,A^2 - A\,P^2 &= On^2 - Pn^2 \\ &= [On - Pn]\cdot[On + Pn] \\ &= OC\cdot OP. \end{aligned}$$

Thus since $O\,A$ and $A\,P$ are both constant $OC\cdot OP$ is always constant, however far or near C and P may be to O. If then the pivot O be fixed to the point O in Fig. 6, and the pivot C be made to describe the circle in the figure by being pivoted to the end of the extra link, the pivot P will

Fig. 7.

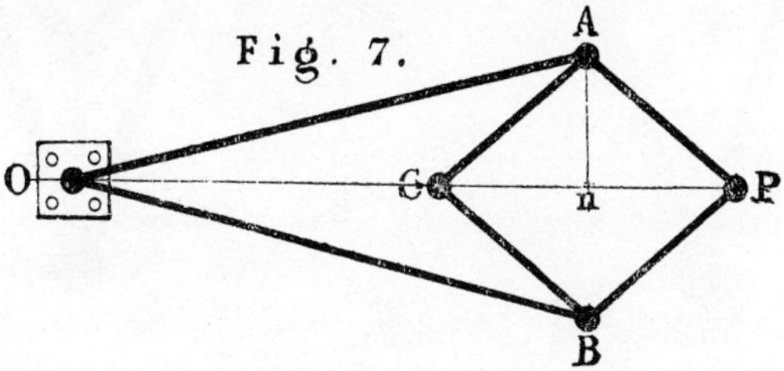

satisfy all the conditions necessary to make it move in a straight line, and if a pencil be fixed at P it will draw a straight line. The distance of the line from the fixed pivots will of course depend on the magnitude of the quantity $OA^2 - OP^2$ which may be varied at pleasure.

I hope you clearly understand the two elements composing the apparatus, the extra link and the cell, and the part each plays, as I now wish to describe to you some modifications of the cell. The extra link will remain the same as before, and it is only the cell which will undergo alteration.

If I take the two linkages in Fig. 8, which are known as the "kite", and the "spear-head," and place one on the other so that the long links of the one coincide with those of the other, and then amalgamate the coincident long links together, we shall get the original cell of Figs. 5 and 7. If then we keep the angles between the long links, or

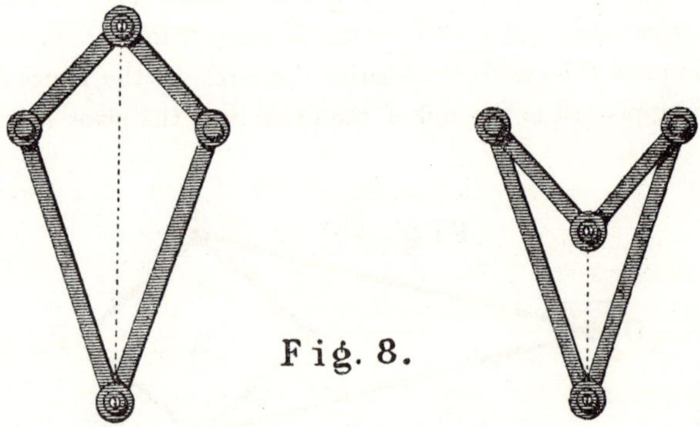

Fig. 8.

that between the short links, the same in the "kite" and "spear-head," we see that the height of the "kite" multiplied by that of the "spear-head" is constant.

Let us now, instead of amalgamating the long links of the two linkages, amalgamate the short ones. We then get the linkage of Fig. 9; and if the pivot where the short links meet is fixed, and one of the other free pivots be made to move in the circle of Fig. 6 by the extra link, the other will describe, not the straight line $P M$, but the straight line $P' M'$. In this form, which is a very compact one, the motion has been applied in a beautiful manner to the air engines which are employed to ventilate the

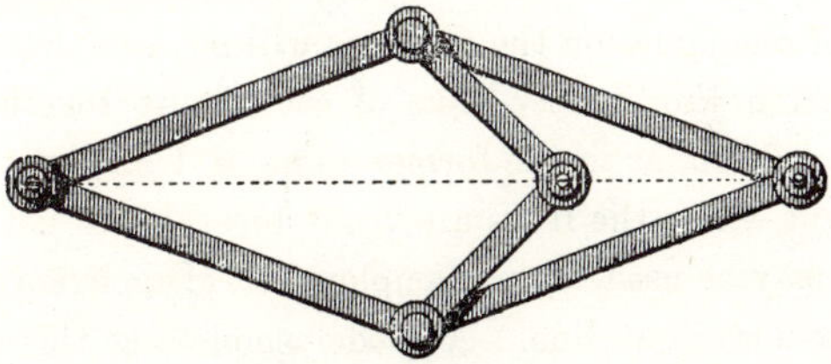

Fig. 9.

Houses of Parliament. The ease of working and absence of friction and noise is very remarkable. The engines were constructed and the Peaucellier apparatus adapted to them by Mr. Prim, the engineer to the Houses, by whose courtesy I have been enabled to see them, and I can assure you that they are well worth a visit.

Another modification of the cell is shown in Fig. 10. If instead of employing a "kite" and "spear-head" of the

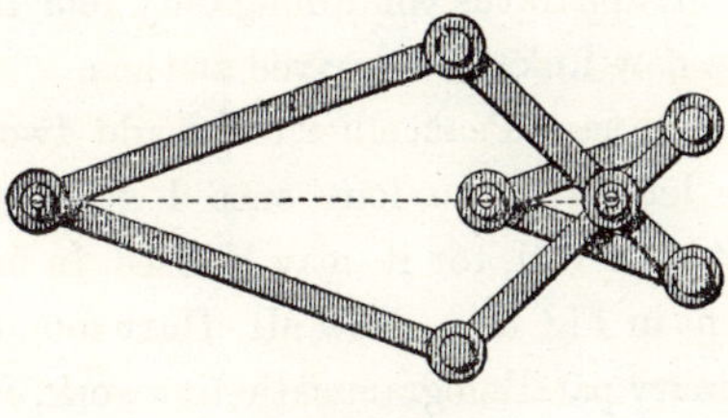

Fig. 10.

same dimensions, I take the same "kite" as before, but use a "spear-head" of half the size of the former one, the angles being however kept the same, the product of the heights of the two figures will be half what it was before, but still constant. Now instead of superimposing the links of one figure on the other, it will be seen that in Fig. 10 I fasten the shorter links of each figure together, end to end. Then, as in the former cases, if I fix the pivot at the point where the links are fixed together, I get a cell which may be used, by the employment of an extra link, to describe a straight line. A model employing this form of cell is exhibited in the Loan Collection by the Conservatoire des Arts et Métiers of Paris, and is of exquisite workmanship; the pencil seems to swim along the straight line.

M. Peaucellier's discovery was introduced into England by Professor Sylvester in a lecture he delivered at the Royal Institution in January, 1874 (5), which excited very great interest and was the commencement of the consideration of the subject of linkages in this country.

In August of the same year Mr. Hart of Woolwich Academy read a paper at the British Association meeting (6), in which he showed that M. Peaucellier's cell could be replaced by an apparatus containing only four links instead of six. The new linkage is arrived at thus.

If to the ordinary Peaucellier cell I add two fresh links of the same length as the long ones I get the double, or rather quadruple cell, for it may be used in four different ways, shown in Fig. 11. Now Mr. Hart found that if he took an ordinary parallelogrammatic linkwork, in which the adjacent sides are unequal, and crossed the links so as to form what is called a contra-parallelogram, Fig. 12, and then took four points on the four links dividing the distances between the pivots in the same proportion, those four points had exactly the same properties as the four points of the double cell. That the four points always lie in a straight line is seen thus: considering the triangle

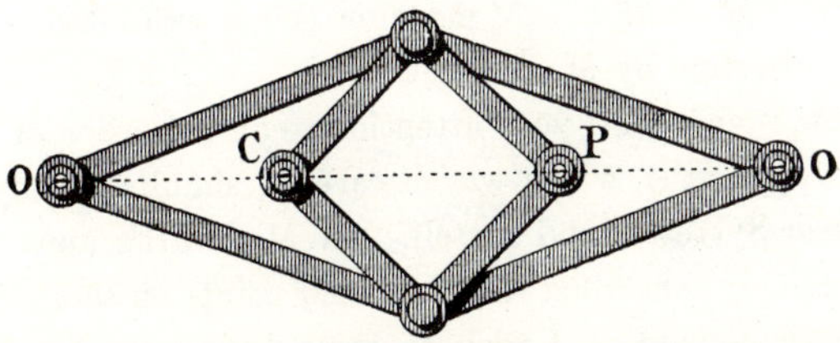

Fig. 11.

abd, since $aO : Ob : : aP : Pd$ therefore OP is parallel to bd, and the perpendicular distance between the parallels is to the height of the triangle *abd* as Ob is to ab; the same reasoning applies to the straight line CO', and since $ab : Ob : : cd : O'd$ and the heights of the triangles *abd*, *cbd*, are clearly

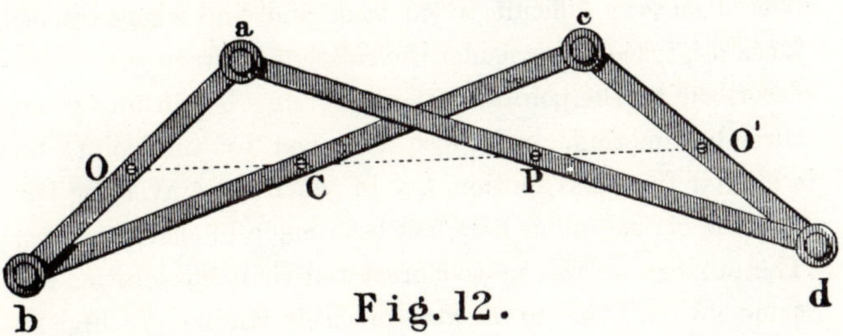

Fig. 12.

the same, therefore the distances of $O\ P$ and $O'C$ from $b\ d$ are the same, and $O\ C\ P\ O'$ lie in the same straight line.

That the product $OC \cdot OP$ is constant appears at once

when it is seen that ObC is half a "spear-head" and OaP half a "kite;" similarly it may be shown that $O'P \cdot O'C$ is constant, as also $OC \cdot CO'$ and $OP \cdot PO'$. Employing then the Hart's cell as we employed Peaucellier's, we get a five-link straight line motion. A model of this is exhibited in the Loan Collection by M. Breguet.

I now wish to call your attention to an extension of Mr. Hart's apparatus, which was discovered simultaneously by Professor Sylvester and myself. In Mr. Hart's apparatus we were only concerned with bars and points on those bars, but in the apparatus I wish to bring before you we have pieces instead of bars. I think it will be more interesting if I lead up to this apparatus by detailing to you its history, especially as I shall thereby be enabled to bring before you another very elegant and very important linkage—the discovery of Professor Sylvester.

When considering the problem presented by the ordinary three-*bar* motion consisting of two radial bars and a traversing bar, it occurred to me—I do not know how or why, it is often very difficult to go back and find whence one's ideas originate—to consider the relation between the curves described by the points on the traversing bar in any given three-bar motion, and those described by the points on a similar three-bar motion, but in which the traversing bar and one of the radial bars had been made to change places. The proposition was no sooner stated than the solution became obvious; the curves were precisely similar. In Fig. 13 let CD and BA be the two radial bars turning about the fixed centres C and B, and let DA be the traversing bar, and let P be any point on it describing a curve depending on the lengths of AB, BC, CD, and DA. Now add to the three-bar motion the bars *CE* and *EAP'*, *CE* being equal to *DA*, and *EA* equal to *CD*. *C D A E* is then a parallelogram, and if an imaginary line *CPP'* be drawn, cutting *EA* produced in *P'*, it will at once be seen that *P'* is a fixed point on *EA* produced, and *CP'* bears always a fixed proportion to *CP*, viz., *CD* : *CE*. Thus the curve described by *P'* is precisely the same as that described by *P*, only it is larger in

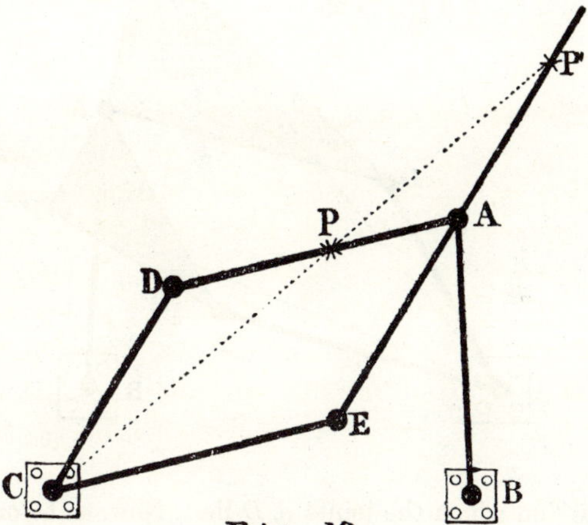

Fig. 13.

the proportion *CE* : *CD*. Thus if we take away the bars *CD* and *DA*, we shall get a three-bar linkwork, describing precisely the same curves, only of different magnitude, as our first three-bar motion described, and this new three-bar linkwork is the same as the old with the radial link *CD* and the traversing link *DA* interchanged (7).

On my communicating this result to Professor Sylvester, he at once saw that the property was one not confined to the particular case of points lying on the traversing bar, in fact to three-*bar* motion, but was possessed by three-*piece* motion. In Fig. 14 $C\ D\ A\ B$ is a three-bar motion, as in Fig. 13, but the tracing point or "graph" does not lie on the line joining the joints $A\ D$, but is anywhere else on a

Fig. 14.

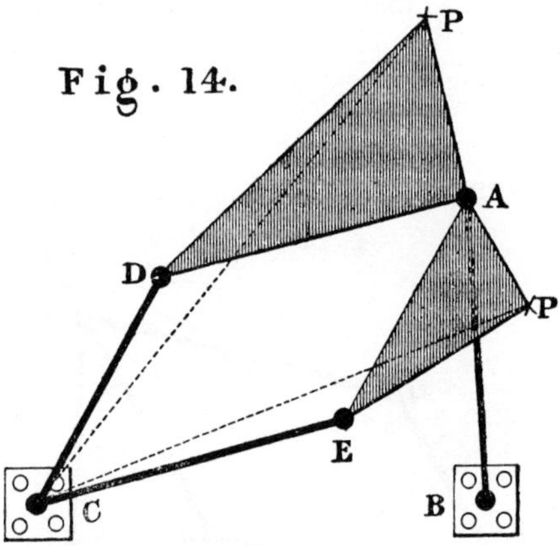

"piece" on which the joints $A\ D$ lie. Now, as before, add the bar $C\ E$, $C\ E$ being equal to $A\ D$, and the piece $A\ E\ P'$, making $A\ E$ equal to $C\ D$, and the triangle $A\ E\ P'$ similar to the triangle $P\ D\ A$; so that the angles $A\ E\ P'$, $A\ D\ P$ are equal, and

$$P'E : E\,A :: A\,D : D\,P.$$

It follows easily from this—you can work it out for yourselves without difficulty—that the ratio $P'C : P\ C$ is constant and the angle $P\ C\ P'$ is constant; thus the paths of P and P', or the "grams" described by the "graphs," P

and P', are similar, only they are of different sizes, and one is turned through an angle with respect to the other.

Now you will observe that the two proofs I have given are quite independent of the bar $A B$, which only affects the particular curve described by P and P'. If we get rid of $A B$, in both cases we shall get in the first figure the ordinary pantagraph, and in the second a beautiful extension of it, called by Professor Sylvester, its inventor, the *Plagiograph* or *Skew Pantagraph.* Like the Pantagraph, it will enlarge or reduce figures, but it will do more, it will turn them through any required angle, for by properly choosing the position of P and P', the ratio of $C P$ to $C P'$ can be made what we please, and also the angle $P C P'$ can be made to have any required value. If the angle $P C P'$ is made equal to 0 or 180°, we get the two forms of the pantagraph now in common use; if it be made to assume successively any value which is a submultiple of 360°, we can, by passing the point P each time over the same pattern make the point P' reproduce it round the fixed centre C after the fashion of a kaleidoscope. I think you will see from this that the instrument, which has, as far as I know, never been practically constructed, deserves to be put into the hands of the designer. I give here a picture of a little model of a possible form for the instrument furnished by me to the Loan Collection by request of Professor Sylvester (8).

After this discovery of Professor Sylvester, it occurred to him and to me simultaneously—our letters announcing our discovery to each other crossing in the post—that the principle of the plagiograph might be extended to Mr. Hart's contra-parallelogram; and this discovery I shall now proceed to explain to you. I shall, however, be more easily able to do so by approaching it in a different manner to that in which I did when I discovered it.

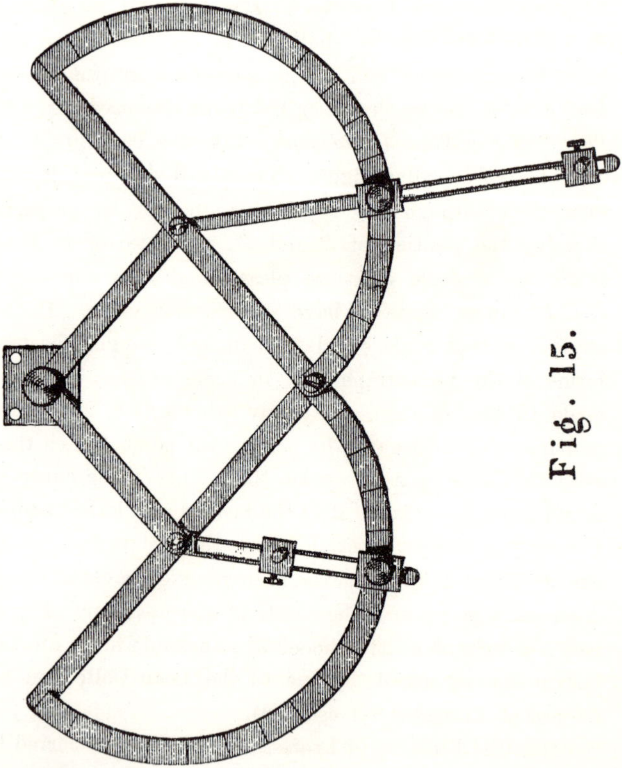

Fig. 15.

If we take the contra-parallelogram of Mr. Hart, and bend the links at the four points which lie on the same straight line, or *foci* as they are sometimes termed, through the same angle, the four points, instead of lying in the same straight line, will lie at the four angular points of a parallelogram of constant angles,—two the angle that the bars are bent through, and the other two their supplements—and of constant area, so that the product of two adjacent sides is constant.

In Fig. 16 the lettering is preserved as in Fig. 12, so that the way in which the apparatus is formed may be at once seen. The holes are taken in the middle of the links,

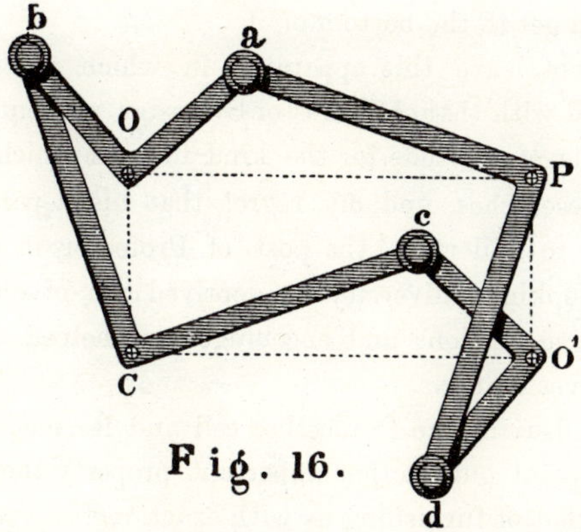

Fig. 16.

and the bending is through a right angle. The four holes *O P O′ C* lie at the four corners of a right-angled parallelogram, and the product of any two adjacent sides, as for example *O C·O P*, is constant. It follows that if *O* be pivoted to the fixed point *O* in Fig. 6, and *C* be pivoted to the extremity of the extra link, *P* will describe a straight line, not *P M*, but one inclined to *P M* at an angle the same as the bars are bent through, *i.e.* a right angle. Thus the straight line will be parallel to the line joining the fixed pivots O and Q. This apparatus, which for simplicity I have described as formed of four straight links which are afterwards bent, is of course strictly speaking formed of four plane links, such as those employed in Fig. 1, on which the various points are taken. This explains the name given to it by Professor Sylvester, the "Quadruplane." Its properties are not difficult to investigate, and when I point out to you that in Fig. 16, as in Fig. 12, Ob, bC form half a "spear-head," and Oa, aP half a "kite," you will very soon get to the bottom of it.

I cannot leave this apparatus, in which my name is associated with that of Professor Sylvester, without expressing my deep gratitude for the kind interest which he took in my researches, and my regret that his departure for America to undertake the post of Professor in the new Johns Hopkins University has deprived me of one whose valuable suggestions and encouragement helped me much in my investigations.

Before leaving the Peaucellier cell and its modifications, I must point out another important property they possess besides that of furnishing us with exact rectilinear motion. We have seen that our simplest linkwork enables us to describe a circle of any radius, and if we wished to describe one of ten miles' radius the proper course would be to have a ten-mile link, but as that would be, to say the least, cumbrous, it is satisfactory to know that we can effect our purpose with a much smaller apparatus. When the Peaucellier cell is mounted for the purpose of describing a straight line, as I told you, the distance between the fixed pivots must be the same as the length of the "extra" link.

If this distance be not the same we shall not get straight lines described by the pencil, but circles. If the difference be slight the circles described will be of enormous magnitude, decreasing in size as the difference increases. If the distance $Q\ O$, Fig. 6, be made greater than $Q\ C$, the convexity of the portion of the circle described by the pencil (for if the circles are large it will of course be only a portion which is described) will be towards O, if less the concavity. To a mathematician, who knows that the inverse of a circle is a circle, this will be clear, but it may not be amiss to give here a short proof of the proposition.

In Fig. 17 let the centres Q, Q' of the two circles be at distances from O proportional to the radii of the circles. If then $O\ D\ C\ P\ S$ be any straight line through O, $D\ Q$ will be parallel to $P\ Q'$, and $C\ Q$ to $S\ Q'$, and $O\ D$ will bear the same proportion to $O\ P$ that $O\ Q$ does to $O\ Q'$. Now considering the proof we gave in connection with Fig. 7, it will be clear that the product $O\ D \cdot O\ C$ is constant, and therefore since $O\ P$ bears a constant ratio to $O\ D$, $O\ P \cdot O\ C$ is constant. That is if $O\ C \cdot O\ P$ is constant and C describes a circle about Q, P will describe one about Q'. Taking then O, C and P as the O, C and P of the Peaucellier cell in Fig. 7, we see how P comes to describe a circle.

It is hardly necessary for me to state the importance of the Peaucellier compass in the mechanical arts for drawing circles of large radius. Of course the various modifications of the "cell" I have described may all be employed for the purpose. The models exhibited by the Conservatoire and M. Breguet are furnished with sliding pivots for the purpose of varying the distance between O and Q, and thus getting circles of any radius.

My attention was first called to these linkworks by the lecture of Professor Sylvester, to which I have referred.

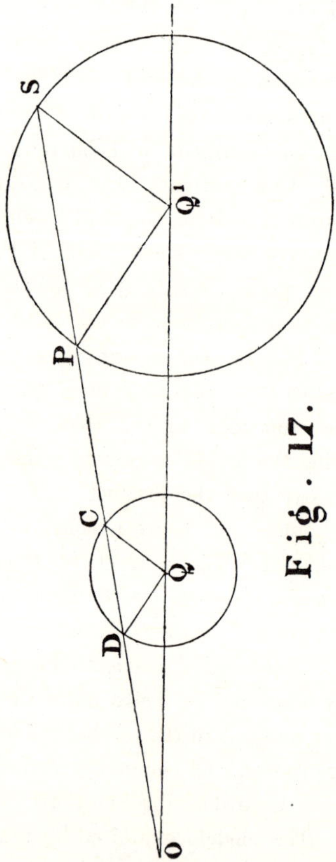

Fig. 17.

A passage in that lecture in which it was stated that there were probably other forms of seven-link parallel motions besides M. Peaucellier's, then the only one known, led me to investigate the subject, and I succeeded in obtaining some new parallel motions of an entirely different character to that of M. Peaucellier (9). I shall bring two of these to your notice, as the investigation of them will lead us to consider some other linkworks of importance.

If I take two kites, one twice as big as the other, such that the long links of each are twice the length of the short ones, and make one long link of the small kite lie on a short one of the large, and a short one of the small on a long one of the large, and then amalgamate the coincident links, I shall get the linkage shown in Fig. 18.

Fig. 18.

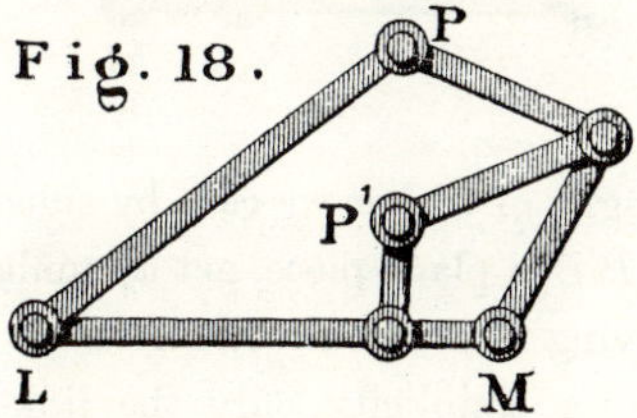

The important property of this linkage is that, although we can by moving the links about, make the points P and P' approach to or recede from each other, the imaginary line joining them is always perpendicular to that drawn through the pivots on the bottom link $L\ M$. It follows that if either of the pivots P or P' be fixed, and the link $L\ M$ be made to move so as always to remain parallel to a fixed line, the other point will describe a straight line perpendicular to the fixed line. Fig. 19 shows you the parallel motion made by fixing P'. It is unnecessary for me to point out how the parallelism of $L\ M$ is preserved by adding the link $S\ L$, it is obvious from the figure. The straight line which is described by the point P is perpendicular to the line joining the two fixed pivots; we can, however, without increasing the number of links, make a point on the linkwork describe a straight line inclined to the line

Fig. 19.

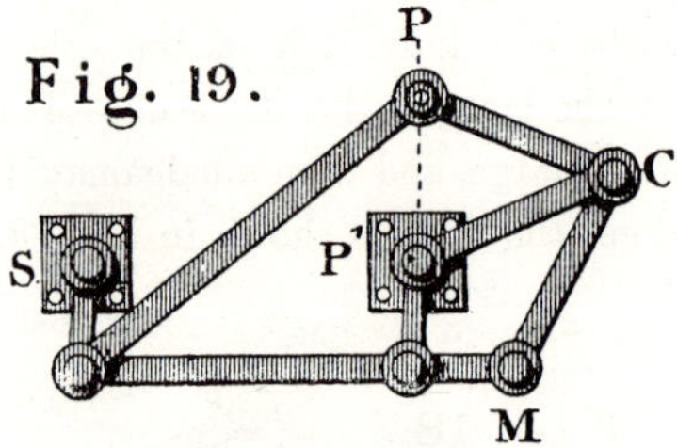

$S\ P$ at any angle, or rather we can, by substituting for the straight link $P\ C$ a plane piece, get a number of points on that piece moving in every direction.

In Fig. 20, for simplicity, only the link $C\ P'$ and the new piece substituted for the link $P\ C$ are shown. The new piece is circular and has holes pierced in it all at the same distance—the same as the lengths $P\ C$ and $P'\ C$—from C. Now we have seen from Fig. 19 that P moves in a vertical straight line, the distance $P\ C$ in Fig. 20 being the same as it was in Fig. 19; but from a well-known property of a circle, if H be any one of the holes pierced in the piece, the angle $H\ P'\ P$ is constant, thus the straight line $H\ P'$ is fixed in position, and H moves along it; similarly all the other holes move along in straight lines passing through the fixed pivot P', and we get straight line motion distributed in all directions. This species of motion is called by Professor Sylvester "tram-motion." It is worth noticing that the motion of the circular disc is the same as it would have been if the dotted circle on it rolled inside the large dotted circle; we have, in fact, White's parallel motion reproduced by linkwork. Of course, if we

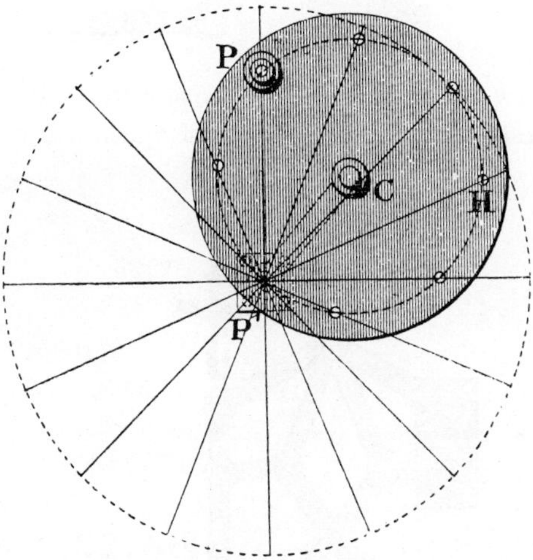

Fig. 20.

only require motion in one direction, we may cut away all the disc except a portion forming a bent arm containing *C*, *P*, and the point which moves in the required direction.

The double kite of Fig. 18 may be employed to form some other useful linkworks. It is often necessary to have, not a single point, but a whole piece moving so that all points on it move in straight lines. I may instance the slide rests in lathes, traversing tables, punches, drills, drawbridges, etc. The double kite enables us

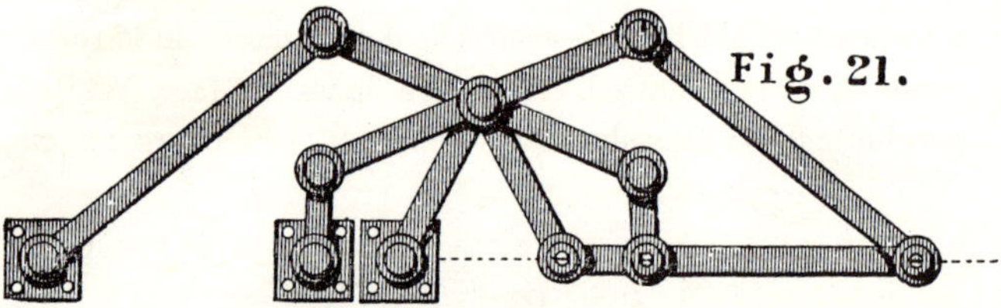

Fig. 21.

to produce linkworks having this property. In the linkwork of Fig. 21, the construction of which will be at once

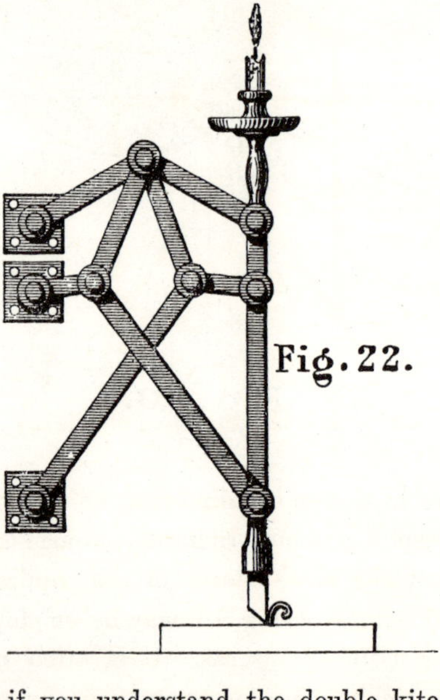

Fig. 22.

appreciated if you understand the double kite, the horizontal link moves to and fro as if sliding in a fixed horizontal straight tube. This form would possibly be useful as a girder for a drawbridge.

In the linkwork of Fig. 22, which is another combination of two double kites, the vertical link moves so that all its points move in horizontal straight lines. There is a modification of this linkwork which will, I think, be found interesting. In the linkage in Fig. 23, which, if the thin links are removed, is a skeleton drawing of Fig 22, let the dotted links be taken away and the thin ones be in-

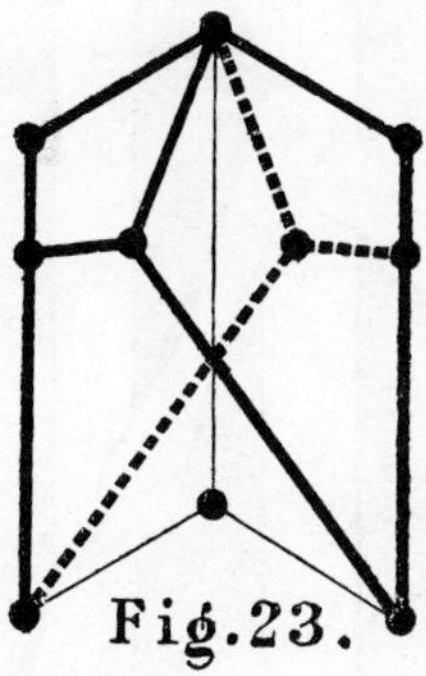

Fig. 23.

serted; we then get a linkage which has the same property as that in Fig. 22, but it is seen in its new form to be the ordinary double parallel ruler with three added links. Fig. 24 is a figure of a double parallel ruler made on this plan with a slight modification. If the bottom ruler be held horizontal the top moves vertically up and down the board, having no lateral movement.

While I am upon this sort of movement I may point out an apparatus exhibited in the Loan Collection by Professor

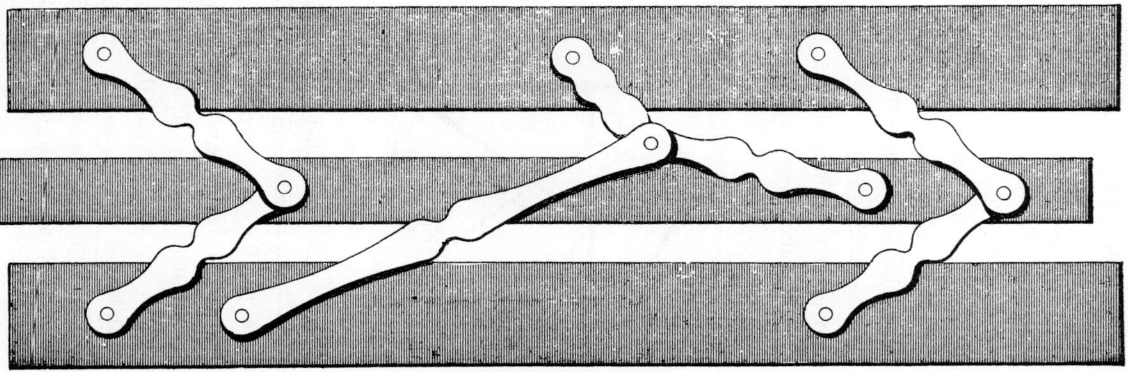

Fig. 24.

Tchebicheff, which bears a strong likeness to a complicated camp-stool, the seat of which has horizontal motion. The motion is not strictly rectilinear; the apparatus being—as will be seen by observing that the thin line in the figure

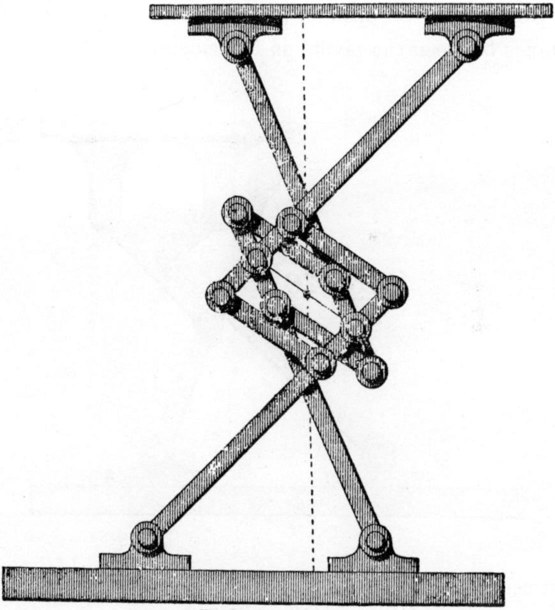

Fig. 25.

is of invariable length, and a link might therefore be put where it is—a combination of two of the parallel motions of Professor Tchebicheff given in Fig. 4, with some links added to keep the seat parallel with the base. The variation of the upper plane from a strictly horizontal movement is therefore double that of the tracer in the simple parallel motion.

Fig. 26 shows how a similar apparatus of much simpler construction, employing the Tchebicheff approximate parallel motion, can be made. The lengths of the links forming the parallel motion have been given before (Fig 4). The distance between the pivots on the moving seat is half that

Fig. 26.

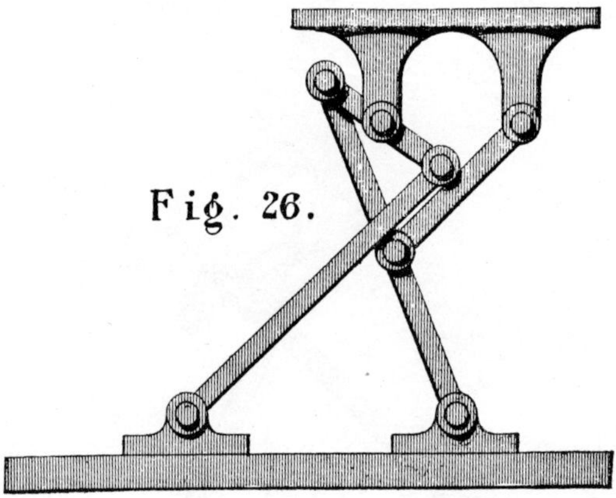

between the fixed pivots, and the length of the remaining link is one-half that of the radial links.

An *exact* motion of the same description is shown in Fig. 27. O, C, O', P are the four *foci* of the quadriplane shown in the figure in which the links are bent through a right angle, so that $O\,C\cdot O\,P$ is constant, and $C\,O\,P$ a right angle. The focus O is pivoted to a fixed point, and C is made by means of the extra link $Q\,C$ to move in a circle of which the radius $Q\,C$ is equal to the pivot distance O

P consequently moves in a straight line parallel to *O Q*, the five moving pieces thus far described constituting the Sylvester-Kempe parallel motion. To this are added the moving seat and the remaining link *R O′*, the pivot distances of which, *P R* and *R O′*, are equal to *O Q*. The seat in consequence always remains parallel to *Q O*, and as *P* moves accurately in a horizontal straight line, every point on

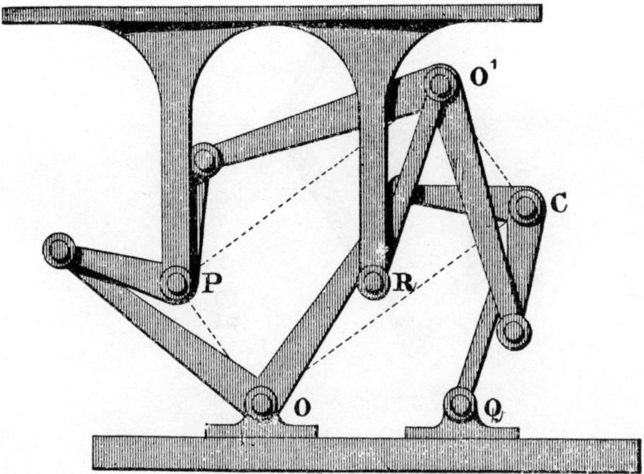

Fig 27.

it will do so also. This apparatus might be used with advantage where a very smoothly-working traversing table is required.

I now come to the second of the parallel motions I said I would show you. If I take a kite and pivot the blunt end to the fixed base, and make the sharp end move up and down in a straight line, passing through the fixed pivot, the short links will rotate about the fixed pivot with equal velocities in opposite directions; and, conversely, if the links rotate with equal velocity in opposite directions, the path of the sharp end will be a straight line, and the same will hold

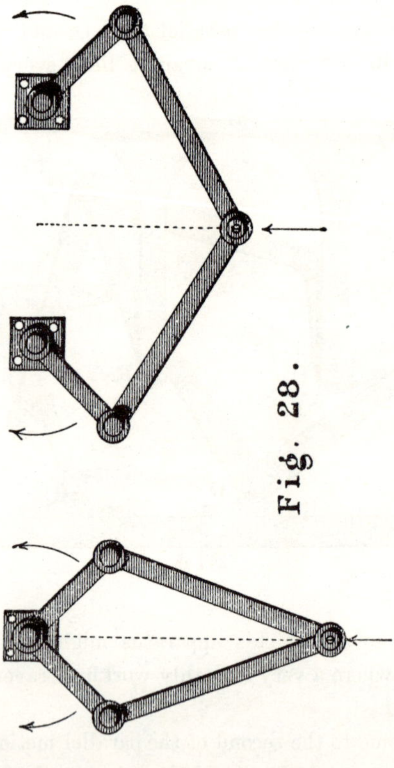

Fig. 28.

good if instead of the short links being pivoted to the same point they are pivoted to different ones.

To find a linkwork which should make two links rotate with equal velocities in opposite directions was one of the first problems I set myself to solve. There was no difficulty in making two links rotate with equal velocities in the same direction,—the ordinary parallelogrammatic linkwork em-

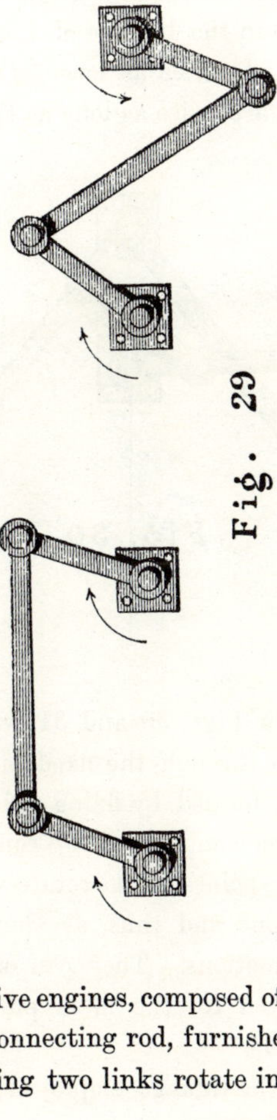

Fig. 29

ployed in locomotive engines, composed of the engine, the two cranks, and the connecting rod, furnished that; and there was none in making two links rotate in opposite directions with *varying* velocity; the contra-parallelogram gave that; but the required linkwork had to be discovered. After some trouble I succeeded in obtaining it by a combination of a large and small contra-parallelogram put together just as the two kites were in the linkage of Fig. 18. One contra-parallelogram is made twice as large as the other, and the long links of each are twice as long as the short (10).

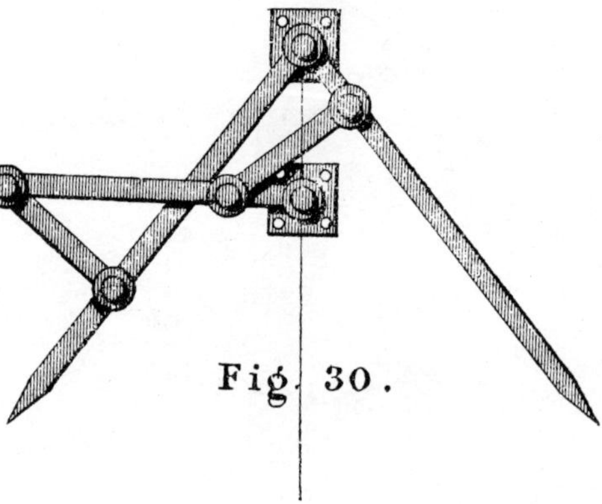

Fig. 30.

The linkworks in Figs. 30 and 31 will, by considering the thin line drawn through the fixed pivots in each as a link, be seen to be formed by fixing different links of the same six-link linkage composed of two contra-parallelograms as just stated. The pointed links rotate with equal velocity n opposite directions, and thus, as shown in Fig. 28, at once give parallel motions. They can of course, however, be usefully employed for the mere purpose of reversing angular velocity (11).

An extension of the linkage employed in these two last

figures gives us an apparatus of considerable interest. If I take another linkage contra-parallelogram of half the size of the smaller one and fit it to the smaller exactly

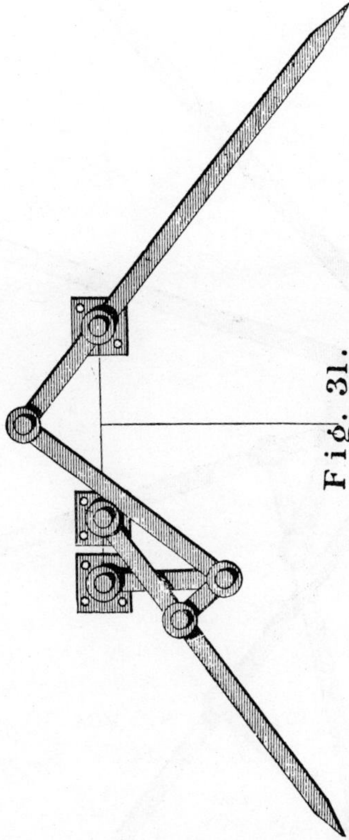

Fig. 31.

I fitted the smaller to the larger, I get the eight-linkage of Fig. 32. It has, you see, four pointed links radiating from a centre at equal angles; if I open out the two extreme ones to any desired angle, you will see that the two intermediate ones will exactly *trisect the angle.* Thus the power

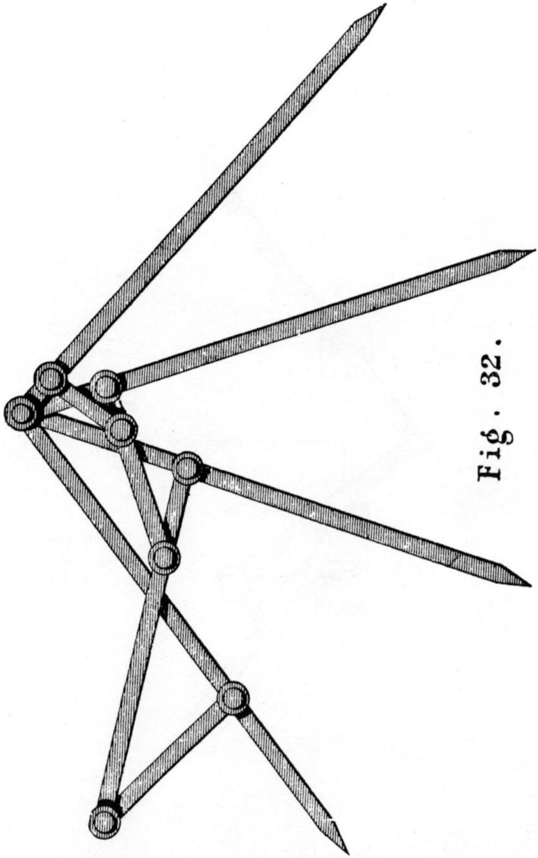

Fig. 32.

we have had to call into operation in order to effect Euclid's first Postulate—linkages—enables us to solve a problem which has no "geometrical" solution. I could of course

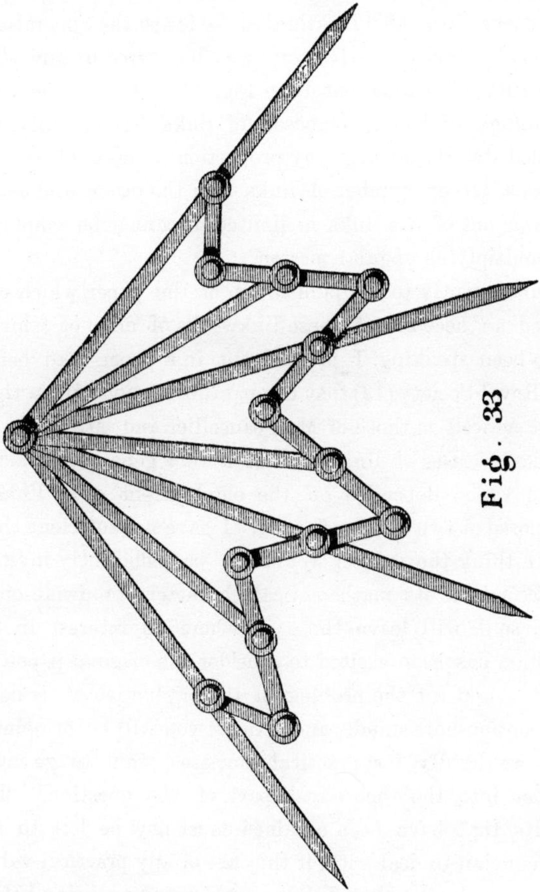

Fig. 33

go on extending my linkage and get others which would divide an angle into any number of equal parts. It is obvious that these same linkages can also be employed as linkworks for doubling, trebling, etc., angular velocity (12.)

Another form of "Isoklinostat"—for so the apparatus is termed by Professor Sylvester—was discovered by him. The construction is apparent from Fig. 33. It has the great advantage of being composed of links having only two pivot distances bearing any proportion to each other, but it has a larger number of links than the other, and as the opening out of the links is limited, it cannot be employed for multiplying angular motion.

Subsequently to the publication of the paper which contained an account of these linkworks of mine of which I have been speaking, I pointed out in a paper read before the Royal Society (13) that the parallel motions given there were, as well as those of M. Peaucellier and Mr. Hart, all particular cases of linkworks of a very general character, all of which depended on the employment of a linkage composed of two similar figures. I have not sufficient time, and I think the subject would not be sufficiently inviting on account of its mathematical character, to dwell on it here, so I will leave those in whom an interest in the question has been excited to consider the original paper.

At this point the problem of the production of straight-line motion now stands, and I think you will be of opinion that we hardly, for practical purposes, want to go much farther into the theoretical part of the question. The results that have been obtained must now be left to the mechanician to deal with, if they are of any practical value. I have, as far as what I have undertaken to bring before you to-day is concerned, come to the end of my tether. I have shown you that we *can* describe a straight line, and *how* we can, and the consideration of the problem has led us to investigate some important pieces of apparatus. But I hope that this is not all. I hope that I have shown you (and your attention makes that hope a belief) that this new field of investigation is one possessing great interest and importance. Mathematicians have no doubt done much more than I have been able to show you to-day (14), but the unexplored fields are still vast, and the earnest investigator can hardly fail to make new discoveries. I hope therefore that you whose duty it is to extend the domain of science will not let the subject drop with the close of my lecture.

NOTES.

(1) The hole through which the pencil passes can be made to describe a circle independently of any surface (see the latter part of Note 3), but when we wish to describe a circle or a given plane surface that surface must of course be assumed to be plane.

(2) "But" (it is carefully added) "not a graduated one." By th use of a ruler with only two graduations, an angle can, as is well known, be readily trisected, thus—Let RST be the angle, and let PP' be the points where the graduations cut the edge of the ruler. Let $2\,RS = PP'$. Draw RU parallel and RV perpendicular to ST. Then if we fit the ruler to the figure $RSTUV$ so that the edge PP' passes through S, P lies on RU and P' on RV, PP' trisects the angle RST. For if Q be the middle point of PP', and RQ be joined, the angle TSP = the angle QPR = the angle QRP = half the angle RQS, that is half the angle RSQ.

This solution is of course not a "geometrical" one in the sense I have indicated, because a graduated ruler and the fitting process are employed. But does Euclid confine himself to his three Postulates of construction? Does he not use a graduated ruler and this fitting process? Is not the side AB of the triangle ABC in Book I. Proposition 4, graduated at A and B, and are we not told to take it up and fit it on to DE?

It seems difficult to see why Euclid employed the second Postulate—that which requires "that a terminated straight line may be produced to any length in a straight line,"—or rather, why he did not put it among the propositions in the First Book as a problem. It is by no means difficult by a rigid adherence to Euclid's methods to find a point outside a terminated straight line which is in the same straight line with it, and to prove it to be so, without the employment of the second Postulate. That point can then, by the first Postulate, be joined to the extremity of the given straight line which is thus produced, and the process can be continued indefinitely, since by the third Postulate circles can be drawn with any centre and radius.

(3) It is important to notice that the fixed base to which the pivots are attached is really one link in the system. It would on that account be perhaps more scientific, in a general consideration of the subject, to commence by calling any combination of pieces (whether those pieces be cranks, beams, connecting-rods, or anything else) jointed or pivoted together, a "*linkage.*" When the motion of the links is confined to one plane or to a number of parallel planes, the system is called a "*plane linkage.*" (It will be seen that this lecture is confined to plane linkages ; a few remarks about solid linkages will be found at the end of the note.) The motion of the links among themselves in a linkage may be determinate or not. When the motion is determinate the number of links must be even, and the linkage is said to be "*complete.*" When the motion is not determinate the linkage is said to have 1, 2, 3, etc. degrees of freedom, according to the amount of liberty the links possess in their relative motion. These linkages may be termed "*primary,*" "*secondary,*" etc. linkages. Thus if we take the linkage composed of four links with two pivots on each, the motion of each link as regards the others is determinate, and the linkage is a "*complete linkage.*" If one link be jointed in the middle the linkage has one degree of liberty and is a "*primary linkage.*" So by making fresh joints "*secondary*" or "*tertiary,*" etc. linkages may be formed. These primary, etc. linkages may be formed in various other ways, but the example given will illustrate the reason for the nomenclature. When one link of a linkage is a fixed base the structure is called a "*linkwork.*" Linkworks, like linkages, may be "*primary,*" "*secondary,*" etc. A "*complete linkwork,*" *i.e.* one in which the motion of every point on the moving part of the structure is definite, is called a "*link-motion.*" The various "grams" described by these link-motions are very difficult to deal with. I have shown, in a paper in the *Proceedings of the London Mathematical Society,* 1876, that a link-motion can be found to describe any given algebraic curve, but the converse problem, "Given the link-motion, what is the curve ?" is one towards the solution of which but little way has been made; and the "tricircular trinodal sextics," which are the "grams" of the simple three-piece motion, are still under the consideration of some of our most eminent mathematicians.

Taking them in their greatest generality, the theoretically simplest form of link-motion is not the flat circle-producing link, but a solid link pivoted to a fixed centre, and capable of motion in all directions about that centre, so that all points on it describe spheres in space; and the most general form a number of such links pivoted together, forming a structure the various points on which describe surfaces. If two simple solid links, turning about two fixed centres, are pivoted together at a common point, that point will describe a circle independently of any plane surface, the other points on the links describing portions of spheres. The form of pivot which would have to be adopted in solid linkages would be the ball-and-socket joint, so that the links could not only move about round the fixed centre, but rotate about any imaginary axis through that centre. It is obvious that it would be impossible to construct any joint which would give the links perfect freedom of motion, as the fixed centre about which any link turned must be fastened to a fixed base in some way, and whatever means were adopted would interfere with the link in some portion of its path. This is not so in plane link-motions. The subject of solid linkages has been but little considered. Hooke's joint may be mentioned as an example of a solid link-motion. (See also Note 11.)

(4) I have been more than once asked to try and get rid of the objectionable term "parallel motion." I do not know how it came to be employed, and it certainly does not express what is intended. The apparatus does not give "parallel motion," but approximate "rectilinear motion." The expression, however, has now become crystallised, and I for one cannot undertake to find a solvent.

(5) See the *Proceedings of the Royal Institution*, 1874.

(6) This paper is printed *in extenso* in the *Cambridge Messenger of Mathematics*, 1875, vol. iv., pp. 82-116, and contains much valuable matter about the mathematical part of the subject.

(7) The interchange of a radial and traversing bar converts Watt's Parallel Motion into the Grasshopper Parallel Motion. The same change shows us that the curves traced by the linkwork formed by fixing one bar of a "kite" are the same as those traced by the linkwork formed by fixing one bar of a contra-parallelogram. This is interesting as showing that there is really only one case in which the sextic curve, the "gram" of three-bar motion, breaks up into a circle and a quartic.

(8) For a full account of this and the piece of apparatus next described, see *Nature*, vol. xii., pp. 168 and 214.

(9) See the *Messenger of Mathematics*, "On Some New Linkages," 1875, vol. iv., p. 121.

(10) A reference to the paper referred to in the last note will show that it is not necessary that the small kites and contra-parallelograms should be half the size of the large ones, or that the long links should be double the short; the particular lengths are chosen for ease of description in lecturing.

(11) By an arrangement of Hooke's joints, pure solid linkages, we can make two axes rotate with equal velocities in contrary directions (See Willis's *Principles of Mechanism*, 2nd Ed. sec. 516, p. 456), and therefore produce an exact parallel motion.

(12) The "kite" and the "contra-parallelogram" are subject to the inconvenience (mathematically very important) of having "dead points." These can be, however, readily got rid of by employin pins and gabs in the manner pointed out by Professor Reuleaux. (See Reuleaux's *Kinematics of Machinery*, translated by Professor Kennedy, Macmillan, pp. 290-294.)

(13) *Proceedings of the Royal Society*, No. 163, 1875, "On a General Method of Obtaining Exact Rectilinear Motion by Linkwork." I take this opportunity of pointing out that the results there arrived at may be greatly extended from the following simple consideration.

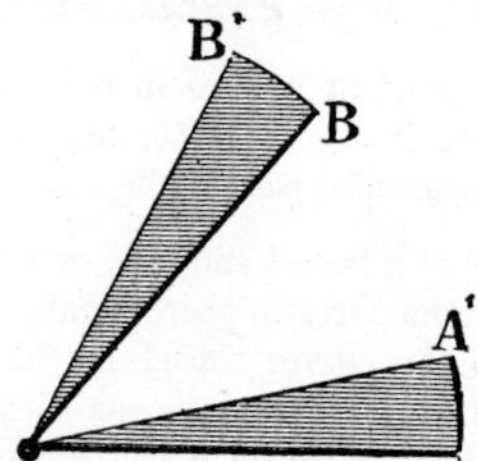

Fig. 34.

If the straight link $O\,B$ makes any angle D with the straight link $O\,A$, and if instead of employing the straight links we employ the pieces $A'OA$, $B'OB$, and if the angle $A'OA$ equals the angle $B'OB$, then the angle $B'OA'$ equals D. The recognition of this very obvious fact will enable us to derive the Sylvester-Kempe parallel motion from that of Mr. Hart.

(14) In addition to the authorities already mentioned, the following may be referred to by those who desire to know more about the mathematical part of the subject of "Linkages." "*Sur les Systèmes de Tiges Articulées,*" par M. V. Liguine, in the *Nouvelles Annales*, December, 1875, pp. 520—560.

Two papers "*On Three-bar Motion,*" by Professor Cayley and Mr. S. Roberts, in the *Proceedings of the London Mathematical Society*, 1876, vol. vii., pp. 14 and 136. Other short papers in the *Proceedings of the London Mathematical Society*, vols. v., vi., vii., and the *Messenger of Mathematics*, Vols. iv. and v.

CHELSEA

SCIENTIFIC

BOOKS

STRING FIGURES, and other monographs

By BALL, CAJORI, CARSLAW, and PETERSEN

FOUR VOLUMES IN ONE:
String Figures, *by W. W. Rouse Ball;*

The Elements of Non-Euclidean Plane Geometry, *by H. S. Carslaw;*

A History of the Logarithmic Slide Rule, *by F. Cajori;*

Methods and Theories for the Solution of Problems of Geometrical Construction, *by J. Petersen*

—528 pp. 5¼x8. [130] Four vols. in one.

THÉORIE DES OPÉRATIONS LINEAIRES

By S. BANACH

—1933. xii + 250 pp. 5¼x8¼. [110]

THEORIE DER FUNKTIONEN MEHRERER KOMPLEXER VERANDERLICHEN

By H. BEHNKE and P. THULLEN

—(Ergeb. der Math.) 1934. vii+115 pp. 5½x8½. [68]

CONFORMAL MAPPING

By L. BIEBERBACH

"The first book in English to give an elementary, readable account of the Riemann Mapping Theorem and the distortion theorems and uniformisation problem with which it is connected. . . . Presented in very attractive and readable form."

—*Math. Gazette.*

". . . thorough and painstaking . . . lucid and clear and well arranged . . . an excellent text."

—*Bulletin of the A. M. S.*

"Engineers will profitably use this book for its accurate exposition."—*Appl. Mechanics Reviews.*

—1952. vi + 234 pp. 4½x6½. [90]

LECTURES ON THE CALCULUS OF VARIATIONS

By O. BOLZA

A standard text by a major contributor to the theory.

—Ready June or July, 1961. Corr. repr. of first ed. xi + 267 pp. 5⅜x8. [145]

THEORIE DER KONVEXEN KORPER

By T. BONNESEN and W. FENCHEL

"Remarkable monograph."
—*J. D. Tamarkin, Bulletin of the A. M. S.*

—1934. 171 pp. 5½x8½. Orig. publ. at $7.50 [54]

THE CALCULUS OF FINITE DIFFERENCES

By G. BOOLE

A standard work on the subject of finite differences and difference equations by one of the seminal minds in the field of finite mathematics.

Some of the topics covered are: *Interpolation, Finite Integration, Summation of Series, General Theory of Difference and Differential Equations of First Order, Linear DEqns with Variable Coefficients, Linear DEqns, Geometrical Applications.*

Numerous exercises with answers.

—Fourth edition. 1958. xii + 336 pp. 5x8. [121] Cloth
[148] Paper

A TREATISE ON DIFFERENTIAL EQUATIONS

By G. BOOLE

Including the Supplementary Volume.

—Fifth edition. 1959. xxiv + 735 pp. 5¼x8. [128]

TEXTBOOK OF ALGEBRA

By G. CHRYSTAL

The usefulness, both as a textbook and as a work of reference, of this charming classic is attested to by the number of editions it has run through—the present being the sixth. Its richness of content can be only appreciated by an examination of the twelve-hundred-page book itself. **Thousands of valuable exercises (with solutions).**

6th ed. 2 Vols. 1235 pages. 5⅜x8. [84] Each volume

HISTORY OF THE THEORY OF NUMBERS

By L. E. DICKSON

"**A monumental work** . . . Dickson always has in mind the needs of the investigator . . . The author has [often] expressed in a nut-shell the main results of a long and involved paper *in a much clearer way than the writer of the article did himself.* The ability to reduce complicated mathematical arguments to simple and elementary terms is highly developed in Dickson."—*Bulletin of A. M. S.*

—Vol. I (Divisibility and Primality) xii + 486 pp. Vol. II (Diophantine Analysis) xxv + 803 pp. Vol. III (Quadratic and Higher Forms) v + 313 pp. [86] Three vol. set

STUDIES IN THE THEORY OF NUMBERS

By L. E. DICKSON

A systematic exposition, starting from first principles, of the arithmetic of quadratic forms, chiefly (but not entirely) ternary forms, including numerous original investigations and correct proofs of a number of classical results that have been stated or proved erroneously in the literature.

—1930. viii + 230 pp. 5⅜x8. [151]

THE INTEGRAL CALCULUS

By J. W. EDWARDS

A leisurely, immensely detailed, textbook of over 1,900 pages, rich in illustrative examples and manipulative techniques and containing much interesting material that must of necessity be omitted from less comprehensive works.

There are forty large chapters in all. The earlier cover a leisurely and a more-than-usually-detailed treatment of all the elementary standard topics. Later chapters include: Jacobian Elliptic Functions, Weierstrassian Elliptic Functions, Evaluation of Definite Integrals, Harmonic Analysis, Calculus of Variations, etc. Every chapter contains many exercises (with solutions).

—2 vols. 1,922 pp. 5x8. Originally published at $31.50 the set. [102], [105]

AUTOMORPHIC FUNCTIONS

By L. R. FORD

"Comprehensive . . . remarkably clear and explicit."—*Bulletin of the A. M. S.*

—2nd ed. (Cor. repr.) x + 333 pp. 5⅜x8. [85]

CURVE TRACING

By P. FROST

This much-quoted and charming treatise gives a very readable treatment of a topic that can only be touched upon briefly in courses on Analytic Geometry. Teachers will find it invaluable as supplementary reading for their more interested students and for reference. The Calculus is not used.

Seventeen plates, containing over 200 figures, illustrate the discussion in the text.

—5th (unaltered) ed. 1960. 210 pp. + 17 fold-out plates. 5⅜x8. [140]

LECTURES ON ERGODIC THEORY

By P. R. HALMOS

CONTENTS: Introduction. Recurrence. Mean Convergence. Pointwise Convergence. Ergodicity. Mixing. Measure Algebras. Discrete Spectrum. Automorphisms of Compact Groups. Generalized Proper Values. Weak Topology. Weak Approximation. Uniform Topology. Uniform Approximation. Category. Invariant Measures. Generalized Ergodic Theorems. Unsolved Problems.

"Written in the pleasant, relaxed, and clear style usually associated with the author. The material is organized very well and painlessly presented. A number of remarks ranging from the serious to the whimsical add insight and enjoyment to the reading of the book."

—*Bulletin of the Amer. Math. Soc.*

—1960. (Repr. of 1956 ed.) viii + 101 pp. 5¼x8. [142]

INTRODUCTION TO HILBERT SPACE AND THE THEORY OF SPECTRAL MULTIPLICITY

By P. R. HALMOS

Prof. Halmos' recent book gives a clear, readable introductory treatment of Hilbert Space. The multiplicity theory of continuous spectra is treated, for the first time in English, in full generality.

—1957. 2nd ed. (c. repr. of 1st ed.). 120 pp. 6x9. [82]

RAMANUJAN: Twelve Lectures on His Life and Works

By G. H. HARDY

The book is somewhat more than an account of the mathematical work and personality of Ramanujan; it is one of the very few full-length books of "shop talk" by an important mathematician.

—viii + 236 pp. 6x9 [136]

GEOMETRY AND THE IMAGINATION

By D. HILBERT and S. COHN-VOSSEN

The theme of this book is *insight.* Not merely proofs, but proofs that offer *insight*—intuitive understanding—into *why they are true.* Not merely properties of the hyperboloid or of Pascal's hexagon, but insight into *why they have these properties.* In this wide-ranging survey, one of the world's greatest and most original mathematicians uses insight as both his technique and his aim. Both the beginner and the mature mathematician will learn much from this fascinating treatise.

Translated from the German by P. NEMENYI.

CHAPTER HEADINGS: I. The Simplest Curves and surfaces. II. Regular Systems of Points. III. Projective Configurations. IV. Differential Geometry. V. Kinematics. VI. Topology.

"A mathematical classic . . . The purpose is to make the reader *see* and *feel* the proofs."—*Science.*

"A fascinating tour of the 20th-century mathematical zoo."—*Scientific American.*

"Students . . . will experience the sensation of being taken into the friendly confidence of a great mathematician and being shown the *real significance* of things."—*Science Progress.*

"A glance down the index (*twenty-five columns of it*) reveal the breadth of range:—

"Annulus; Atomic structure; Automorphic functions; Bubble, soap; Caustic Curve; Color problem; Density of packing, of circles; Four-dimensional space; Gears, hyperboloidal; Graphite; Lattices; Mapping; "Monkey Saddle"; Table salt; Zinc.

"*These are but a few of the topics* . . . The title evokes the imagination and the text must surely capture it."—*Math. Gazette.*

—1952. 358 pp. 6x9. [87]

SQUARING THE CIRCLE, and other Monographs

By HOBSON, HUDSON, SINGH, and KEMPE

FOUR VOLUMES IN ONE.

SQUARING THE CIRCLE, by *Hobson.* A fascinating account of one of the three famous problems of antiquity, its significance, its history, the mathematical work it inspired in modern times, and its eventual solution in the closing years of the last century.

RULER AND COMPASSES, by *Hudson.* "An analytical and geometrical investigation of how far Euclidean constructions can take us. It is as thoroughgoing as it is constructive."—*Sci. Monthly.*

THE THEORY AND CONSTRUCTION OF NON-DIFFERENTIABLE FUNCTIONS, by *Singh.* I. Functions Defined by Series. II. Functions Defined Geometrically. III. Functions Defined Arithmetically. IV. Properties of Non-Differentiable Functions.

HOW TO DRAW A STRAIGHT LINE, by *Kempe.* An intriguing monograph on linkages. Describes, among other things, a linkage that will trisect any angle.

"Intriguing, meaty."—*Scientific American.*

—388 pp. 4½x7½. [95] Four vols. in one

SPHERICAL AND ELLIPSOIDAL HARMONICS

By E. W. HOBSON

"A comprehensive treatise . . . and the standard reference in its field."—*Bulletin of the A. M. S.*

—1930. 512 pp. 5⅜x8. Orig. pub. at $13.50. [104]

ERGODENTHEORIE

By E. HOPF

—(Ergeb. der Math.) 1937. 89 pp. 5½x8½. [43]

FOUNDATIONS OF ANALYSIS

By E. LANDAU

"Certainly no clearer treatment of the foundations of the number system can be offered. . . . One can only be thankful to the author for this fundamental piece of exposition, which is alive with his vitality and genius."—*J. F. Ritt, Amer. Math. Monthly.*

—2nd ed. 1960. 6x9. [79]

HANDBUCH DER LEHRE VON DER VERTEILUNG DER PRIMZAHLEN

By E. LANDAU

To Landau's monumental work on prime-number theory there has been added, in this edition, two of Landau's papers and an up-to-date guide to the work: an Appendix by Prof. Paul T. Bateman.

—2nd ed., 1953. 1,028 pp. 5½x8½. [96] Two vol. set

DIFFERENTIAL AND INTEGRAL CALCULUS

By E. LANDAU

Landau's sparkling *Einführung* in English translation. Completely rigorous, completely self-contained, borrowing not even the fundamental theorem of algebra (of which it gives a rigorous elementary proof), it develops the entire calculus including Fourier series, starting only with the properties of the number system. A masterpiece of rigor and clarity.

—2nd ed. 1960. 372 pp. 6x9. [78]

ELEMENTARE ZAHLENTHEORIE

By E. LANDAU

"Interest is enlisted at once and sustained by the accuracy, skill, and enthusiasm with which Landau marshals . . . facts and simplifies . . . details."

—*G. D. Birkhoff, Bulletin of the A. M. S.*

—1927. vii + 180 + iv pp. 5½x8¼. [26]

KNOTENTHEORIE

By K. REIDEMEISTER

—(Ergeb. der Math.) 1932. 78 pp. 5½x8½. [40]

FOURIER SERIES

By W. ROGOSINSKI

Translated by H. COHN. Designed for beginners with no more background than a year of calculus, this text covers, nevertheless, an amazing amount of ground. It is suitable for self-study courses as well as classroom use.

"The field covered is extensive and the treatment is thoroughly modern in outlook . . . An admirable guide to the theory."—*Mathematical Gazette.*

—Second ed. 1959. vi + 176 pp. 4½x6½. [67]

CONIC SECTIONS

By G. SALMON

"The classic book on the subject, covering the whole ground and full of touches of genius."

—*Mathematical Association.*

—6th ed. xv + 400 pp. 5¼x8¼. [99] Cloth
[98] Paper

HIGHER PLANE CURVES

By G. SALMON

CHAPTER HEADINGS: I. Coordinates. II. General Properties of Algebraic Curves. III. Envelopes. IV. Metrical Properties. V. Cubics. VI. Quartics. VII. Transcendental Curves. VIII. Transformation of Curves. IX. General Theory of Curves.

—3rd ed. xix + 395 pp. 5⅜x8. [138]